MRSA
Current Perspectives

Edited by

Ad C. Fluit and Franz-Josef Schmitz

*Eijkman-Winkler Institute, 3508 GA Utrecht,
The Netherlands*

Copyright © 2003
Caister Academic Press
32 Hewitts Lane
Wymondham
Norfolk NR18 0JA
England

www.caister.com

British Library Cataloguing-in-Publication Data

A catalogue record for this book is available from the British Library

ISBN:0-9542464-5-4

Description or mention of instrumentation, software, or other products in this book does not imply endorsement by the author or publisher. The author and publisher do not assume responsibility for the validity of any products or procedures mentioned or described in this book or for the consequences of their use.

Printed and bound in Great Britain
by Cromwell Press Ltd, Trowbridge, Wiltshire BA14 0XB, UK

Contents

Books of Related Interest

For further information on these books contact:

Caister Academic Press
32 Hewitts Lane, Wymondham
Norfolk
NR18 0JA England

Tel: +44(0)1953-601106
Fax: +44(0)1953-603068
Email: mail@caister.com
Internet: www.caister.com

Our Web site has details of all our books including full chapter abstracts, book reviews, and ordering information:

Contributors

Debby Ben-David
Infectious Diseases Unit
Sheba Medical Center
Tel Hashomer
Israel
Email:
bendaviddebby@hotmail.com

B. Berger-Bächi
Institute of Medical Microbiology
University of Zürich
Gloriastr. 32
CH 8028 Zürich
Switzerland
Email: bberger@immv.unizh.ch

Karsten Becker
Institute of Medical Microbiology
Westfälische Wilhelms-Universität
Münster
Domagkstraße 10
48149 Münster
Germany

Markus Bischoff
Institute of Medical Microbiology
University of Zürich
Gloriastr. 32
CH 8028 Zürich
Switzerland

Derek F.J. Brown
Health Protection Agency
Clinical Microbiology
and Public Health Laboratory
Addenbrooke's Hospital
Hills Road
Cambridge CB2 2QW
UK
Email: DFJB2@CAM.AC.UK

Barry Cookson
Laboratory of Healthcare
-Associated Infection
Specialist and Reference Division
Health Protection Agency
London
UK

Longzhu Cui MD
Dept.of Bacteriology
Medical Shchool
Juntendo University
2-1-1 Hongo
Bunkyo-Ku
Tokyo
Japan
Email:
longzhu@med.juntendo.ac.jp

David J. Farrell
GR Micro Ltd.
7-9 William Road
London NW1 3ER
UK
Email: d.farrell@grmicro.co.uk

Ross J. Fitzgerald
Department of Microbiology
Moyne Institute of Preventive
Medicine
University of Dublin
Trinity College
Dublin 2
Ireland
Email: jrfitzge@tcd.ie

Ad C. Fluit
Eijkman-Winkler Center
UMCU
room G04.614
PO box 85500
3508 GA Utrecht
The Netherlands
Email: A.C.Fluit@lab.azu.nl

Uwe Frank
Institute for Environmental Medicine
and Hospital Epidemiology
Hugstetter Str. 55
79106 Freiburg
Germany
Email:
ufrank@iuk3.ukl.uni-freiburg.de

Keiichi Hiramatsu
Department of Bacteriology
Juntendo University
2-1-1 Hongo
Bunkyo-Ku
Tokyo
Japan 113-8421

Ian Morrissey
GR Micro Ltd.
7-9 William Road
London NW1 3ER
UK
Email: i.morrissey@grmicro.co.uk

James M. Musser
Rocky Mountain Laboratories
National Institute of Allergy
and Infectious Diseases
National Institutes of Health
903 South 4th Street
Hamilton
MT 59840
USA
Email:jmusser@niaid.nih.gov

Richard P. Novick
Skirball Institute
Departments of Microbiology and
Medicine
New York University School of
Medicine
540 First Ave.
New York
NY 10016
USA
Email: novick@saturn.med.nyu.edu

Susanne Rohrer
Institute of Medical Microbiology
University of Zürich
Gloriastr. 32
CH 8028 Zürich
Switzerland

Jutta Rossi
Institute of Medical Microbiology
University of Zürich
Gloriastr. 32
CH 8028 Zürich
Switzerland

Ethan Rubinstein
Infectious Diseases Unit
Sheba Medical Center
Tel Hashomer
Israel

Franz-Josef Schmitz
Eijkman-Winkler Center
University Medical Center Utrecht
Utrecht
The Netherlands

Willem B. van Leeuwen
Erasmus MC
Department of Medical Microbiology
and Infectious Diseases (room L333)
Dr. Molewaterplein 40
3015 GD Rotterdam
The Netherlands
Email:
w.vanleeuwen@erasmusmc.nl

Christof von Eiff
Institute of Medical Microbiology
Westfälische Wilhelms-Universität
Münster
Domagkstraße 10
48149 Münster
Germany
Email: eiffc@uni-muenster.de

Jesse S. Wright III
Skirball Institute
Departments of Microbiology and
Medicine
New York University School of
Medicine
540 First Ave.
New York
NY 10016
USA
Email: wright@saturn.med.nyu.edu

From: *MRSA: Current Perspectives*
Edited by: A.C. Fluit and F.-J. Schmitz

Chapter 1

Introduction

Barry Cookson,
Franz-Josef Schmitz, and Ad C. Fluit

Staphylococci: Discovery

Staphylococci were first observed and cultured by Pasteur and Koch, but the initial detailed studies on staphylococci were made by Ogston in 1881 and Rosenbach in 1884 (Ogston, 1881; Rosenbach, 1884). The genus *Staphylococcus* was given its name by Ogston in 1881 when he observed grape-like clusters of bacteria in pus from human abscesses (Ogsten, 1881). Three years later, Rosenbach was able to isolate and grow these micro-organisms in pure culture. He gave these bacteria the specific epithet *Staphylococcus aureus* because of the yellow-to-orange pigmented appearance of their colonies. Rosenbach showed that *S. aureus* was responsible for wound infections and furunculosis, and that *Staphylococcus epidermidis* was a normal colonizer of the skin (Rosenbach, 1884). Ever since Rosenbach first described the growth of this 'golden' coccus, surgeons have feared staphylococcal wound infections after surgery. Staphylococci also caused life-threatening disease after trauma and fatal pneumonia during the influenza season. Therefore, in the pre-antibiotic era, *S. aureus* was known as a major life-threatening pathogen.

Staphylococcal Infections

Staphylococcal infections are characterized by intense suppurative inflammation of local tissues with a tendency for the infected area to become encapsulated, leading to abscess formation. The most common staphylococcal infection is the furuncle or boil, which is a localized painful superficial skin infection that develops in a hair follicle or gland. Similar infections at the base of the eyelashes are the common styes. The most usual sites for boils are the neck and the buttocks, often as a result of wearing tight clothes. The infecting organism is usually identical to that carried in the patients' anterior nares. Approximately 2–3% of the population have chronic furunculosis. Staphylococcal infection may spread from a furuncle to the deeper subcutaneous tissues resulting in the development of one or more abscesses known as carbuncles. These are now less common, but such patients were often suffering from diabetes mellitus, and a carbuncle may have been the initial presentation of the condition. These abscesses occur mostly on the back of the neck, but may involve other skin sites. They may result in bloodstream invasion.

Certain strains of *S. aureus* can cause bullous impetigo. This is characterized by small bullae, which form and burst. When this occurs in infants it may be described as pemphigus neonatorum. Bullous impetigo is a highly contagious superficial skin infection characterized by large blisters containing many staphylococci in the superficial layers of the skin. Bullous impetigo is seen most often in infants and children under conditions where direct spread can occur (e.g. sharing of contaminated towels). Impetigo mainly occurs on the face and limbs. Extensive atypical bullous impetigo is associated with HIV infection.

Staphylococcus aureus can cause a wide variety of deep tissue, often metastatic, infections. These infections include osteomyelitis, arthritis, endocarditis and cerebral, pulmonary, renal and breast abscesses (the latter in the nursing mother). *Staphylococcus aureus* can also cause pneumonia, which is almost always secondary (e.g. after influenza or other viral infections). In many of these situations, diabetes mellitus, leukocyte defects or a general reduction of host defenses by alcoholism, malignancy, old age, or corticosteroid or cytotoxic therapy are predisposing factors. Severe *S. aureus* infections, including endocarditis, are particularly common in intravenous drug abusers. *Staphylococcus aureus* is a notorious cause of wound infection. Infections can be major complications following surgery. The source may be the patient's own carrier state, other carriers (e.g. physicians or nurses) or other infected patients. Staphylococcal infections at the site of intravenous lines can result in bacteremia with metastatic infection (Weatherall *et al.*, 1995).

The pathogenesis of *S. aureus* infections is determined by the interaction of the *S. aureus* surface-associated and extracellular proteins with host cell determinants. The pathogenesis of staphylococci is multifactorial and therefore it is difficult to determine precisely the role of each individual staphylococcal component in the etiology of staphylococcal disease (Chapter 9). The role of small colony variants (SCV) in the pathogenesis of *S. aureus* infections and MRSA infections in particular has hardly been studied. However, the difficulty in detecting these isolates may have led to an underestimation of their importance (Chapter 10).

Antibiotic Resistance

The Second World War, with its commensurate cases of *S. aureus* sepsis, provided a good testing ground for the recently invented sulphonamides and a spur to the development of penicillin (Shanson, 1982), which in the 1940's reversed the appalling prognosis of *S. aureus* infections.

However, resistance to sulphonamides and then to penicillin (Shanson, 1982, Cookson and Phillips, 1990) soon followed. Indeed, by 1948 about 60% of *S. aureus* in the UK were resistant to penicillin, and penicillinase-producing *S. aureus* strains of phage group I were the cause of world-wide nosocomial infection in maternity and neonatal units, and similar strains of phage group III were the cause of infection in non-maternity areas (Altemeier *et al.*, 1981). This pattern was repeated when the versatile staphylococcus developed, in turn, resistance to streptomycin, tetracycline, chloramphenicol and erythromycin (Cookson and Phillips, 1990).

Multiple Resistant *S. aureus*

Background

Multiple resistant *S. aureus* became a world-wide problem in the 1950's (Shanson, 1982; Cookson and Phillips, 1990). During this time a new virulent strain of penicillin- resistant *S. aureus* appeared; firstly in Australia, where it reacted with phage 80 and then in Canada, where it reacted with phage 81. These strains have since been found to be closely related (Richardson *et al.*, 1994). Although much of the early history of nosocomial infection control is anecdotal (Haley, 1985), it seems fairly clear that this staphylococcal pandemic was the major stimulus to an organised infection control effort, although throughout the '60s and '70s most clinicians were blissfully unaware of the magnitude of nosocomial infection in their hospitals.

However, after 1960 the "hospital staphylococcus" as the 80/81 strain had become known, was much less common. The isolation of 6-aminopenicillinic acid in 1959 resulted in the subsequent synthesis of a large number of compounds with different radicals on the side-chain of the penicillinic acid nucleus. Methicillin was one such compound, and it proved of value in the treatment of penicillin-resistant staphylococci because of the molecule's stability to staphylococcal penicillinase. Barber reported artificially induced methicillin-resistant strains in 1960, but it was Jevons (1961) who first described naturally occurring methicillin-resistant *S. aureus*, three of which were found amongst 5,440 strains examined shortly after the therapeutic agent had been introduced. The strains were thought to have emerged without exposure to the agent because the infected patients had not received the antibiotic. There is of course the possibility that they may have been cross-infected by carriers who had previously received methicillin treatment. The term "natural resistance" has in fact been used in two ways: either occurring without exposure to the agent (perhaps the term spontaneous would be clearer) as opposed to laboratory selected or unnatural resistance.

Evolution of Resistance in the Absence of Methicillin Selection

The evolution of MRSA strains in India, without selection by exposure to methicillin was described by Pal and Ghosh Ray (1964). A total of 128 MRSA strains were isolated from 98 patients, in three hospitals. However the possibility that these strains emerged as a result of methicillin treatment cannot be excluded. All of the new MRSA strains were nosocomial isolates, and none were community associated. Only 24 of the strains were typeable and these typed with phages in group III (one also reacted with phage 29). All strains were resistant to mercury salts, penicillin, streptomycin, tetracycline and 25% were chloramphenicol resistant. Methicillin had been used only for one week by 9 patients, and so it was assumed that the MRSA strains had not arisen due to exposure to the agent. However, it is not stated whether MRSA was detected on any of the treated patients, or even if they had been screened. One hundred and forty-nine of their isolates were detected in doctors, nurses and sweepers in the hospital wards, perhaps a reflection of widespread colonisation of patients or the environment. Methicillin had also been sprayed into the nares of 50 hospital staff, presumably an attempt to clear them of MRSA carriage, rather than prophylaxis. It seems most probable that the MRSA had extensively spread between patients, possibly by staff, and the workers claim to the spontaneous evolution of MRSA was not proven.

However, MRSA strains were reported in countries, such as Poland and Turkey, before the introduction of methicillin (Borowski *et al.*, 1964). It is tempting to speculate that this was either due to an earlier (or perhaps an alternative form) of resistance to naturally occurring penicillin in the environment,

Alternatively, it may have resulted from exposure to an unknown source of naturally occurring methicillin-like antimicrobial agent. There are also genetic relationships between the *mec* and beta-lactamase genes (see Chapter 3), but prime candidate for *mecA* ancestor is *Staphylococcus sciuri* (Chapters 3, 6 and 7). In 1963 the first nosocomial MRSA outbreak was described over a 14 month period when 37 children were affected on eight wards. In the same year, Jevons and co-workers found 102 MRSA strains amongst 27,000 *S. aureus* strains sent to Colindale PHLS (Cookson and Phillips, 1990). All of these strains showed a similar antibiotic and mercurial salt resistogram, and most belonged to phage group III.

MRSA Resistance to Other Antimicrobials

During the early half of the 1970's (Shanson, 1982), many UK centres experienced a decrease in MRSA (~9% to 0%), and the same was true in Switzerland (from 20% to 3%) and Denmark (from 19% to 6%). The geographical differences in experience with MRSA were also evident in the USA. In the late 1960's, MRSA comprised less than 1% of *S. aureus* isolates (Barrett *et al.*, 1968). However, problems did arise in the 1970's and these have since continued (Cookson and Philips, 1990). Resistance to other antimicrobials, many occurring very shortly after they were introduced clinically, has been a recurrent theme in MRSA evolution. Erythromycin, tetracycline and chloramphenicol resistance were documented in the 1960s (Shanson, 1982). In the 1970s gentamicin-resistance was reported for the first time in the UK and is now a global phenomenon. Many plasmids are described conveying resistance to antimicrobials and antiseptics (Lyon and Skurray, 1987). However, the resistance elements can be readily incorporated into the chromosome. More recent examples of resistance include the emergence of mupirocin resistance (Noble *et al.*, 1988) and linezolid resistance (Tsiodras *et al.*, 2001; Wilson *et al.,* 2003).

Monitoring the spread MRSA was made possible by the development of typing techniques. Until a decade ago, phage typing was the most commonly applied technique. However with the development of molecular biology, a whole array of new typing techniques have become available (Chapter 5).

Genetics of MRSA Resistance

MRSA strains contain an extra DNA fragment of approximately 30 to 50kb in that is absent in sensitive isolates. This extra DNA segment has been given various names as follows; *mec, mec* determinant or associated DNA, and staphylococcal cassette chromosome *mec* (SCC*mec*). The essential methicillin resistance gene *mecA,* encoding the extra penicillin-binding protein PBP2a or 2', (Hackbarth *et al.*, 1994) is located within this element. SCC*mec* may contain up to ~100 open reading frames. Regulatory genes (*mecRI* and *mecI*) may or may not be present and there is considerable dynamism with transposons such as Tn*554* (encoding inducible resistance to erythromycin) and IS*431*. The IS*431* element is also commonly found in plasmids and probably related to the propensity for MRSA to trap resistance elements in the chromosome near to *mecA* (Chambers, 1977). In addition to IS*431* and Tn*554,* another mobile element that can be present is SCC*mec* (Chapter 6 and 7). More recently several strains have or are in the process of being sequenced. This, together with other approaches such as MLST and SCC*mec* typing are providing important insights into the evolution and detailed resistance mechanisms of MRSA (Chapters 6 and 7). There is evidence that the methicillin resistance genes have been introduced into *S. aureus* on at least five occasions. Different elements have appeared in the same genetic background and the same element into different backgrounds.

Resistance to Vancomycin

The methicillin-resistance of *S. aureus* mediates clinically inadequate susceptibility to all currently available beta-lactam antibiotics. In addition, as already mentioned, MRSA is typically resistant to several other antimicrobial agents including; aminoglycosides, chloramphenicol, clindamycin, fluoroquinolones and macrolides (Figure 1 and Chapter 4). Multiple resistance varies greatly geographically, so in planning treatment it is imperative to carry out susceptibility testing (Chapter 2). To date, most clinical isolates have been susceptible only to vancomycin, which consequently has become the agent of choice for the treatment of MRSA infections. However, recently, several investigators have described clinical isolates of *S. aureus* with diminished sensitivity to vancomycin. In the laboratory, vancomycin resistance has been transferred from *Enterococcus faecalis* to *S. aureus* and is stably expressed. In addition, the first two vancomycin resistant *S. aureus* isolates were detected in the USA. Probably the *van*A gene, coding for vancomycin resistance, was transferred from enterococci to *S. aureus* (Anonymous, 2002a and 2002b; Centers for Disease Control and Prevention, 2002).

S. *aureus* isolates with reduced vancomycin susceptibility had already been described worldwide. These isolates, called GISA isolates, have reduced susceptibility to glycopeptides. GISA isolates are also sometimes called VRSA (i.e. vancomcyin-resistant *S. aureus*). Thus far, all studied isolates contain thickened cell walls, and all, with the exception of an isolates from Illinois, showed a reduced cross-linking when compared to isogenic revertants (Boyle *et al.*, 2001). Interestingly, only some strains showed a reduction in D-glutamic acid amidation. The data described so far on GISA strains seem to indicate that, depending on the strain studied, several independent mutations have been accumulated in these strains, which in various combinations lead to the

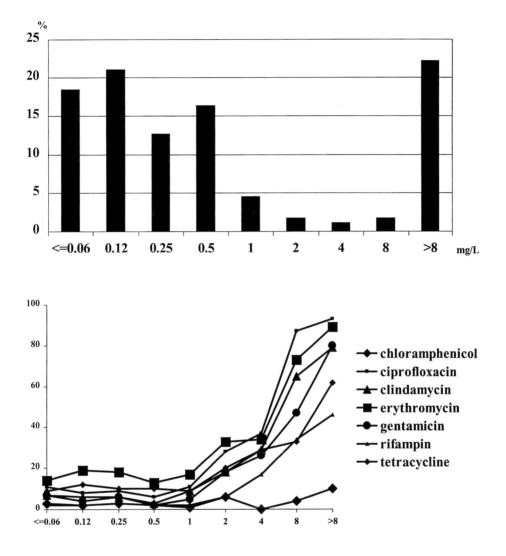

Figure 1. **Top**. MIC distribution of 3,051 *S. aureus* isolates. The isolates were obtained from blood cultures, respiratory tract, urinary tract, wound and soft tissue infections from 25 European university hospitals during 1997 and 1998 as part of the SENTRY Antimicrobial Surveillance Program (Fluit *et al.*, 2001). **Bottom**. Resistance of the 3,051 *S. aureus* isolates for different antibiotics as function of their MIC for oxacillin. The resistance is expressed as percentage resistant for each MIC-value for oxacillin (Fluit *et al.*, 2001).

observed resistance phenotype. The emergence of the GISA strain is probably related to the accelerated cell wall synthesis leading to thickened cell wall; the thickened cell-wall is capable of affinity trapping large amounts of vancomycin and thus shielding the membrane associated lipid II molecules, the target for glycopeptide antimicrobial activity (Chapter 8).

Should vancomycin-resistance and reduced susceptibility to glycopeptides become widespread, alternative MRSA therapies will be urgently needed. Careless prescription of drugs active against MRSA will inevitably lead to the emergence of resistance and the permanent loss of a valuable agent. Responsible antibiotic prescription therefore mandates close and frequent consultation between the prescribing physician and medical microbiologist (Chapter 11). However, the use of prudent hospital hygiene measures will also be required to battle both MRSA and VRSA (Chapter 12).

From this introduction it is clear that MRSA presents an unique challenges to the clinician, microbiologist, and the molecular biologist. In the following chapters a comprehensive review of the knowledge on all aspects of MRSA is presented. The data may help guide future research and efforts for the treatment of patients infected with MRSA.

References

Altemeier, W.A., Lewis, S., and Brackett, K. 1981. The versatile *Staphylococcus*. In: The Staphylococci. Proceedings of the Alexander Ogston Centennial Conference. Macdonald, A., and Smith, G., eds. Aberdeen University Press. p. 125-148.

Anonymus. 2002a. *Staphylococcus aureus* resistant to vancomycin - United States, 2002. Morb. Mortal. Wkly. Rep. 51: 565.

Anonymus. 2002b. Vancomycin-resistant *Staphylococcus aureus* - Pennsylvania, 2002. MMWR Morb. Mortal Wkly Rep. 51: 902.

Barrett, F.F., McGehee, R.F., and Finland, M. 1968. Methicillin-resistant *Staphylococcus aureus* at Boston City Hospital: bacteriologic and epidemiologic observations. N. Engl. J. Med. 279: 441-448.

Borowski, J., Kamienska, K., and Rutecka, I. 1964. Methicillin- resistant staphylococci. Brit. Med. J. I: 983.

Boyle-Vavra, S., Carey, R.B., and Daum, R.S. 2001. Development of vancomycin and lysostaphin resistance in a methicillin-resistant *Staphylococcus aureus* isolate. J. Antimicrob. Chemother. 48: 617-625.

Centers for Disease Control and Prevention. 2002. Vancomycin resistant *Staphylococcus aureus*-Pennsylvania. 2002. J.A.M.A. 288: 2116.

Cookson, B.D., and Phillips, I. 1990. Methicillin-resistant staphylococci. Soc. Appl. Bacteriol. Symp. Ser. 19: 55S-70S.

Fluit, A.C., Wielders, C.L., Verhoef, J., Schmitz, F.-J. 2001. Epidemiology and susceptibility of 3,051 *Staphylococcus aureus* isolates from 25 university hospitals participating in the European SENTRY study. J. Clin. Microbiol. 39: 3727-3732.

Hackbarth, C.J., Miick, C., Chambers, H.F. 1994. Altered production of penicillin-binding protein 2a can affect phenotypic expression of methicillin resistance in *Staphylococcus aureus*. Antimicrob. Agents Chemother. 38: 2568-2571.

Haley, R.W. 1985. Surveillance by objective: A new priority-directed approach to the control of nosocomial infections. Am. J. Infect. Contr. 13: 78-89.

Lyon, B.R. and Skurray, R., 1987. Antimicrobial resistance of *Staphylococcus aureus*: genetic basis. Microbiol. Rev. 51: 88-134

Noble, W.C., Rahman, M., Cookson, B.D., and Phillips, I. 1988. Transferable mupirocin resistance. J. Antimicrob. Chemother. 22: 771-772.

Ogsten, A. 1881. Report upon microorganisms in surgical diseases. Br. Med. J. 1:369-375.

Richardson, J.F., Aparicio, P., Marples, R.R., and Cookson, B.D. 1994. Ribotyping of *Staphylococcus aureus*: an assessment using well-defined strains. Epidemiol. Infect. 112: 93-101.

Rosenbach, F.J. 1884. Mikro-Organismen bei den Wund-Infektions-Krankenheiten des Menschen. Wiebaden. J.F. Bergmann's Verlag.

Shanson, D.C. 1982. Antibiotic-resistant *Staphylococcus aureus*. J. Hosp. Infect. 2: 11-36.

Tsiodras, S., Gold, H.S., Sakoulas, G., Elliopoulos, G.M., Wennersten, C., Venkataraman, L., Moellering, R.C., and Ferraro, M.J. 2001. Linezolid resistance in a clinical isolate of *Staphylococcus aureus*. Lancet 358: 207-208.

Weatherall, D.J., Ledingham, J.G.G., and Warrell, D.A. eds. 1995. Oxford Textbook of Medicine, 3rd Ed. Oxford University Press.

Wilson, P., Andrews, J.A., Charlesworth, R., Walesby, R., Singer, M., Farrell, D.J., and Robbins, M. 2003. Linezolid resistance in clinical isolates of *Staphylococcus aureus*. J. Antimicrob. Chemother. 51: 186-188.

From: *MRSA: Current Perspectives*
Edited by: A.C. Fluit and F.-J. Schmitz

Chapter 2

Detection of MRSA

Derek Brown and Barry Cookson

Abstract

The detection of MRSA in clinical specimens is a challenge to any microbiology laboratory. We describe the various aspects of this process: isolation media, confirmation of species, determination of methicillin-resistance and the many factors that can affect this. The phenotypic and genotypic laboratory techniques, their advantages and various pitfalls are outlined as well as current and promising rapid methods. We also discuss the cost effectiveness of these methods and propose a categorisation of MRSA screening strategies. Mathematical modelling to evaluate cost effectiveness is needed so that informed choices of the most appropriate technique for each screening strategy can be made.

Introduction

The discovery of methicillin resistance in *Staphylococcus aureus* shortly after the agent entered into clinical use in 1961 was by chance (Jevons, 1961). There was no room on the antibiotic susceptibility testing agar plate for another antibiotic disc so a methicillin disc was placed on the phage typing agar plate, which was incubated at 30°C. This was fortunate as the early methicillin resistant *S. aureus* (MRSA) were susceptible at 37°C (Hewitt *et al.,* 1969). This was because of the heterogeneous expression of resistance to methicillin and other penicillinase-stable isoxazolyl penicillins (e.g. cloxacillin, oxacillin), which is seen as large differences in the expression of resistance among individual cells in a population (Knox, 1961). Varying the test conditions (e.g. the temperature, salt concentration, pH, chelating agents, trace metals, atmosphere of incubation) could have major effects on the detection of resistance (Barber, 1964; Annear, 1968; Sutherland and Rolinson, 1964). These effects can vary for different strains, although the genetic basis is still unclear. The fact that strains could appear susceptible to methicillin at 37°C stimulated an extensive debate about the significance of the resistance before a consensus was achieved that these isolates were indeed clinically resistant to isoxazolyl penicillins. More homogeneously resistant MRSA were subsequently described (Cookson and Phillips, 1990).

Most methicillin resistance depends on the production of an additional penicillin-binding protein, PBP2a or PBP2' (Hartman and Tomasz, 1984; Reynolds and Brown, 1985), encoded by the additional *mecA* gene (Matsuhashi *et al.*, 1986), which has no allelic equivalent in methicillin susceptible staphylococci. There are also additional repressor and inducer genes (*mecR* and *mecI*), which can be associated with *mecA*. These and several other genes known as *fem* (factor essential for methicillin resistance) or *aux* (auxilliary) genes, which are found in susceptible as well as resistant strains, affect the expression of methicillin resistance in *S. aureus* (de Lencastre and Tomasz, 1994; Labischinski *et al.,* 1998).

Low-level resistant strains with alterations to existing PBPs ("moderately" resistant *S. aureus,* MODSAs) have been described (Montanari *et al.*, 1990; de Lencastre *et al.*, 1991; Bignardi *et al.*, 1996). Their clinical importance is unknown and they are not often isolated. Some strains which hyper-produce penicillinase may appear low-level or "borderline" resistant (BORSAs) under some conditions (Chambers *et al.*, 1989; McDougal and Thornsberry, 1986). Their clinical significance is doubtful, in that there are no reports of failure of treatment with penicillinase-resistant penicillins in infections with BORSAs and, in an experimental rat endocarditis model, oxacillin was equally effective in treatment of infections with fully susceptible and penicillinase hyper-producing strains (Thauvin-Eliopoulos *et al.*, 1990). Interestingly, BORSAs are often of Phage group V (Phage 94/96 reacting). BORSAs and MODSAs can be very difficult to distinguish from genuine MRSA, which carry the *mecA* gene and are highly heterogeneous in the expression of resistance. The test conditions may be modified to increase expression of resistance in strains with the *mecA* gene, but distinguishing penicillinase hyper-producing strains from true resistant strains remains a problem.

Screening for MRSA

MRSA are readily isolated on routine culture media. However, they can be difficult to detect when mixed with other microbial flora and many different media have been recommended for the isolation of MRSA from screening specimens. Ideally, a screening method should allow the growth of all MRSA, inhibit or differentiate other organisms, and allow direct identification tests to be performed on colonies. Unfortunately, some of these requirements conflict, and compromise is necessary. Enrichment broth (nutrient broth or cooked meat medium) containing 7% added sodium chloride was recommended by the British Society for Antimicrobial Chemotherapy, the Hospital Infection Society and the Infection Control Nurses Association working party (Ayliffe *et al.*, 1998), although several other media have been used (Davies and Zadik, 1997; Jones *et al.*, 1997; Sauter *et al.*, 1988; Wagenvoort *et al.*, 1996). The broths recommended by the working party were designed to detect the original UK Epidemic MRSA (EMRSA-1), but they may inhibit the growth of some isolates of MRSA if present in small numbers (Jones *et al.*, 1997). Selective agents suggested as alternatives to sodium chloride (NaCl) include aztreonam (Wood *et al.*, 1993), or nalidixic acid plus colistin (Morton and Holt, 1989), although these have been evaluated in solid media, not in enrichment broth.

In order to select resistant *S. aureus* on direct plating or plating from enrichment broth, the BSAC, HIS, ICNA working party (Ayliffe *et al.*, 1998) suggested using plain blood agar with 4 mg/L methicillin (methicillin is not available now as manufacture has been stopped but oxacillin 2mg/L can be substituted). However, problems may be experienced with contamination, and there may be difficulties distinguishing coagulase-negative staphylococci, which are likely to be present in many screened sites. The alternative is to use a more selective medium. Mannitol salt agar (MSA; Allen *et al.*, 1994; Lally *et al.*, 1985, Van Enk and Thompson, 1992) with 7% NaCl, and variations of MSA (La Zonby and Styarzyk, 1986; Martinez *et al.*, 1992; Merlino *et al.*, 1996) are used widely. They have the disadvantages of poor growth of some MRSA, growth of some coagulase-negative staphylococci and unreliable direct agglutination tests for *S. aureus* identification (see below; Davies and Zadik, 1997; Van Enk and Thompson, 1992). Baird Parker agar with 8 mg/L ciprofloxacin has been used where the majority of MRSA are known to be ciprofloxacin resistant (Davis and Zadik, 1997).

In recent years there has been discussion about models that may help evaluate the cost-effectivess of MRSA screening strategies. This is a very complex field and requires dynamic modelling (Cooper, 2000). Screening for MRSA may be done for different purposes, which can be categorised into five groups (Table 1). Improved speed and sensitivity of screening tests will have different implications for different screening categories (Table 1). Rapidly confirming that a result is negative has a far greater potential for immediate

Table 1. Proposed categorisation of screening tests

Category of Screening (see text)	Effects of improved sensitivity and speed of detection
Category 1: Patient introductions to hospitals from other Hospitals, Re-admissions, Nursing homes*	Control of spread and bed management improved
Category 2: Patient and staff carriage resulting in spread	Control of spread and bed management improved
Category 3: Detection of MRSA acquisition related to interventions e.g. isolation, hand hygiene	Greater confidence in results, improved control of spread and bed management
Category 4: Clearance of MRSA from treated patient	Greater confidence in results and faster establishment of clearance, improved control and bed management
Category 5: Clinical "screening"	Earlier correct treatment: fewer "reserve" antimicrobials used and less selection pressure for resistance

* Practice varies regarding policies for this depending on local MRSA prevalence in homes (see Ayliffe *et al.*, 1998: Cookson, 2000).

economic impact than rapid confirmation of positives as negative results will free-up isolation beds, may re-open wards more quickly, reduce waiting lists and isolation costs, and facilitate the change to cheaper antimicrobials. Kunori *et al.* (2002) attempted to use mathematical modelling to explore the cost effectiveness of MRSA screening methods in patients admitted to an intensive care unit. Although there was a scarcity of published data on some methods, it was concluded that taking a sample from the nose alone and directly inoculating onto Baird-Parker agar with ciprofloxacin and confirmation by a latex test for *S. aureus* without any test to confirm methicillin resistance was the most cost-effective approach. However, reporting without confirmation of resistance is not recommended. Detection of a presumptive MRSA strain should be followed by its full identification as *S.aureus,* confirmation of methicillin resistance and testing susceptibility to other antimicrobial agents. The susceptibility profile may be helpful in identifying EMRSA strains or local types (Ayliffe *et al.*, 1998).

Confirmation of Identification of *S. aureus*

When isolating a staphylococcus from a clinical or screening specimen, it is of the utmost importance to ensure that it is in fact a *S. aureus* rather than a coagulase-negative staphylococcus, although the latter can be opportunistic pathogens. There is a variety of biochemical tests that can help in the speciation of staphylococci. Some of these tests, e.g. urease, can be used to biotype strains of *S. aureus* (Ayliffe *et al.*, 1998). However, the most useful identifying biochemical features comprise the presence of extra-cellular and cell-wall bound protein A, cell-bound clumping factor, extra-cellular coagulase and heat-stable nuclease. Protein A is responsible for the agglutination of most strains by normal human serum and is due to non-specific combination with the Fc portion of some human IgGs, IgAs and IgM. Clumping factor provides a useful and rapid means of identifying *S. aureus*, as long as organisms are easily emulsified, although up to 12% of strains, including MRSA, are negative. Clumping factor can be hidden by large capsules and differs from free coagulase in that it requires only fibrinogen (Cookson, 1997).

Coagulase

Free coagulase clots plasma in the absence of calcium, but it does not clot purified fibrinogen unless a coagulase reacting factor (CRF) is present. CRF varies in type and specificity, and is similar but not identical to prothrombin. The gold standard for identification of *S. aureus* is the tube coagulase test with rabbit sera and examination of results at 4h and 24h (Cookson, 1997; Wichelhaus *et al.*, 1999). This is longer than desirable for routine testing and other more rapid tests are being used increasingly. Several rapid identification tests are available and have been assessed in many studies (see Wichelhaus *et al.*, 1999; Mason *et al.*, 2001). A problem encountered when comparing the results of the various studies is that different strains of *S. aureus* have been used. In addition the strains are not described adequately to enable reliable comparison of results.

Latex Agglutination Tests

So-called "first generation" tests detected protein A and/or clumping factor (slide coagulase) e.g. Staphaurex (Murex Diagnostic Limited, Kent, England). These tests were usually less sensitive, although more specific, than so-called "second generation" tests, e.g. Pastorex Staph-Plus (Sanofi Diagnostic Pasteur, Paris, France), Staph Latex (Am. Microsan, Mahwah, N.J, USA), Staphaurex Plus (Murex Diagnostic Limited, Kent, England), Dryspot Staphytect (Oxoid, UK), which detect protein A, clumping factor and various surface antigens. The lower specificity of these second generation tests may be due to cross-reactivity of surface antigens in coagulase-negative staphylococci. In addition, *Staphylococcus lugdunensis* may be mistaken for *S aureus* by first and second generation tests which include clumping factor (Wichelhaus *et al.*, 1999). The situation is made more complex by the fact that some strains of MRSA are negative for clumping factor and protein A and require second generation tests and perhaps confirmation by tube coagulase (Kuusela *et al.*, 1994).

Heat-stable Nuclease Test

Heat-stable nuclease tests have been used to identify *S. aureus* (Barry *et al.*, 1973), particularly in direct tests on blood cultures (Madison and Baselski, 1983). However, some rarer coagulase-negative staphylococci can also be positive for this enzyme and the test is medium dependent (Faruki and Murray, 1986). A latex test, which detects this enzyme, is also available (Remmal Lab., Kansas, USA).

Methicillin/Oxacillin Susceptibility Testing

The problems with methicillin susceptibility testing have been the subject of multiple publications with many, sometimes conflicting, recommendations regarding the most reliable method for routine use. There are several reasons for this. Studies have included different strains, which have marked differences in heterogeneity (Tomasz *et al.*, 1991), and may respond differently to changes in test conditions. There is also variable interaction between multiple factors affecting the expression of resistance e.g. the effect of modifying the NaCl concentration may depend on the medium and the temperature of incubation. The MIC determined by a dilution method has been the "gold standard" for susceptibility testing. MIC methods, however, are affected by variation in test conditions as much as diffusion tests for methicillin susceptibility and it is likely that some "incorrect" reports in studies of diffusion tests are due to errors in the reference MIC tests. More recently, MIC methods have been replaced by molecular methods, which detect the *mecA* gene as the "gold standard" for methicillin resistance in *S. aureus*, although resistance unrelated to *mecA* will not be detected (see below). Although some commercial systems for detection of methicillin resistance are available (see below), most laboratories continue to use diffusion methods for routine tests.

Effect of Test Conditions on the Detection of Resistance

Test conditions have a significant effect on the expression of resistance. Optimal conditions for susceptibility testing vary among strains although some "epidemic" strains are less affected than others. No single set of test conditions is ideal for all resistant strains.

Agent Tested

There is cross-resistance between methicillin, oxacillin and other β-lactam agents with *S. aureus*. Tests with methicillin or oxacillin are most reliable and are representative of all β-lactam agents (Acar *et al.*, 1970; Chambers *et al.*, 1984). In discs, oxacillin is less labile than methicillin (Drew *et al.*, 1972), although this is not a problem if manufacturers' instructions for storage and handling are followed. Methicillin is more resistant to staphylococcal β-lactamases than oxacillin so there are fewer problems with hyper-producers of penicillinase when testing with methicillin. However, methicillin is not currently being manufactured and discs will not be available in the near future.

Media

Susceptible and resistant populations of *S. aureus* are more clearly distinguished on Mueller-Hinton and Columbia media than on IsoSensitest and DST media (Brown and Kothari, 1974). There have been reports of variation in the performance of Mueller-Hinton medium from different manufacturers (Hindler and Inderlied, 1985; Hindler and Warner, 1987) and of different batches of Mueller-Hinton medium from the same manufacturer (Coombs *et al.*, 1996). Differences in Columbia agar from different manufacturers have yet to be investigated. Comparative reports on the performance of different liquid media are not available.

Adding NaCl to Mueller-Hinton, Columbia and DST media is beneficial for detection of resistance (Brown and Yates, 1986, Milne *et al.*, 1987a, 1987b), but with Iso-Sensitest agar such addition is detrimental for some strains (Brown and Yates, 1986). Up to 5% NaCl has been widely used but the effect is dependent on the combination of NaCl concentration, medium, incubation temperature, inoculum density, and test format. The growth of some MRSA is reduced by 5% NaCl, resulting in false sensitive reports. The United States National Committee for Clinical Laboratory Standards (NCCLS, 2000) currently recommend 2% NaCl for dilution tests only, while BSAC recommend 2% for both disc diffusion and dilution tests (Brown, 2001).

Table 2. Oxacillin MICs by agar dilution

Medium	1. Prepare Mueller-Hinton or Columbia agar according to the manufacturer's instructions and add 2% NaCl. 2. Autoclave medium.
Oxacillin stock	1. Obtain standard powder from a chemical supplier or pharmaceutical company. 2. Allowing for the potency of the powder, prepare stock solutions of 10^4, 10^3 and 10^2 mg/L in sterile water. 3. Solutions may be stored at –20°C or below for up to one month but should not be thawed and refrozen.
Antibiotic dilutions	1. Label a series of universal containers from 0.125 to 128 in a twofold series. Also include an antibiotic-free control (0). 2. Use micropipettes to transfer the following volumes of stock solution: 256 µL 10^4 stock to universal labelled 128 128 µL 10^4 stock to universal labelled 64 64 µL 10^4 stock to universal labelled 32 32 µL 10^4 stock to universal labelled 16 160 µL 10^3 stock to universal labelled 8 80 µL 10^3 stock to universal labelled 4 40 µL 10^3 stock to universal labelled 2 200 µL 10^2 stock to universal labelled 1 100 µL 10^2 stock to universal labelled 0.5 50 µL 10^4 stock to universal labelled 0.25 25 µL 10^4 stock to universal labelled 0.125
Pouring plates	1. Cool medium to 50°C and mix well. 2. For each universal in turn, add 20 mL molten L-agar, mix and pour into prelabelled Petri dish. 3. Plates should be stored at 4°C and used within 24h of preparation.
Inoculum	1. Touch at least four colonies and inoculate into Mueller-Hinton or Iso-Sensitest broth. Incubate overnight at 37°C. Alternatively suspend colonies to an even turbidity in broth or water. 2. Adjust the density of the suspension (eg against a McFarland standard) so that a multipoint inoculator delivers an inoculum of 10^4 cfu/spot.
Inoculation	1. Transfer inoculum to the wells of a multipoint inoculator and inoculate plates, including an antibiotic-free control, with 1-2µL spots on the surface of the media. 2. Allow the inoculum spots to dry before incubation of plates.
Incubation	24h at 30°C.

Reading	The MIC is read as the point of complete inhibition of growth. Trailing endpoints or reduced grown (other than a single colony in the spot) should not be ignored.
Interpretation	Susceptible ≤2 mg/L, Resistant ≥4 mg/L
Control strains	1. Include an oxacillin susceptible control strain (*S aureus* ATCC 29213 or NCTC 6571) with each batch of tests. 2. With both strains oxacillin MICs should be within one two-fold dilution step of 0.25 mg/L. 3. *S. aureus* NCTC 12493 may be included as a resistant control.

Inoculum Density

Increasing the inoculum with strains of MRSA which are heterogeneous increases the chances of detecting the minority resistant sub-population, although this also depends on other test conditions. Increasing inoculum will not always lead to increased reliability of testing, especially if other test conditions are optimal for detection of resistance. A heavy inoculum also creates problems: on media containing higher concentrations of NaCl it may make endpoints for susceptible strains in dilution tests very difficult to read and, in disc diffusion tests with higher concentrations of NaCl, it may lead to increased false resistant reports, particularly with oxacillin (Huang *et al.*, 1993).

Incubation

Lower incubation temperature generally improves the detection of resistance, (Annear, 1968; Brown and Kothari, 1974). A few strains of *S. aureus* do not grow well at 30°C, particularly with 5% NaCl in the medium, and this can cause false sensitive results. Resistant sub-populations of some heterogeneous strains grow slowly and resistance is more reliably detected by incubation for 48h, although this may be influenced by other test conditions.

Reading Tests

Trailing endpoints are not uncommon in agar dilution MIC tests, making discrimination of endpoints difficult. In disc diffusion tests, most strains show no zone of inhibition around methicillin 5μg or oxacillin 1μg discs, but resistance may be indicated by reduced zone sizes, zones containing various numbers of different sized colonies, partial inhibition with colonies reducing in size towards the disc, or concentric rings of inhibition of growth around the disc.

Minimum Inhibitory Concentration Determination

MIC by Agar Dilution

A variety of test conditions has been recommended but Mueller-Hinton or Columbia agars with 2% NaCl and an inoculum of 10^4 cfu/mL will distinguish most resistant from susceptible strains (Huang *et al.,* 1993; Watson and Brown, 1998). The method recommended by the BSAC (Andrews, 2001; Brown, 2001) is shown in Table 2.

MIC by Broth Microdilution

The method defined by NCCLS is the only widely used broth dilution method. It involves the use of Mueller-Hinton broth with 2% NaCl, an inoculum of 5 x 10^5 cfu/mL, and incubation at 35°C for 24h (NCCLS, 2000).

MIC by Etest

The Etest (AB Biodisk, Solna, Sweden) is affected by variation in test conditions in a similar way to other MIC methods but has the advantage that it is as easy to set up disc diffusion test. Comparisons of the Etest with dilution and Polymerase Chain Reaction (PCR) methods are generally favourable (Huang *et al.,* 1993; Novak *et al.,* 1993; Petersson *et al.*, 1996; Weller *et al.,* 1997). The manufacturer recommends Mueller-Hinton agar with 2% NaCl, an inoculum in the range of 0.5-1 McFarland standards, inoculation with a swab, and incubation at 35°C for 24h.

Agar Breakpoint Methods

These are similar to agar dilution MIC methods but test only the breakpoint concentration, 2 mg/L oxacillin. Test details are otherwise the same as for agar dilution MICs.

MRSA Susceptibility Screening Methods

These methods have been recommended for screening suspicious colonies isolated from clinical specimens on routine media and confirmation of doubtful resistant strains from disc diffusion tests. The screening method documented by the NCCLS (NCCLS, 2000) involves inoculation of Mueller-Hinton agar containing 4% NaCl and 6 mg/L oxacillin with a spot or a streak of the test

Table 3. Disc diffusion tests for methicillin and oxacillin susceptibility testing

Medium	Columbia agar with 2% NaCl or Mueller-Hinton agar with 2% NaCl.
Inoculum preparation	1. Touch at least four colonies and inoculate into Mueller-Hinton or IsoSensitest broth. Grow organisms for 4 h at 37°C. 2. Alternatively make a suspension of the organisms in water. 3. Adjust the density of the inoculum so that semi-confluent growth is obtained after incubation.
Inoculation	Use a sterile swab to inoculate the surface of the medium evenly with the standardized culture.
Antimicrobial discs	1. Methicillin 5µg or oxacillin 1µg. 2. Place a disc on the surface of the medium, ensuring that there is close contact with the medium.
Incubation	24h at 30°C in air.
Reading	1. Measure zones to the nearest mm. 2. Examine zones carefully in good light to detect colonies, some of which may be very small. 3. Any colonies within zones may be indicative of resistance. 4. Colonies growing within zones may indicate heteroresistance but to exclude the possibility of mixed cultures, colonies within zones should be identified and retested.
Interpretation	Susceptible ≥ 15mm Resistant ≤14mm
Control strains	Include a sensitive control *S. aureus* strain (ATCC 25923 or NCTC 6571) with each batch of tests. Zone diameters should be within the following ranges:

		Columbia	Mueller-Hinton
ATCC 25923	methicillin 5µg	18-27	19-28
	oxacillin 1µg	19-28	20-29
NCTC 6571	methicillin 5µg	18-30	19-30
	oxacillin 1µg	19-30	20-30

S. aureus NCTC 12493 is methicillin resistant and should give no zone with either agent.

organism on a swab dipped in a suspension adjusted to the density of a 0.5 McFarland standard. Plates are incubated at no higher than 35°C for 24h and any growth other than a single colony indicates resistance.

Disc Diffusion

As for agar dilution MICs, there are several different combinations of conditions recommended for disc diffusion. The conditions recommended by the BSAC (Andrews, 2001; Brown, 2001) are shown in Table 3. Hyper-producers of penicillinase may have zones of inhibition <15mm in diameter, whereas most true methicillin/oxacillin resistant isolates give no zone. Resistance may be confirmed by PCR or latex methods. Some hyper-producers of penicillinase give no zone and will therefore be falsely reported as MRSA.

Polymerase Chain Reaction (PCR) Tests for Detecting MRSA

There is a plethora of PCR methods that have been developed to detect methicillin resistance in *S. aureus*. The target comprises the *mecA* gene (Murakami *et al.*, 1991; Bignardi *et al.*, 1996*)* found in almost all current MRSA and methicillin resistant coagulase-negative staphylococci. Borderline resistance, which is not mediated by *mecA*, will not be detected (Bignardi *et al.*, 1996; Araj *et al.*, 1999), but such resistance is currently uncommon and has yet to be shown to be clinically significant. PCR methods are regarded by many as the gold standard, although strains containing *mecA* can rarely be methicillin susceptible and *mecA* positive isolates can also be missed (Mason *et al.*, 2001). Low-cost kits based on PCR methods are not yet available and the method is currently most likely to be used as a reference technique. Indeed, it is interesting that some workers in the field see the test as a screen to augment rather than replace conventional susceptibility testing (Mason *et al.*, 2001). This is not surprising given the complexities of the phenotypic expression of methicillin resistance.

PCR tests have also been developed to identify *S. aureus*. Early tests required the southern blotting of products to confirm their identity (Greisen *et al.*, 1994). They have been replaced with more specific multiplex PCR methods that combine the *mecA* gene detection with a *S. aureus* species-specific target gene. These genes have included nuclease (*nuc*), coagulase (*coa*), protein A (*spa*), surface-associated fibrinogen-binding protein (Mason *et al.*, 2001), an undefined DNA fragment from the *S. aureus* chromosome, Sa442 (Martineau *et al.*, 1998) and *femB* (Jonas *et al.*, 1999). The original tests were used on possible MRSA colonies taken from culture media, but several workers have now applied these to clinical samples with promising results. These have included synovial fluid (Mariani *et al.*, 1996) and blood cultures (Mason *et al.*, 2001; Thean *et al.*, 2001). The use of the faster PCR amplification methods has made it possible to identify MRSA even more rapidly (Thean *et al.*, 2001). The introduction of multiplex-PCR machines may even enable the detection of mixtures of methicillin-resistant coagulase-negative staphylococci and methicillin-sensitive *S. aureus*. Many other approaches are being assessed (e.g. Oliveira *et al.*, 2002) and the field is moving very rapidly. Several biotechnology companies are working on microarray (chip) systems to make the approach more practical.

Automated Methods for Detecting Methicillin/Oxacillin Resistance

Automated systems such as the Vitek (bioMérieux), Phoenix (Becton Dickinson) and Microscan (Dade Behring) include tests for methicillin/oxacillin susceptibility and are generally reported to be reliable for testing methicillin susceptibility of *S. aureus*, although problems have been reported with some strains (Frebourg *et al.*, 1998; Ribero *et al.*, 1999).

Other Methods for Detecting Methicillin/Oxacillin Resistance

Same-day results may be provided by the Crystal MRSA method (Becton Dickinson), which is based on the quenching of fluorescence of an oxygen-sensitive fluorescent indicator by oxygen remaining in broth when growth of an organism is inhibited by oxacillin. However, this method still requires several hours of incubation (Knapp *et al.*, 1994; Wallet *et al.*, 1996).

A slide latex agglutination test based on detection of PBP2a has been developed (Nakatomi and Sugiyama, 1998) and is commercially available (Mast, Oxoid). In this method, PBP2a is extracted from colonies and detected by agglutination with latex particles coated with monoclonal antibodies to PBP2a. The method is rapid (ten minutes) and is highly sensitive and specific for *S. aureus* grown on blood agar (Van Griethuysen *et al.*, 1999), but may not be reliable for colonies grown on screening media containing NaCl (Brown and Walpole, 2001).

References

Acar, J.F., Courvalin, P., and Chabbert, Y.A. 1970. Methicillin-resistant staphylococci: bacteriological failure of treatment with cephalosporins. Antimicrob. Agents Chemother. 1: 280-285.

Allen, J.L., Cowan, M.E., and Cockroft, P.M. 1994. A comparison of three semi-selective media for the isolation of methicillin-resistant *Staphylococcus aureus*. J. Med. Microbiol. 40: 2: 98-101.

Andrews, J.M. 2001. Determination of minimum inhibitory concentrations. J. Antimicrob. Chemother. Suppl. S1. 48: 516.

Andrews, J.M. for the BSAC Working Party on Susceptibility Testing. 2001. BSAC standardized disc susceptibility testing method. J. Antimicrob. Chemother. Suppl. S1: 43-57.

Annear, D.I. 1968. The effect of temperature on resistance of *Staphylococcus aureus* to methicillin and some other antibiotics. Med. J. Australia. 1: 444-446.

Araj, G.F., Talhouk, R.S., Simaan, C.J., and Maasad, M.J. 1999. Discrepancies between *mecA* PCR and conventional tests used for detection of methicillin-resistant *Staphylococcus aureus*. Int. J. Antimicrob. Agents. 11: 47-52.

Ayliffe, G.A.J., Buckles, A., Casewell, M.S., Cookson, B.D., Cox, A., Duckworth, G.J., French, G.L., Griffith-Jones, A., Heathcock, R., Humphreys, H., Keane, C.T., Marples, A.A., Shanson, D.C., Slack, R., and Tebbs E. 1998. Revised guidelines for the control of methicillin-resistant *Staphylococcus aureus* infections in hospitals. Report of a combined working party of the British Society of Antimicrobial Chemotherapy, the Hospital Infection Society and the Infection Control Nurses' Association. J. Hosp. Infect. 39: 253-290.

Barber, M. 1964. Naturally occurring methicillin-resistant staphylococci. J. Gen. Microbiol. 35: 183-90.

Barry, A.L., Lachica, R.V.F. and Atchison, F.W. 1973. Identification of *Staphylococcus aureus* by simultaneous use of tube coagulase and thermostable nuclease tests. Appl. Microbiol. 25: 496-497.

Bignardi, G.E., Woodford, N., Chapman, A., Johnson, A.P., and Speller, D.C.E. 1996. Detection of the *mec-A* gene and phenotypic detection of resistance in *Staphylococcus aureus* isolates with borderline or low-level methicillin resistance. J. Antimicrob Chemother. 37: 53-63.

Brown, D.F.J., and Kothari, D. 1974. The reliability of methicillin sensitivity tests on four culture media. J. Clin. Pathol. 27: 420-426.

Brown, D.F.J., and Yates, V.S. 1986. Methicillin susceptibility testing of *Staphylococcus aureus* on media containing five percent sodium chloride. European J. Clin. Microbiol. Infect. Diseases. 5: 726-728.

Brown, D.F.J., and Walpole, E. 2001. Evaluation of the Mastalex latex agglutination test for methicillin resistance in *Staphylococcus aureus* on different screening media. J. Antimicrob. Chemother. 47: 187-189.

Brown, D.F.J. 2001. Detection of methicillin/oxacillin resistance in staphylococci. J. Antimicrob. Chemother. 48, Suppl. S1: 65-70.

Chambers, H.F., Hackbarth, C.J., Drake, T.A., Rusnack, M.G., and Sande, M.A. 1984. Endocarditis due to methicillin-resistant *Staphylococcus aureus* in rabbits: Expression of resistance to beta-lactam antibiotics *in vitro* and *in vivo*. J. Infect. Diseases. 149: 894-903.

Chambers, H.F., Archer, G., and Matsuhashi, M. 1989. Low-level methicillin resistance in strains of *Staphylococcus aureus*. Antimicrob. Agents Chemother. 33: 424-8.

Cookson, B.D. 1997. *Staphylococcus aureus*. In: Principles in Clinical Bacteriology. Emmerson, M., Kibbler, C., Hawkey, P., eds. John Wiley, Oxford. p.109-130.

Cookson, B.D., and Phillips, I. 1990. Methicillin-resistant staphylococci. J. Appl. Bacteriol. 69: Suppl 19: 55S-70S.

Cookson, B.D. 2000. Methicillin-resistant *Staphylococcus aureus* in the community: New battlefronts, or are the battles lost? Infect. Control Hosp. Epidemiol. 21: 398-403.

Coombs, G.W., Pearman, J.W., Khinsoe, C.H., and Boehm, J.D. 1996. Problems in detecting low-expression-class methicillin resistance in *Staphylococcus aureus* with batches of Oxoid Mueller-Hinton agar. J. Antimicrob. Chemother. 38: 551-555.

Cooper, B.S. 2000. The transmission dynamics of methicillin-resistant *Staphylococcus aureus* and vancomycin-resistant enterococci in hospital wards. PhD thesis, University of Warwick, UK.

Davies, S., and Zadik, P.M. 1997. Comparison of methods for the isolation of methicillin-resistant *Staphylococcus aureus*. J. Clin. Pathol. 50: 257-258.

de Lencastre, H., Sa Figueiredo, A.M., Urban, C., Rahal, J., and Tomasz, A. 1991. Multiple mechanisms of methicillin resistance and improved methods of detection in clinical isolates of *Staphylococcus aureus*. Antimicrob. Agents Chemother. 35: 632-639.

de Lencastre, H., and Tomasz, A. 1994. Reassessment of the number of auxillary genes essential for expression of high-level methicillin resistance in *Staphylococcus aureus*. Antimicrob. Agents Chemother. 38: 2590-2598.

Drew, W.L., Barry, A.L., O'Toole, R., and Sherris, J.C. 1972. Reliability of the Kirby-Bauer disc diffusion method for detecting methicillin-resistant strains of *Staphylococcus aureus*. Appl. Microbiol. 24: 240-247.

Faruki, H., and Murray, P. 1986. Medium dependence for rapid detection of thermonuclease in blood culture broths. J. Clin. Microbiol. 24: 482-483.

Frebourg, N.B., Nouet, D., Lemee, L., Martin, E., and Lemeland, J.F. 1998. Comparison of ATB staph, rapid ATB staph, Vitek, and E-test methods for detection of oxacillin heteroresistance in staphylococci possessing *mecA*. J. Clin. Microbiol. 36: 52-7.

Greisen, M.K., Loeffelholz, M., Purohit, A., and Leong, D. 1994. PCR primers and probes for the 16S rRNA gene of most species of pathogenic bacteria, including bacteria found in cerebrospinal fluid. J. Clin. Microbiol. 32:335-351.

Hartman, B.J., and Tomasz, A. 1984. Low-affinity penicillin-binding protein associated with β-lactam-resistance in *Staphylococcus aureus*. J. Bacteriol. 158: 513-516.

Hewitt, J., Coe, A.W., and Parker, M.T. 1969. The detection of methicillin resistance in *Staphylococcus aureus*. J. Med. Microbiol. 2: 443-456.

Hindler, J.A., and Inderlied, C.B. 1985. Effect of the source of Mueller-Hinton agar and resistance frequency on the detection of methicillin-resistant *Staphylococcus aureus*. J. Clin. Microbiol. 21: 205-210.

Hindler, J.A., and Warner, N.L. 1987. Effect of the source of Mueller-Hinton agar on the detection of oxacillin resistance in *Staphylococcus aureus* using a screening methodology. J. Clin. Microbiol. 25: 734-735.

Huang, M.B., Gay, T.E., Baker, C.N., Banerjee, S.N., and Tenover, F.C. 1993. Two percent sodium chloride is required for susceptibility testing of staphylococci with oxacillin when using agar-based dilution methods. J. Clin. Microbiol. 31: 2683-2688.

Jevons, M.P. 1961. "Celbenin"-resistant staphylococci. B. Med. J. i, 124-125.

Jonas, D., Grundmann, H., Hartung, D., Daschner, F.D., and Towner, K.J. 1999. Evaluation of the *mecA fem B* duplex polymerase chain reactions for detection of methicillin-resistant *Staphylococcus aureus*. Eur. J. Clin. Microbiol. Infect. Dis. 18: 643-647.

Jones, E.M., Bowker, K.E., Cooke ,R,. Marshall, R.J., Reeves, D.S., and MacGowan A.P. 1997. Salt tolerance of EMRSA-16 and its effect on the sensitivity of screening cultures. J. Hosp. Infect. 35: 59-62.

Knapp, C.C., Ludwig, M.D., and Washington, J. 1994. Evaluation of BBL Crystal MRSA ID system. J. Clin. Microbiol. 32: 2588-2589.

Knox, R. 1961. "Celbenin"-resistant staphylococci. Br. Med. J. i: 126.

Kunori, T., Cookson, B., Kibbler, C., Stone, S., and Roberts, J. 2002. The cost effectiveness of different MRSA screening methods. J. Hosp. Infect. 51:189-200.

Kuusela, P., Hilden, P., Savolainen, K., Vuento, M., Lyytikaim, O., and Vuopio-Varkila, J. 1994. Rapid detection of methicillin-resistant *Staphylococcus aureus* strains not identified by slide agglutination tests. J. Clin. Microbiol. 32: 144-147.

Labischinski, H., Ehlert, K., and Berger-Bächi, B. 1998. The targeting of factors necessary for expression of methicillin resistance in staphylococci. J. Antimicrob. Chemother. 41: 581-584.

Lally, R.T., Ederer, M.N., and Woolfrey, B.F. 1985. Evaluation of mannitol salt agar with oxacillin as a screening medium for methicillin-resistant *Staphylococcus aureus*. J. Clin. Microbiol. 22: 501-504.

La Zonby, J.G., and Styarzyk, M.J. 1986. Screening method for recovery of methicillin-resistant *Staphylococcus aureus* from primary plates. J. Clin. Microbiol. 24: 186-188.

Madison, B.M. and Baselski, V.S. 1983. Rapid identification of *Staphylococcus aureus* in blood cultures by thermonuclease testing. J. Clin. Microbiol. 18:502-5.

Mariani, B.D., Martin, D.S., Levine, M.J., Booth Jr., R.E., and Tuan RS. 1996. The Coventry Award. Polymerase chain reaction detection of bacterial infection in total knee arthroplasty. Clin. Orthop. 331: 11-22.

Martineau, F., Picard, F.J., Roy, P.H., Ouellette, M., and Bergeron, M.G.1998. Species-specific and ubiquitous-DNA-based assays for rapid identification of *Staphylococcus aureus*. J. Clin. Microbiol. 36: 618–623.

Martinez, O.V., Cleary, T., Baker, M., and Civetta, J. 1992. Evaluation of a mannitol-salt-oxacillin-tellurite medium for the isolation of methicillin-resistant *Staphylococcus aureus* from contaminated sources. Diag. Microbiol. Infect. Dis.15: 207-211.

Mason, W.J., Blevins, J.S., Beenken, K., Wibowo, N., Ojha, N., and Smeltzer, M.S. 2001. Multiplex PCR protocol for the diagnosis of staphylococcal infection. J. Clin. Microbiol. 39: 3332-3338.

Matsuhashi, M., Song, M.D., Ishino, F., Wachi, M., Doi, M., Inoue, M., Ubukata K, Yamashita N, and Konno M. 1986. Molecular cloning of the gene of a penicillin-binding protein supposed to cause high resistance to beta-lactam antibiotics in *Staphylococcus aureus*. J. Bacteriol. 167: 975-980.

McDougal, L., and Thornsberry, C. 1986. The role of β-lactamase in staphylococcal resistance to penicillinase-resistant penicillins and cephalosporins. J. Clin. Microbiol. 23: 832-839.

Merlino, J., Gill, R., and Robertson, G.J. 1996. Application of lipovitellin-salt-mannitol agar for screening, isolation and presumptive identification of *Staphylococcus aureus* in a teaching hospital. J. Clin. Microbiol. 34: 3012-3015.

Milne, L.M., Curtis, G.D.W., Crow, M., Kraak, W.A.G., and Selkon, J.B. 1987a. Comparison of culture media for detecting methicillin resistance in *Staphylococcus aureus* and coagulase-negative staphylococci. J. Clin. Pathol. 40: 1178-1181.

Milne, L.M., Curtis, G.D.W., Crow, M., Kraak, W.A.G., and Selkon, J.B. 1987b. Effects of culture media on detection of methicillin resistance in *Staphylococcus aureus* and coagulase negative staphylococci by disc diffusion methods. J. Clin. Pathol. 46: 394-397.

Montanari, M.P., Tonin, E., Biavasco, F., and Varaldo, P.E. 1990. Further characterization of borderline methicillin-resistant *Staphylococcus aureus* and analysis of penicillin-binding proteins. Antimicrob. Agents and Chemother. 34: 911-913.

Morton, C.E.G., and Holt, H.A. 1989. A problem encountered using Staphylococcus/Streptococcus supplement. Med. Lab. Sci. 46: 72-73.

Murakami, K., Minamide, W., Wada, K., Nakamura, E., Teraoka, H., and Watanabe, S. 1991. Identification of methicillin-resistant strains of staphylococci by polymerase chain reaction. J. Clin. Microbiol. 29: 2240-2244.

Nakatomi, Y., and Sugiyama, J. 1998. A rapid latex agglutination assay for the detection of penicillin-binding protein 2'. Microbiol. Immunol. 42: 739-743.

National Committee for Clinical Laboratory Standards 2000. Methods for dilution antimicrobial susceptibility tests for bacteria that grow aerobically – Fifth Ed. Approved Standard M7-A4. NCCLS, Wayne, PA.

Novak, S.M., Hindler, J., and Bruckner, D.A. 1993. Reliability of two novel methods, Alamar and E Test, for detection of methicillin-resistant *Staphylococcus aureus*. J. Clin. Microbiol. 31: 3056-3057.

Oliveira, K., Procop, G.W., Wilson, D., Coull, J., and Stender1, H. 2002. Rapid identification of *Staphylococcus aureus* directly from blood cultures by fluorescence *in situ* hybridization with peptide nucleic acid probes. J Clin Microbiol. 40: 247-251.

Petersson, A.C., Mirner, H., and Kamme, C. 1996. Identification of *mecA*-related oxacillin resistance in staphylococci by the Etest and the broth microdilution method. J. Antimicrob. Chemother. 37: 445-456.

Reynolds, P.E., and Brown, D.F.J. 1985. Penicillin-binding proteins of β-lactam-resistant strains of *Staphylococcus aureus*. Effect of growth conditions. FEBS Letters. 192: 28-32.

Ribero, J., Vieira, F.D., King, T., D'Arezzo, J.B., and Boyce, J.M. 1999. Misclassification of susceptible strains of *Staphylococcus aureus* as methicillin-resistant *S. aureus* by a rapid automated susceptibility testing system. J. Clin. Microbiol. 37: 1619-1620.

Sauter, R.L., Brown, W.J., and Mattman, L.H. 1988. The use of a selective staphylococcal broth v direct plating for the recovery of *Staphylococcus aureus*. Infect. Control Hosp. Epidemiol. 9: 204-205.

Sutherland, R., and Rolinson, G.N. 1964. Characteristics of methicillin-resistant staphylococci. J. Bacteriol. 87: 887-899.

Thauvin-Eliopoulos, C., Rice, L.B., Elioupolos, G.M., and Moellering, R.C. 1990. Efficacy of oxacillin and ampicillin-sulbactam combination in experimental endocarditis caused by β-lactamase-hyperproducing *S. aureus*. Antimicrob. Agents Chemother. 34: 728-732.

Thean, Y.T., Corden, S., Barnes, S. R., and Cookson, B. 2001. Rapid identification of methicillin-resistant *Staphylococcus aureus* from positive blood cultures using real-time fluorescence PCR. J. Clin. Microbiol. 39: 4529-4531.

Tomasz, A., Nachman, S., and Leaf, H. 1991. Stable classes of phenotypic expression in methicillin-resistant clinical isolates of staphylococci. Antimicrob. Agents Chemother. 35: 124-129.

Van Enk, R.A., and Thompson, K.D. 1992. Use of a primary isolation medium for recovery of methicillin-resistant *Staphylococcus aureus*. J. Clin. Microbiol. 30: 504-505.

Van Griethuysen, A., Pouw, M., van Leeuwen, N., Heck, M., Willemse, P., Buiting, A., and Kluytmans, J. 1999. Rapid slide latex agglutination test for detection of methicillin resistance in *Staphylococcus aureus*. .J. Clin. Microbiol. 37: 2789-2792.

Wagenvoort , J.H.T., Werink, T.J., Gronenschild, J.M.H., and Davies, B.I. 1996. Optimization of detection and yield of methicillin-resistant *Staphylococcus aureus*. Infect. Control Hosp. Epidemiol. 17: 208-209.

Wallet, F., Roussel-Delvallez, M., and Courcol, R.J. 1996. Choice of a routine method for detecting methicillin-resistance in staphylococci. J. Antimicrob. Chemother. 37: 901-909.

Watson, A., and Brown, D.F.J. 1998. Methods for determining susceptibility of *Staphylococcus aureus* to methicillin and oxacillin. In: Abstracts of the

38th Interscience Conference on Antimicrobial Agents and Chemotherapy, San Diego, CA, 1998. Abstr. D-53. Am. Soc. Microbiol. Washington, DC. p. 144.

Weller, T.M.A., Crook, D.W., Crow, M.R., Ibrahim, W., Pennington, T.H., and Selkon, J.B. 1997. Methicillin susceptibility testing of staphylococci by Etest and comparison with agar dilution and *mecA* detection. J. Antimicrob. Chemother. 39: 251-253.

Wichelhaus, T.A. , Kern, S., Schafer, V., Brade, V., and Hunfeld, K.-P. 1999. Evaluation of modern agglutination tests for identification of methicillin-susceptible and methicillin-resistant *Staphylococcus aureus*. Eur. J. Clin. Microbiol. Infect. Dis.18: 756-758.

Wood, W., Harvey, G., Plsen, E.S., and Reid, T.M.S. 1993. Aztreonam selective agar for Gram positive bacteria. J. Clin. Pathol. 46: 769-771.

From: *MRSA: Current Perspectives*
Edited by: A.C. Fluit and F.-J. Schmitz

Chapter 3

Mechanisms of Methicillin Resistance

Susanne Rohrer, Markus Bischoff, Jutta Rossi, and Brigitte Berger-Bächi

Abstract

Methicillin-resistant staphylococci have been on the rise since shortly after the introduction of penicillinase-resistant β-lactams in 1960. The methicillin resistance is due to the acquisition of the low-affinity penicillin binding protein, PBP2', encoded by *mecA*. This gene resides on a large, chromosomally inserted element termed SCC*mec* (staphylococcal cassette chromosome *mec*) from an extraspecies source. It has been possible to trace the evolution and dissemination of methicillin resistance within the genus *Staphylococcus*, but the origin of SCC*mec* is still unclear. Although *mecA* is the sole prerequisite for methicillin resistance, the resulting phenotype is highly complex. A characteristic of methicillin resistance is its heterogeneous expression – highly resistant subclones arising under antibiotic pressure at a frequency that lies above that of spontaneous mutation. This phenomenon has puzzled clinicians and researchers for the past 30 years. Although different approaches have yielded a great number of genetic factors that influence the expression of methicillin resistance and have given us insight into a wide variety of processes in the staphylococcal cell, the underlying mechanism causing heteroresistance has not been found. It is to be hoped that modern techniques enabling the analysis of cellular metabolism on a global scale will unravel the "mystery" of heterogeneous resistance to methicillin.

Introduction

Staphylococci are a major cause of both nosocomial and community-acquired infections (Diekema *et al.*, 2001). Strains that have acquired methicillin resistance pose the greatest problems for treatment and eradication. Methicillin resistance in staphylococci is an intrinsic resistance to virtually all β-lactams and their derivatives, including cephalosporins and carbapenems. The first methicillin-resistant *Staphylococcus aureus* (MRSA) clinical isolates were reported shortly after the introduction of methicillin, the first semisynthetic penicillinase resistant β-lactam, into clinical use (Jevons, 1961). Epidemics that occurred in European hospitals in the early 1960s were of clonal origin. Since then, MRSA and methicillin-resistant coagulase-negative staphylococci have spread worldwide and have diversified. MRSA are mainly a problem in hospital and tertiary care home settings. Under antibiotic pressure, they have proven efficient in acquiring multiple resistance determinants to other, non-related antibiotics. The incidence of methicillin resistance in staphylococci is rising, but varies greatly depending on the country, region, and hospital. The recent emergence of community–acquired MRSA (cMRSA) has added a new dimension to the dissemination of MRSA (Chambers, 2001). Some of the characteristics of cMRSA resemble those of early MRSA, with a relatively low level of methicillin resistance and few or no additional resistance determinants. The increasing rate of methicillin resistance among staphylococci, the changing epidemiology of MRSA, and the imminent threat of vancomycin-resistant strains calls for a global control of multiple resistant staphylococci.

The Genetic Basis of Methicillin Resistance

SCC*mec*, A Mosaic, Mobile "Resistance Island"

Methicillin resistance in staphylococci is due to the acquisition of a large DNA element that ranges from 20 to more than 100 kb in size, termed staphylococcal cassette chromosome *mec* (SCC*mec*) (Katayama *et al.*, 2000) (Figure 1), which integrates site- and orientation-specifically close to the origin of replication of the *S. aureus* chromosome. A characteristic feature of the SCC*mec* element are terminal inverted and direct repeats, reminiscent of the mechanism of cassette gene insertion of integrons. Integration occurs within the 3' end of

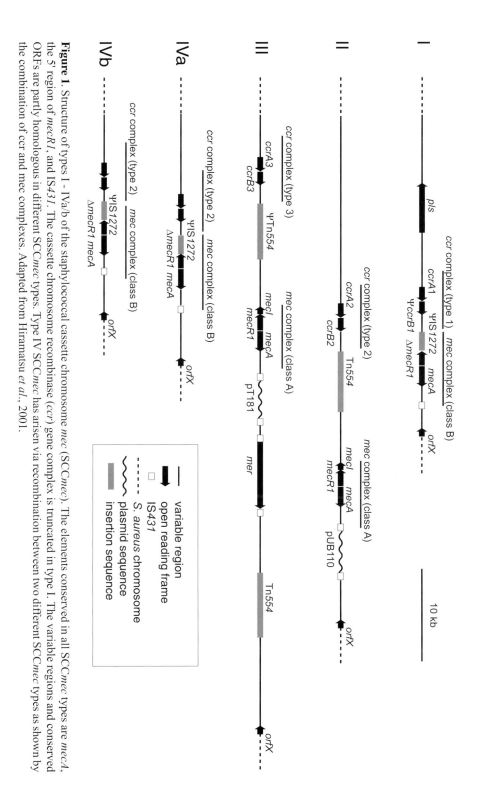

Figure 1. Structure of types I - IVa/b of the staphylococcal cassette chromosome *mec* (SCC*mec*). The elements conserved in all SCC*mec* types are *mecA*, the 5' region of *mecR1*, and IS*431*. The cassette chromosome recombinase (*ccr*) gene complex is truncated in type I. The variable regions and conserved ORFs are partly homologous in different SCC*mec* types. Type IV SCC*mec* has arisen via recombination between two different SCC*mec* types as shown by the combination of ccr and mec complexes. Adapted from Hiramatsu *et al.*, 2001.

orfX, a gene of unknown function, while preserving the integrity of the ORF (Ito *et al.*, 1999). The ubiquitous presence of *orfX* in over 50 *S. aureus* sequenced has led to the hypothesis that it may have an important function. SCC*mec* contains a set of recombinase genes, *ccrA* and *ccrB*, of the invertase/resolvase family, that catalyze the precise excision and site-specific integration of SCC*mec* (Katayama *et al.*, 2000). The gene responsible for methicillin resistance is *mecA*, which codes for a penicillin-binding protein termed PBP2' (synonym PBP2a). At least one copy of the insertion element IS*431* (synonym IS*257*) is located downstream of *mecA* within SCC*mec* and acts as a receptor site for the integration of IS*431*-associated plasmids and transposons by homologous recombination, leading to the accumulation of resistance determinants in this element. Attachment sites for transposons, such as Tn*554*, promote the capturing of additional resistance genes. Since the precise excision of SCC*mec* and site- and orientation-specific insertion into the chromosome has been demonstrated (Katayama *et al.*, 2000), SCC*mec* is considered a novel type of mobile element that resembles neither transposons nor bacteriophages, and has been termed a "resistance island" in analogy to pathogenicity islands (Ito *et al.*, 1999).

Three distinct types of hospital-acquired SCC*mec* have been identified, which are characterized by different alleles of each of the *ccr* genes (Ito *et al.*, 2001). Type I SCC*mec*, which disseminated among the early MRSA of the 1960s, carries only the methicillin resistance determinant (*mec* complex). Type II and type III SCC*mec*, which became predominant in the 1980s, carry multiple additional antibiotic resistance genes. cMRSA contain one of two variants of the type IV SCC*mec*, with methicillin resistance as the only resistance determinant. At 20-24 kb, type IV SCC*mec* is considerably shorter than the hospital-acquired SCC*mec* types, and arose by recombination between the *ccr* and *mec* gene complexes of two different types of SCC*mec*. Whether its smaller size improves mobility remains to be tested.

While the *mecA* gene remained intact, many of the genes localized on the various SCC*mec* are incomplete or disrupted, and may have originated from housekeeping genes of a bacterial chromosome. An example of an intact gene is *pls*, which occurs only in type I SCC*mec*. The gene codes for a plasmin-sensitive cell surface protein and reduces adhesion to fibronectin, fibrinogen and IgG in its uncleaved form (Hilden *et al.*, 1996).

As judged by the GC content, it is assumed that SCC*mec* is of non-staphylococcal origin. Methicillin resistance has found its way into at least five genetically distinct lineages of *S. aureus*, demonstrating that methicillin resistant strains have evolved on several independent occasions (Fitzgerald *et al.*, 2001). For yet unknown reasons, coagulase-negative staphylococci have a higher incidence of methicillin resistance than *S. aureus*. They carry the same SCC*mec* elements as MRSA, which has led to the suggestion that they are the reservoir for methicillin resistance in *S. aureus* (Archer and Niemeyer, 1994). This hypothesis is supported by the recent observation that a methicillin-resistant coagulase-negative strain was the source for transfer of methicillin resistance in a patient with an initially methicillin-susceptible *S. aureus* infection (Wielders *et al.*, 2001). Comparison of the *mecA* gene and its framing sequences provides additional evidence for transfer across species barriers (Katayama *et al.*, 2001). An unusually high *mecA* carrier rate was recently found in coagulase negative staphylococci isolated from frogs, suggesting that amphibians may be a reservoir for methicillin resistance (Slaughter *et al.*, 2001); however, an intermediate reservoir or vector that humans frequently come in contact with would need to be identified.

The closest homologue to the *mecA* antibiotic resistance gene is the *Staphylococcus sciuri sscA* gene. SscA does not confer methicillin resistance upon its host, but exposure of *S. sciuri* to increasing concentrations of methicillin selects for methicillin resistance and overexpression of *sscA* due to a mutation in the promoter region. When transferred into *S. aureus* on a plasmid, the overexpressed *sscA* gene increases resistance of the recipient to methicillin from three- to more than tenfold (Wu *et al.*, 2001).

The stability of SCC*mec* is variable, presumably depending on the SCC*mec* type, the number of insertion sequences and additional resistance cassettes present, the genetic background, and external conditions. This is exemplified by several lines of evidence. While in a collection of early MRSA of the 1960s, methicillin resistance was stably maintained, multiresistant MRSA strains of the 1980s more readily lost the *mec* determinant during storage (Hürlimann-Dalel *et al.*, 1992; Poston and Li Saw Hee, 1991; Wada *et al.*, 1991). Stress conditions such as UV irradiation, starvation, and elevated temperature, can induce deletions in the *mec* region resulting in methicillin susceptibility (Inglis *et al.*, 1990). Loss of methicillin resistance *in vivo* has been observed in MRSA isolated from two patients (Deplano *et al.*, 2000).

The Physiological Role of PBP2'

Synthesis of the low-affinity penicillin-binding protein PBP2' is sufficient for methicillin resistance. Its action is closely linked to the endogenous machinery of cell wall synthesis. Typical for the staphylococcal peptidoglycan is the high degree of cross-linkage, enabled by the long and flexible pentaglycine side chain which branches off the lysine of the peptidoglycan stem peptide (Labischinski, 1992) (Figure 2). The transpeptidation reaction is catalysed by penicillin-binding proteins (PBPs). *S. aureus* possesses three high molecular mass PBPs, PBP1 to 3, and one low molecular mass PBP, PBP4, which has a secondary transpeptidase activity (Wyke *et al.*, 1981). While the transglycosylases of *S. aureus* have been shown to be monofunctional proteins without penicillin-binding activity (Park and Matsuhashi, 1984), a

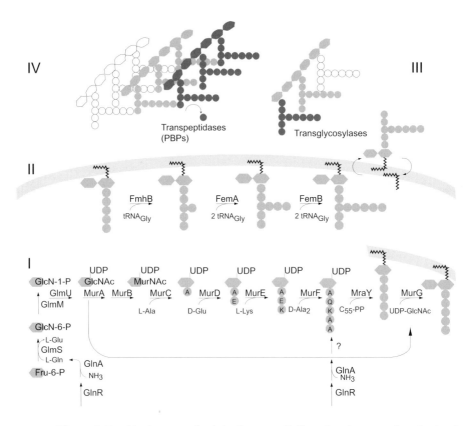

Figure 2. Peptidoglycan synthesis in *S. aureus*. I) Cytoplasmic steps of synthesis of the bactoprenol-linked disaccharide-pentapeptide precursor (lipid II). II) Synthesis of the pentaglycine interpeptide by FmhB, FemA and FemB using tRNA$_{Gly}$ as glycine donor. III) Transglycosylation on the outer face of the cytoplasmic membrane. IV) Transpeptidation by penicillin-binding proteins (PBPs).

transglycosylase-like activity was also demonstrated for PBP2 (Pinho *et al.*, 2001b). This transglycosylase domain of PBP2 is required by PBP2' for expression of methicillin resistance (Pinho *et al.*, 2001a). This study confirms the hypothesis that high molecular mass PBPs of class B1, such as PBP2', which have a low affinity for β-lactams, perform basic functions required for cell wall assembly in conjunction with a monofunctional transglycosylase or the transglycosylase module of a class A PBP (Goffin and Ghuysen, 1998).

Although PBP2' was shown to have transpeptidase activity (Gaisford and Reynolds, 1989; Pinho *et al.*, 2001b), it does not seem to affect cell wall composition and cross-linking in absence of β-lactams (de Jonge *et al.*, 1992) and seems to be a poor transpeptidase; since in the presence of otherwise inhibitory concentrations of β-lactams it is only able to form muropeptide dimers (de Jonge and Tomasz, 1993).

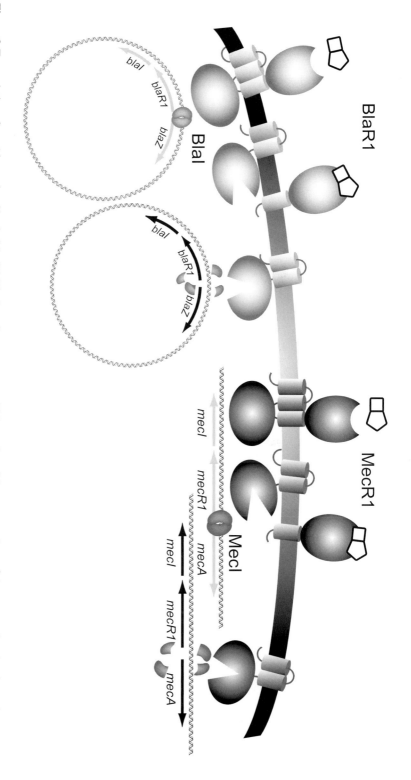

Figure 3. Transcription of *mecA* is regulated by the MecR1-MecI regulatory system, which consists of the transmembrane β-lactam-sensing signal transducer, MecR1, and the repressor, MecI. Plasmid-encoded β-lactamases are regulated by the analogous BlaR1-BlaI system. Upon induction, the sensor-transducer is autocatalytically cleaved and a metalloprotease domain located in the cytoplasmic part of the protein is activated. This protease then cleaves the repressor, allowing transcription of both *mecA* and *mecR1-mecI*, or *blaZ* and *blaR1-blaI*, respectively.

Regulation of *mecA* by *mecR1-mecI*

Transcription of *mecA* is regulated by the divergently transcribed *mecR1-mecI* regulatory region (Figure 3), which codes for a transmembrane β-lactam-sensing signal transducer, MecR1, and a repressor, MecI (Hiramatsu *et al.*, 1992). In the absence of an inducer, MecI represses transcription of *mecA* as well as *mecR1-mecI*. The close sequence similarity between the *mec* element and the regulatory elements of staphylococcal penicillinases (Song *et al.*, 1987) also allows control of *mecA* by the penicillinase repressor BlaI. Upon induction, the sensor-transducer is autocatalytically cleaved, thereby activating a metalloprotease domain located in the cytoplasmic part of the protein (Zhang *et al.*, 2001). This protease then cleaves the repressor either directly or via yet unknown intermediates, allowing transcription of both *mecA* and *mecR1-mecI*. While the *mecA* operator is recognized by either MecI or BlaI, cleavage of the repressor molecules is only possible by their cognate sensor-transducers (Lewis and Dyke, 2000; McKinney *et al.*, 2001). Induction by the *bla* system is more rapid than by the *mec* system (Clarke and Dyke, 2001). Since methicillin and oxacillin, which are routinely used in susceptibility tests, are not recognized as inducers by MecR1, PBP2' production in presence of these drugs is too low to confer resistance in strains carrying *mecR1-mecI*. MRSA with an intact *mec* regulatory region are therefore often phenotypically susceptible despite the presence of SCC*mec*. It has been proposed that additional factors may be involved in the induction pathway, since mutations in a yet unidentified chromosomal locus, termed *blaR2*, confers microconstitutive synthesis of inducible penicillinases (Cohen and Sweeney, 1968).

The regulatory elements of the *mec* complex are often truncated by insertion of IS*431* or ψIS*1272* or inactivated by mutations (Suzuki *et al.*, 1993), resulting in derepression of *mecA* (Katayama *et al.*, 2001). Surprisingly, some strains with an apparently intact regulator region can express high and homogenous resistance. A biochemical explanation for this behaviour has not been found so far, but constitutive cleavage of the repressor is a possible mechanism.

Heterogeneous Expression of Methicillin Resistance

A characteristic of methicillin resistance is its usually heterogeneous expression. The minimal inhibitory concentrations (MIC) of methicillin can vary widely, from values below the susceptibility breakpoint to greater than 1000 µg/ml, depending on which genetic background a specific SCC*mec* element has integrated into. Growth in the presence of β-lactams selects resistant subclones from an MRSA population with low methicillin MICs. The frequency at which these subclones arise is a reproducible, strain-specific characteristic (Tomasz *et al.*, 1991) and usually lies clearly above the rate of spontaneous mutation due to a yet unidentified mechanism, which is not likely a mutator phenotype

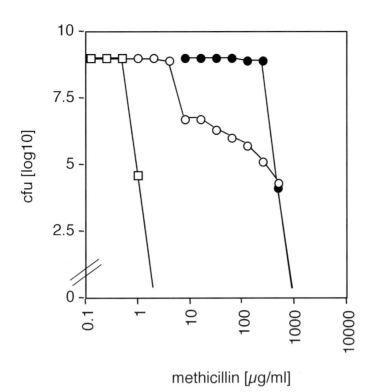

Figure 4. Methicillin resistance profile of *S. aureus* strains. Overnight cultures were plated on increasing concentrations of methicillin. Squares, susceptible strain BB255; circles, heterogenously methicillin-resistant MRSA strain BB270; full circles, homogeneously, highly methicillin-resistant MRSA strain COL. Both MRSA strains are constitutive PBP2' producers.

(Finan *et al.*, 2002). With few exceptions (de Lencastre *et al.*, 1993), once high-level resistance has been selected, it remains high (Finan *et al.*, 2002).

Although a minimal amount of PBP2' is needed for expression of the resistance, constitutive, high level expression of PBP2' does not necessarily correlate with high, homogeneous resistance. Strain pairs with essentially identical *mecA* sequences and high PBP2' production can vary considerably with respect to their methicillin resistance levels as shown in Figure 4. The presence of unrelated genetic factors such as a β-lactamase in a homogeneously and highly resistant MRSA clone can lead to heterogeneous resistance (Boyce *et al.*, 1990; Hackbarth *et al.*, 1994). Methicillin resistance levels depend strongly on external conditions such as temperature, osmolarity, availability of divalent cations, oxygen pressure, and light (Matthews and Stewart, 1984), which suggests that specific regulatory systems control the resistance mechanism.

Mutations Promoting High Level Resistance

The events leading to high resistance have not yet been satisfactorily described. Genetic analysis has shown that chromosomal loci unlinked to SCC*mec*, termed *chr** , are involved in the switch to high level methicillin resistance (Ryffel *et al.*, 1994). Only a few mutations, accounting for just a fraction of the highly resistant subclones, have been identified. This suggests that many pathways lead to high resistance, or that an unknown global regulator controls its expression.

Two genes, termed *hmrA* and *hmrB* (for high methicillin resistance) were shown to confer high level resistance to a heterogeneously resistant strain and to enhance resistance when overexpressed. The *hmrA* gene codes for a putative aminohydrolase, and *hmrB* for a homolog of an acyl carrier protein (Kondo *et al.*, 2001). However, their function and relevance *in vivo* are not clear. *hmrA* was found to be induced in one of ten highly resistant MRSA strains analysed, whereas no evidence for *hmrB* induction has been found in any highly resistant strains.

Cell growth and lysis are accompanied by autolytic activities that must be tightly controlled. β-lactams uncouple septum formation from autolytic activities, leading to lytic death. At β-lactam concentrations corresponding to the MIC, crosslinking of the peptidoglycan and new septum initiation are inhibited, new cell wall polymer is deposited as loose material in the septal plane, while the timing of the autolytic activities involved in the separation of the daughter cells continues normally. Lysis occurs in the next division cycle by murosome-induced puncturing of the cell wall. Very high concentrations of penicillin, in contrast, lead to non-lytic death by interaction with a yet unknown target (Giesbrecht *et al.*, 1998). Multiple autolytic enzymes have been identified in staphylococci (Sugai, 1997) and it is conceivable that resistance to β-lactams may be influenced by their activity as is illustrated below.

High-level methicillin resistance resulting from inactivation of *lytH*, encoding a putative lytic enzyme with similarity to an *N*-acetylmuramyl-L-alanine amidase, suggests that loss of autolytic activities contributes to resistance (Fujimura and Murakami, 1997). However, this effect was found only in a fraction of highly methicillin resistant strains.

Inactivation of the *dlt* operon, which catalyzes D-alanine substitution of the wall teichoic acids and lipoteichoic acids, leads to an increase in methicillin resistance (Nakao *et al.*, 2000). Loss of D-alanine substitution increases the negative charge of teichoic acids and improves their ability to sequester the cationic autolytic enzymes (Fischer *et al.*, 1981). However, the details of posttranslational control of autolysins and their impact on methicillin resistance remain to be elucidated, since contradicting levels of autolytic activities were reported in strains of different genetic backgrounds by two groups (de Jonge *et al.*, 1991; Gustafson *et al.*, 1992).

Mutations Reducing Methicillin Resistance

Transposon-mediated inactivation of methicillin resistance in MRSA has revealed a multitude of chromosomal loci that affect resistance levels. The genes identified were initially termed *fem* (factor essential for methicillin resistance) (Berger-Bächi *et al.*, 1989a) or *aux* (auxiliary) factors (de Lencastre and Tomasz, 1994). Their names are being gradually substituted by functional designations. Many of these insertions still allow a reduced expression of the target gene, either by exerting a polar effect on transcription of a downstream gene, or a partial truncation of the ORF, because complete inactivation would be lethal. None of the *fem* or *aux* genes analysed so far appear to directly influence the production of PBP2'. Instead, most of them are directly or indirectly involved in peptidoglycan precursor formation or affect overall peptidoglycan composition (Figure 2).

Inactivation of phosphoglucomutase, GlmM (Jolly *et al.*, 1997) (originally FemD (Glanzmann *et al.*, 1999)), reduces resistance by decreasing the conversion of glucosamine-6-phosphate into glucosamine-1-phosphate, which is the substrate for the synthesis of UDP-*N*-acetylglucosamine (Jolly *et al.*, 1997). By growing *femD* mutants in presence of methicillin, compensatory mutations, that restore resistance can arise at a low frequency (Glanzmann *et al.*, 1999). Their nature has not yet been identified, but there may be an alternative pathway for the production of glucosamine-1-phosphate. Inactivation of FemF (MurE), which catalyzes addition of lysine to the UDP-linked muramyl-dipeptide (Ludovice *et al.*, 1998; Ornelas-Soares *et al.*, 1994), has a similarly negative effect on the resistance level.

The glutamine synthetase repressor, GlnR (formerly FemC) (Gustafson *et al.*, 1994), is an example of a factor with an indirect effect on the peptidoglycan structure. Inactivation of *glnR* has a polar effect on the *glnA* gene, encoding glutamine synthetase, thus reducing the amount of glutamine in the cell. This, in consequence, affects the rate of amidation of iD-glutamate in the peptidoglycan stem peptide. Reduced amidation of the stem peptide prevents the efficient crosslinking of peptidoglycan strands, which results in reduced methicillin resistance in MRSA strains. PBP2' presumably does not recognize the non-amidated stem peptide, indicating that the precise chemical structure of the substrate is essential for its activity (Gustafson *et al.*, 1994). It is conceivable that transposon insertions in *glnA* itself were not recovered because they are lethal. Extracellular addition of glutamine restores both amidation of iD-glutamate and methicillin resistance. Similar to *glmM* mutants, *glnR* mutants can restore resistance by an alternative pathway, which is different from the *femD* compensatory mutation.

The peptidoglycan structure with the highest impact on methicillin resistance is the long and flexible pentaglycine interpeptide, which allows the high rate of peptidoglycan cross-linking characteristic for staphylococci. It is synthesized by the sequential addition of glycine from glycyl-tRNA to the bactoprenol-linked peptidoglycan precursor, lipid II (Matsuhashi *et al.*, 1965). The synthesis of the interpeptide is catalysed by a novel class of nonribosomal peptide synthetases (Hegde and Shrader, 2001). FmhB is needed for the addition of the first glycine (Rohrer *et al.*, 1999), FemA for the glycine residues in positions two and three (Stranden *et al.*, 1997), and FemB for the last two glycyl residues (Henze *et al.*, 1993). Inactivation of *fmhB* is lethal for the cells. Inactivation of *femB* results in a shortened interpeptide of only three glycine residues, leading to reduced peptidoglycan cross-linking and loss of methicillin resistance. Deletion of the *femAB* operon results in a single glycine in the interpeptide and a weakened, poorly crosslinked cell wall and abnormal septum localization, and needs a compensatory mutation for survival, which considerably reduces the growth rate (Ling and Berger-Bächi, 1998). The effect of the *femAB* deletion is pleiotropic; not only is methicillin resistance abolished, but the mutants are also hypersusceptible to all other classes of antibiotics. FemAB-like proteins have therefore been considered to be potential targets for novel therapeutic agents that may restore the efficacy of β-lactams in MRSA (Kopp *et al.*, 1996).

A number of membrane and cell wall-associated proteins of yet unknown function have been shown to modulate methicillin resistance. Llm is a 38 kDa hydrophobic protein of unknown function whose loss leads to increased autolysis in the presence of Triton X-100 and to higher susceptibility to methicillin (Maki *et al.*, 1994). Interestingly, *llm* revertants selected for increased resistance to methicillin were found to have an increased transcription of *llm* due to transposition of IS*256*, creating a new hybrid promoter (Maki and Murakami, 1997). Transposition of IS elements may therefore also be a mechanism contributing to higher resistance.

FmtA (for factor affecting methicillin resistance in presence of triton) is a membrane protein that contains two of the three conserved motifs found in low molecular mass PBPs, but does not bind penicillin. *fmtA* mutants have a reduction in crosslinking and partially reduced amidation of glutamate residues in the peptidoglycan similar to *femC* mutants. *fmtA* transcription is induced by subinhibitory concentrations of β-lactams, fosfomycin and bacitracin (Komatsuzawa *et al.*, 1999).

Mrp (for multiple repeat polypeptide), identical to FmtB, is a surface protein which was identified independently by two different groups (Komatsuzawa *et al.*, 2000; Wu and De Lencastre, 1999) Methicillin resistance is reduced upon transposon insertion into *mrp/fmtB*, but this phenotype may be due to a polar effect on an unknown target. Resistance is restored by adding N-acetyl glucosamine or glucosamine to the medium, or by trans-complementation with the phosphoglucosamine mutase GlmM, which maps upstream of *mrp/fmtB* (Komatsuzawa *et al.*, 2000).

The *fmtC* gene (synonymous with *mprF*) was identified simultaneously by two different approaches. While *fmtC* mutants were characterized by decreased methicillin and bacitracin resistance in the presence of Triton X-100 (Komatsuzawa *et al.*, 2001), MprF was described as a factor influencing staphylococcal virulence and determining resistance to defensins and protegrins. MprF mutants are unable to modify the membrane lipid phosphatidylglycerol with lysine, thus increasing the negative charge of the membrane surface (Peschel *et al.*, 2001). However, a model correlating increased negative charge of the membrane with reduced methicillin resistance has not yet been proposed.

The global regulators *agr* and *sar* control expression of cell wall and extracellular proteins involved in virulence. Inactivation of either *sar* and/or *agr* in a typical heterogeneously resistant MRSA results in a small decrease in the number of cells in the subpopulation expressing high methicillin resistance. Decrease in resistance may be due to an observed reduction in PBP1 and PBP3, suggesting that these PBPs somehow cooperate with PBP2' in high-level methicillin resistance (Duran *et al.*, 1996).

The stress-inducible alternative transcription factor σ^B is interlinked with the regulation by *agr* and *sar* and controls the synthesis of a set of at least 30 proteins, among them the *ica* genes involved in biofilm production (Bischoff *et al.*, 2001; Gertz *et al.*, 2000). Mutations reducing the level of σ^B activity in staphylococci results in a reduction of methicillin resistance (Giachino, 2000; Wu *et al.*, 1996) whereas σ^B overexpression increases the resistance level, concomitantly with the overexpression of cell division and *pbp* genes as well as *mecA* (Morikawa *et al.*, 2001).

A correlation between biofilm formation and methicillin resistance was observed in a biofilm-producing *S. epidermidis* strain (Christensen *et al.*, 1990), suggesting a positive effect of slime production on the resistance level. However, the impact of biofilm on the development of methicillin resistance in *S. epidermidis* still needs to be clarified, as different classes of mutations were recently reported that resulted in the loss of slime production without affecting resistance levels of the mutants (classes 1, 2) or even increasing them (class 4) (Mack *et al.*, 2000).

Analysis of an extensive library of Tn*551* insertions inactivating methicillin resistance in a highly and homogeneously resistant MRSA has yielded a series of genes that affect methicillin resistance. Included in this set of genes are putative sensor transducers, ABC transporters and a catabolite control protein (de Lencastre *et al.*, 1999). Confirmation and characterization of their functions will be a tremendous but exciting task, but we might draw the preliminary conclusion that the general metabolic state of the cell has an impact on the level of methicillin resistance.

Non-*mec*-mediated Methicillin Resistance

In addition to the acquisition of *mecA*, staphylococci are able to achieve high level methicillin resistance via intrinsic mechanisms. Methicillin resistant staphylococci can be obtained *in vitro* by stepwise selection for growth on increasing concentrations of methicillin. In contrast to *mecA*-dependent methicillin resistance, this type of resistance is associated with slower growth (unpublished results), and resistance is expressed homogeneously. Overexpression of PBP2 and/or PBP4, and changes in their penicillin affinity were shown to increase resistance to β-lactams (Berger-Bächi *et al.*, 1989b; Hackbarth *et al.*, 1995; Henze and Berger-Bächi, 1996). This phenomenon does not only occur in the laboratory, as clinical isolates resembling first step *in vitro* selected methicillin-resistant strains have indeed been identified. They were termed MODSA in reference to their modified PBPs (Tomasz *et al.*, 1989). Decreased binding capacity of PBP3 was found to be responsible for low-level methicillin resistance in *S. epidermidis* (Petinaki *et al.*, 2001). Overexpression of a penicillinase in a specific genetic background, the so-called BORSA (borderline resistant *S. aureus*) may also mimic methicillin resistance (Barg *et al.*, 1991). Strains with borderline methicillin resistance have less clinical impact than true MRSA, since their methicillin resistance has not yet been shown to attain the high levels of MRSA.

Outlook

After more than 30 years of research into the mechanism of methicillin resistance, many questions remain unanswered. While multiresistant MRSA have historically been mainly hospital-associated, the emerging cMRSA are a new phenomenon and raises new questions. The reservoir and the selective forces in the community need to be found. The success behind cMRSA, seemingly spreading in humans without any selective pressure, may be due to an isolated event combining SCC*mec* with a powerful epidemic strain. Evolution appears to be streamlining SCC*mec* into a smaller element retaining only the crucial *mec* and *ccr* complexes. An extrapolation of this process makes it likely that a more mobile, transposable element may arise that carries *mecA* and may even end up located on a plasmid, which would be a frightening vision.

It is clear that PBP2' plays the key role in methicillin resistance, but the genetic background into which PBP2' has entered is obviously equally important. Methicillin resistance appears to reduce the overall fitness of MRSA. The cell wall synthesis complex of naive, susceptible *S. aureus* is probably not optimised to accommodate PBP2', and introduction of PBP2' may require some compensatory mutations in the *S. aureus* genome for efficient expression of methicillin resistance, as the presence of PBP2' may not be entirely beneficial to the cell. This assumption is supported by the finding, that in *in vitro* selected methicillin resistant strains, or MODSA, heterogeneous resistance does not occur. Since the frequency at which resistant subclones arise lies far above the rate of spontaneous mutation, it is conceivable that epigenetic changes, such as DNA methylation, possibly leading to gene silencing, may alter the regulation of cell wall metabolism sufficiently to trigger high resistance.

The structural requirements of PBP2' on the peptidoglycan precursor identified so far are the amidation of the iD-glutamate of the stem peptide and a full-length pentaglycine interpeptide. The protein structure of PBP2' may be optimized for the chemical composition of the precursor, and the low penicillin-binding activity may imply structural restraints that are incompatible with a broader substrate range. The crystal structures of a streptococcal PBP2x (Pares *et al.*, 1996) and PBP2' from a methicillin resistant strain of *S. aureus*, have both been published.

Peptidoglycan turnover is tightly regulated by autolytic enzymes whose inactivation increases methicillin resistance levels. The control of autolytic enzymes occurs at the transcriptional, translational, and postranslational levels. Sequestration of the enzymes by charged cell wall molecules such as teichoic acids is one example of post-translational control of autolytic activity, but the multiple pathways controlling autolytic activities have not been fully explored in regard to methicillin resistance.

While insertional inactivation of methicillin resistance turned out to be a useful tool to identify genes involved in cell wall biosynthesis, it has been unable to resolve the mechanisms behind high level methicillin resistance satisfactorily. All evidence points towards multiple mechanisms or redundant pathways leading to high level resistance. Conditions able to abolish heteroresistance without affecting the basal resistance level have not yet been found, which impairs the detailed genetic analysis of this phenomenon. Newer techniques such as differential analysis of genome-wide expression patterns of high versus low level resistant strains by DNA microarray techniques, or proteome analysis by 2-D gel electrophoresis, may give us new insights into this complex regulatory network. Using such novel technologies may open up new directions for successful research.

Acknowledgement

Research in the laboratory of B. Berger-Bächi is supported by the Swiss National Science Foundation grant 3200-063552.00. The authors thank Dr. Teruyo Ito, Juntendo University, Tokyo, Japan for helpful comments regarding Figure 1.

References

Archer, G.L., and Niemeyer, D.M. 1994. Origin and evolution of DNA associated with resistance to methicillin in staphylococci. Trends Microbiol. 2: 343-347.

Barg, N., Chambers, H., and Kernodle, D. 1991. Borderline susceptibility to anti-staphylococcal penicillins is not conferred exclusively by the overproduction of beta-lactamase. Antimicrob. Agents Chemother. 35: 1975-1979.

Berger-Bächi, B., Barberis-Maino, L., Strässle, A., and Kayser, F.H. 1989a. FemA, a host-mediated factor essential for methicillin resistance in *Staphylococcus aureus*: molecular cloning and characterization. Mol. Gen. Genet. 219: 263-269.

Berger-Bächi, B., Strässle, A., and Kayser, F.H. 1989b. Natural methicillin resistance in comparison with that selected by *in vitro* drug exposure in *Staphylococcus aureus*. J. Antimicrob. Chemother. 23: 179-188.

Bischoff, M., Entenza, J.M., and Giachino, P. 2001. Influence of a functional *sigB* operon on the global regulators *sar* and *agr* in *Staphylococcus aureus*. J. Bacteriol. 183: 5171-5179.

Boyce, J.M., Medeiros, A.A., Papa, E.F., and O Gara, C.J. 1990. Induction of beta-lactamase and methicillin resistance in unusual strains of methicillin-resistant *Staphylococcus aureus*. J. Antimicrob. Chemother. 25: 73-81.

Chambers, H.F. 2001. The changing epidemiology of *Staphylococcus aureus*? Emerg. Infect. Dis. 7: 178-182.

Christensen, G.D. Baddour, L.M., Madison, B.M., Parisi, J.T., Abraham, S.N., Hasty, D.L., Lowrance, J.H., Josephs, J.A., and Simpson, W.A. 1990. Colonial morphology of staphylococci on memphis agar - phase variation of slime production, resistance to beta-lactam antibiotics, and virulence. J. Infect. Dis.161: 1153-1169.

Clarke, S.R., and Dyke, K.G.H. 2001. Studies of the operator region of the *Staphylococcus aureus* beta-lactamase operon. J. Antimicrob. Chemother. 47: 377-389.

Cohen, S., and Sweeney, H.M. 1968. Constitutive penicillinase formation in *Staphylococcus aureus* owing to a mutation unlinked to the penicillinase plasmid. J. Bacteriol. 95: 1368-1374.

de Jonge, B.L., de Lencastre, H., and Tomasz, A. 1991. Suppression of autolysis and cell wall turnover in heterogeneous Tn*551* mutants of a methicillin-resistant *Staphylococcus aureus* strain. J. Bacteriol. 173: 1105-1110.

de Jonge, B.L.M., Chang, Y.S., Gage, D., and Tomasz, A. 1992. Peptidoglycan composition of a highly methicillin-resistant *Staphylococcus aureus* strain - the role of penicillin binding protein-2A. J. Biol. Chem. 267: 11248-11254.

de Jonge, B.L.M., and Tomasz, A. 1993. Abnormal peptidoglycan produced in a methicillin-resistant strain of *Staphylococcus aureus* grown in the presence of methicillin - functional role for penicillin-binding protein-2A in cell wall synthesis. Antimicrob. Agents Chemother. 37: 342-346.

de Lencastre, H., Figueiredo, A.M.S., and Tomasz, A. 1993. Genetic control of population structure in heterogeneous strains of methicillin resistant *Staphylococcus aureus*. Eur. J. Clin. Microbiol. Infect. Dis. 12(Suppl. 1): 13-18.

de Lencastre, H., and Tomasz, A. 1994. Reassessment of the number of auxiliary genes essential for expression of high-level methicillin resistance in *Staphylococcus aureus*. Antimicrob. Agents Chemother. 38: 2590-2598.

de Lencastre, H. Wu, S. W., Pinho, M. G., Ludovice, A. M., Filipe, S., Gardete, S., Sobral, R., Gill, S., Chung, M., and Tomasz, A.1999. Antibiotic resistance as a stress response: Complete sequencing of a large number of chromosomal loci in *Staphylococcus aureus* strain COL that impact on the expression of resistance to methicillin. Microb. Drug Res. 5: 163-175.

Deplano, A., Tassios, P.T., Glupczynski, Y., Godfroid, E., and Struelens, M.J. 2000. *In vivo* deletion of the methicillin resistance *mec* region from the chromosome of *Staphylococcus aureus* strains. J. Antimicrob. Chemother. 46: 617-619.

Diekema, D.J., Pfaller, M. A., Schmitz, F. J., Smayevsky, J., Bell, J., Jones, R. N., and Beach, M. 2001. Survey of infections due to *Staphylococcus* species: Frequency of occurrence and antimicrobial susceptibility of isolates collected in the United States, Canada, Latin America, Europe, and the Western Pacific region for the SENTRY Antimicrobial Surveillance Program, 1997-1999. Clin Infect Diseases, 32: S114-S132.

Duran, S.P., Kayser, F.H., and Berger-Bächi, B. 1996. Impact of *sar* and *agr* on methicillin resistance in *Staphylococcus aureus*. FEMS Microbiol. Lett. 141: 255-260.

Finan, J.E., Rosato, A.E., Dickinson, T.M., Ko, D., and Archer, G.L. 2002. Conversion of oxacillin-resistant staphylococci from heterotypic to homotypic resistance expression. Antimicrob. Agents Chemother. 46: 24-30.

Fischer, W., Rösel, P., and Koch, H.U. 1981. Effect of alanine ester substitution and other structural features of lipoteichoic acids on their inhibitory activity against autolysins of *Staphylococcus aureus*. J. Bacteriol. 146: 467-475.

Fitzgerald, J.R., Sturdevant, D.E., Mackie, S.M., Gill, S.R., and Musser, J.M. 2001. Evolutionary genomics of *Staphylococcus aureus*: Insights into the origin of methicillin-resistant strains and the toxic shock syndrome epidemic. Proc. Natl. Acad. Sci. USA. 98: 8821-8826.

Fujimura, T., and Murakami, K. 1997. Increase of methicillin resistance in *Staphylococcus aureus* caused by deletion of a gene whose product is homologous to lytic enzymes. J. Bacteriol. 179: 6294-6301.

Gaisford, W.C., and Reynolds, P.E. 1989. Methicillin resistance in *Staphylococcus epidermidis*. Relationship between the additional penicillin-binding protein and an attachment transpeptidase. Eur. J. Biochem. 185: 211-218.

Gertz, S. Engelmann, S., Schmid, R., Ziebandt, A-K., Tischer, K., Scharf, C., Hacker, J., and Hecker, M. 2000. Characterization of the σ^B regulon in *Staphylococcus aureus*. J. Bacteriol. 182: 6983-6991.

Giachino, P. 2000. Der alternative Sigma-Faktor σ^B in *Staphylococcus aureus*: Einfluss des intakten σ^B Operons auf die Virulenz des Bakteriums. PhD Thesis, University of Zürich, Zürich. p.115.

Giesbrecht, P., Kersten, T., Maidhof, H., and Wecke, J. 1998. Staphylococcal cell wall: Morphogenesis and fatal variations in the presence of penicillin. Microbiol. Mol. Biol. Rev. 62: 1371-1414.

Glanzmann, P., Gustafson, J., Komatsuzawa, H., Ohta, K., and Berger-Bächi, B. 1999. *glmM* operon and methicillin-resistant *glmM* suppressor mutants in *Staphylococcus aureus*. Antimicrob. Agents Chemother. 43: 240-245.

Goffin, C., and Ghuysen, J.M. 1998. Multimodular penicillin binding proteins: An enigmatic family of orthologs and paralogs. Microbiol. Mol. Biol. Rev. 62: 1079-1093.

Gustafson, J., Strässle, A., Hächler, H., Kayser, F.H., and Berger-Bächi, B. 1994. The *femC* locus of *Staphylococcus aureus* required for methicillin resistance includes the glutamine synthetase operon. J. Bacteriol. 176: 1460-1467.

Gustafson, J.E., Berger-Bächi, B., Strässle, A., and Wilkinson, B.J. 1992. Autolysis of methicillin-resistant and -susceptible *Staphylococcus aureus*. Antimicrob. Agents Chemother. 36: 566-572.

Hackbarth, C.J., Kocagoz, T., Kocagoz, S., and Chambers, H.F. 1995. Point mutations in *Staphylococcus aureus* PBP 2 gene affect penicillin-binding kinetics and are associated with resistance. Antimicrob. Agents Chemother. 39: 103-106.

Hackbarth, C.J., Miick, C., and Chambers, H.F. 1994. Altered production of penicillin-binding protein 2a can affect phenotypic expression of methicillin resistance in *Staphylococcus aureus*. Antimicrob. Agents Chemother. 38: 2568-2571.

Hegde, S.S., and Shrader, T.E. 2001. FemABX family members are novel nonribosomal peptidyltransferases and important pathogen-specific drug targets. J. Biol. Chem. 276: 6998-7003.

Henze, U., Sidow, T., Wecke, J., Labischinski, H., and Berger-Bächi, B. 1993. Influence of *femB* on methicillin resistance and peptidoglycan metabolism in *Staphylococcus aureus*. J. Bacteriol. 175: 1612-1620.

Henze, U., and Berger-Bächi, B. 1996. Penicillin-binding protein 4 overproduction increases beta-lactam resistance in *Staphylococcus aureus*.

Antimicrob. Agents Chemother. 40: 2121-2125.

Hilden, P., Savolainen, K., Tyynela, J., Vuento, M., and Kuusela, P. 1996. Purification and characterisation of a plasmin-sensitive surface protein of *Staphylococcus aureus*. Eur. J. Biochem. 236: 904-10.

Hiramatsu, K., Asada, K., Suzuki, E., Okonogi, K., and Yokota, T. 1992. Molecular cloning and nucleotide sequence determination of the regulator region of *mecA* gene in methicillin-resistant *Staphylococcus aureus* (MRSA). FEBS Lett. 298: 133-136.

Hiramatsu, K., Cui, L., Kuroda, M., and Ito, T. 2001. The emergence and evolution of methicillin-resistant *Staphylococcus aureus*. Trends Microbiol. 9:486-493.

Hürlimann-Dalel, R.L., Ryffel, C., Kayser, F.H., and Berger-Bächi, B. 1992. Survey of the methicillin resistance-associated genes *mecA*, *mecR1-mecI*, and *femA-femB* in clinical isolates of methicillin-resistant *Staphylococcus aureus*. Antimicrob. Agents Chemother. 36: 2617-2621.

Inglis, B., Matthews, P.R., and Stewart, R.A. 1990. Induced deletions within a cluster of resistance genes in the *mec* region of the chromosome of *Staphylococcus aureus*. J. Gen. Microbiol. 136: 2231-2239.

Ito, T. Katayama, Y., Asada, K., Mori, N., Tsutsumimoto, K., Tiensasitorn, C., and Hiramatsu, K. 2001. Structural comparison of three types of staphylococcal cassette chromosome *mec* integrated in the chromosome in methicillin-resistant *Staphylococcus aureus*. Antimicrob. Agents Chemother. 45: 1323-1336.

Ito, T., Katayama, Y., and Hiramatsu, K. 1999. Cloning and nucleotide sequence determination of the entire *mec* DNA of pre-methicillin-resistant *Staphylococcus aureus* N315. Antimicrob. Agents Chemother. 43: 1449-1458.

Jevons, M.P. 1961. "Celbenin" resistant staphylococci. British Med. J. i: 124-125.

Jolly, L. Wu, S. W., Van Heijenoort, J., De Lencastre, H., Mengin Lecreulx, D., and Tomasz, A.1997. The *femR*315 gene from *Staphylococcus aureus*, the interruption of which results in reduced methicillin resistance, encodes a phosphoglucosamine mutase. J. Bacteriol. 179: 5321-5325.

Katayama, Y., Ito, T., and Hiramatsu, K. 2000. A new class of genetic element, staphylococcus cassette chromosome *mec*, encodes methicillin resistance in *Staphylococcus aureus*. Antimicrob. Agents Chemother. 44: 1549-1555.

Katayama, Y., Ito, T., and Hiramatsu, K. 2001. Genetic organization of the chromosome region surrounding *mecA* in clinical staphylococcal strains: Role of IS431-mediated *mecI* deletion in expression of resistance in *mecA*-carrying, low-level methicillin-resistant *Staphylococcus haemolyticus*. Antimicrob. Agents Chemother. 45: 1955-1963.

Komatsuzawa, H. Ohta, K., Fujiwara, T., Choi, G. H., Labischinski, H., and Sugai, M.2001. Cloning and sequencing of the gene, *fmtC*, which affects oxacillin resistance in methicillin-resistant *Staphylococcus aureus*. FEMS

Microbiol. Lett. 203: 49-54.

Komatsuzawa, H., Ohta, K., Labischinski, H., Sugai, M., and Suginaka, H. 1999. Characterization of *fmtA*, a gene that modulates the expression of methicillin resistance in *Staphylococcus aureus*. Antimicrob. Agents Chemother. 43: 2121-2125.

Komatsuzawa, H. Ohta, K., Sugai, M., Fujiwara, T., Glanzmann, P., Berger-Bächi, B., and Suginaka, H.2000. Tn*551*-mediated insertional inactivation of the *fmtB* gene encoding a cell wall-associated protein abolishes methicillin resistance in *Staphylococcus aureus*. J. Antimicrob. Chemother. 45: 421-431.

Kondo, N., Kuwahara-Arai, K., Kuroda-Murakami, H., Tateda-Suzuki, E., and Hiramatsu, K. 2001. Eagle-type methicillin resistance: New phenotype of high methicillin resistance under *mec* regulator gene control. Antimicrob. Agents Chemother. 45: 815-824.

Kopp, U., Roos, M., Wecke, J., and Labischinski, H. 1996. Staphylococcal peptidoglycan interpeptide bridge biosynthesis: a novel anti-staphylococcal target? Microb. Drug Resist. 2: 29-41.

Labischinski, H. 1992. Consequences of the interaction of beta-lactam antibiotics with penicillin binding proteins from sensitive and resistant *Staphylococcus aureus* strains. Med. Microbiol. Immunol. 181: 241-265.

Lewis, R.A., and Dyke, K.G.H. 2000. MecI represses synthesis from the beta-lactamase operon of *Staphylococcus aureus*. J. Antimicrob. Chemother. 45: 139-144.

Lim, D., and Strynadka, N.C. 2002. Structural basis for the beta lactam resistance of a PBP2a from methicillin resistant *Staphylococcus aureus*. Nature Structural Biology 9: 870-876.

Ling, B.D., and Berger-Bächi, B. 1998. Increased overall antibiotic susceptibility in *Staphylococcus aureus femAB* null mutants. Antimicrob. Agents Chemother. 42: 936-938.

Ludovice, A.M., Wu, S.W., and De Lencastre, H. 1998. Molecular cloning and DNA sequencing of the *Staphylococcus aureus* UDP-N-acetylmuramyl tripeptide synthetase (*murE*) gene, essential for the optimal expression of methicillin resistance. Microb. Drug Resist. 4: 85-90.

Mack, D. Rohde, H., Dobinsky, S., Riedewald, J., Nedelmann, M., Knobloch, J. K. M., Elsner, H. A., and Feucht, H. H. 2000. Identification of three essential regulatory gene loci governing expression of *Staphylococcus epidermidis* polysaccharide intercellular adhesin and biofilm formation. Infect. Immun. 68: 3799-3807.

Maki, H., and Murakami, K. 1997. Formation of potent hybrid promoters of the mutant *llm* gene by IS*256* transposition in methicillin-resistant *Staphylococcus aureus*. J. Bacteriol. 179: 6944-6948.

Maki, H., Yamaguchi, T., and Murakami, K. 1994. Cloning and characterization of a gene affecting the methicillin resistance level and the autolysis rate in *Staphylococcus aureus*. J. Bacteriol. 176: 4993-5000.

Matsuhashi, M., Dietrich, C.P., and Strominger, J.L. 1965. Incorporation of glycine into the cell wall of glycopeptide in *Staphylococcus aureus*: role of sRNA and lipid intermediates. Proc. Natl. Acad. Sci. USA. 54: 587-594.

Matthews, P.R., and Stewart, P.R. 1984. Resistance heterogeneity in methicillin-resistant *Staphylococcus aureus*. FEMS Microbiol. Lett. 22: 161-166.

McKinney, T.K., Sharma, V.K., Craig, W.A., and Archer, G.L. 2001. Transcription of the gene mediating methicillin resistance in *Staphylococcus aureus* (*mecA*) is corepressed but not coinduced by cognate *mecA* and beta-lactamase regulators. J. Bacteriol. 183: 6862-6868.

Morikawa, K. Maruyama, A., Inose, Y., Higashide, M., Hayashi, H., and Ohta, T.2001. Overexpression of sigma factor σ^B, urges *Staphylococcus aureus* to thicken the cell wall and to resist β-lactams. Biochem. Biophys. Res. Commun. 288: 385-389.

Nakao, A., Imai, S., and Takano, T. 2000. Transposon-mediated insertional mutagenesis of the D-alanyl-lipoteichoic acid (*dlt*) operon raises methicillin resistance in *Staphylococcus aureus*. Res. Microbiol. 151: 823-829.

Ornelas-Soares, A., de Lencastre, H., de Jonge, B.L.M., and Tomasz, A. 1994. Reduced methicillin resistance in a new *Staphylococcus aureus* transposon mutant that incorporates muramyl dipeptides into the cell wall peptidoglycan. J. Biol. Chem. 269: 27246-27250.

Pares, S., Mouz, N., Petillot, Y., Hakenbeck, R., and Dideberg, O. 1996. X-ray structure of *Streptococcus pneumoniae* PBP2x, a primary penicillin target enzyme. Nature Struct. Biol. 3: 284-289.

Park, W., and Matsuhashi, M. 1984. *Staphylococcus aureus* and *Micrococcus luteus* peptidoglycan transglycosylases that are not penicillin-binding proteins. J. Bacteriol. 157: 538-544.

Peschel, A. Jack, R. W., Otto, M., Collins, L. V., Staubitz, P., Nicholson, G., Kalbacher, H., Nieuwenhuizen, W. F., Jung, G., Tarkowski, A., van Kessel, K. P. M., and van Strijp, J. A. G.2001. *Staphylococcus aureus* resistance to human defensins and evasion of neutrophil killing via the novel virulence factor MprF is based on modification of membrane lipids with L-lysine. J. Exp. Med. 193: 1067-1076.

Petinaki, E., Dimitracopoulos, G., and Spiliopoulou, I. 2001. Decreased affinity of PBP3 to methicillin in a clinical isolate of *Staphylococcus epidermidis* with borderline resistance to methicllin and free of the *mecA* gene. Microb. Drug Resistance. 7: 297-300.

Pinho, M.G., de Lencastre, H., and Tomasz, A. 2001a. An acquired and a native penicillin-binding protein cooperate in building the cell wall of drug-resistant staphylococci. Proc. Nat. Acad. Sci. USA. 98: 10886-10891.

Pinho, M.G., Filipe, S.R., de Lencastre, H., and Tomasz, A. 2001b. Complementation of the essential peptidoglycan transpeptidase function of penicillin-binding protein 2 (PBP2) by the drug resistance protein PBP2A in *Staphylococcus aureus*. J. Bacteriol. 183: 6525-6531.

Poston, S.M., and Li Saw Hee, F.L. 1991. Genetic characterisation of resistance to metal ions in methicillin- resistant *Staphylococcus aureus*: elimination of resistance to cadmium, mercury and tetracycline with loss of methicillin resistance. J. Med. Microbiol. 34:193-201.

Rohrer, S., Ehlert, K., Tschierske, M., Labischinski, H., and Berger-Bächi, B. 1999. The essential *Staphylococcus aureus* gene *fmhB* is involved in the first step of peptidoglycan pentaglycine interpeptide formation. Proc. Natl. Acad. Sci. USA. 96: 9351-9356.

Ryffel, C., Strässle, A., Kayser, F.H., and Berger-Bächi, B. 1994. Mechanisms of heteroresistance in methicillin-resistant *Staphylococcus aureus*. Antimicrob. Agents Chemother. 38: 724-728.

Slaughter, D.M., Patton, T.G., Sievert, G., Sobieski, R.J., and Crupper, S.S. 2001. Antibiotic resistance in coagulase-negative staphylococci isolated from Cope's gray treefrogs (*Hyla chrysoscelis*). FEMS Microbiol. Lett. 205: 265-270.

Song, M.D., Wachi, M., Doi, M., Ishino, F., and Matsuhashi, M. 1987. Evolution of an inducible penicillin-target protein in methicillin-resistant *Staphylococcus aureus* by gene fusion. FEBS Lett. 221: 167-171.

Strandén, A.M., Ehlert, K., Labischinski, H., and Berger-Bächi, B. 1997. Cell wall monoglycine cross-bridges and methicillin hypersusceptibility in a *femAB* null mutant of methicillin-resistant *Staphylococcus aureus*. J. Bacteriol.179: 9-16.

Sugai, M. 1997. Peptidoglycan hydrolases of the staphylococci. J. Infect. Chemother. 3: 113-127.

Suzuki, E., Kuwahara-Arai, K., Richardson, J.F., and Hiramatsu, K. 1993. Distribution of *mec* regulator genes in methicillin-resistant *Staphylococcus* clinical strains. Antimicrob. Agents Chemother. 37: 1219-26.

Tomasz, A., Drugeon, H.B., de Lencastre, H.M., Jabes, D., and McDougall, L. 1989. New mechanism for methicillin resistance in *Staphylococcus aureus*: clinical isolates that lack the PBP 2a gene and contain normal penicillin-binding proteins with modified penicillin-binding capacity. Antimicrob. Agents Chemother. 33: 1869-1874.

Tomasz, A., Nachman, S., and Leaf, H. 1991. Stable classes of phenotypic expression in methicillin-resistant clinical isolates of staphylococci. Antimicrob. Agents Chemother. 35: 124-129.

Wada, A., Katayama, Y., Hiramatsu, K., and Yokota, T. 1991. Southern hybridization analysis of the *mecA* deletion from methicillin-resistant *Staphylococcus aureus*. Biochem. Biophys. Res. Commun. 176: 1319-1325.

Wielders, C.L.C., Vriens, M. R., Brisse, S., de Graaf-Miltenburg, L.A.M., Troelstra, A., Fleer, A., Schmitz, F. J., Verhoef, J., and Fluit, A. C. 2001. Evidence for *in-vivo* transfer of *mecA* DNA between strains of *Staphylococcus aureus*. Lancet 357: 1674-1675.

Wu, S.W., de Lancastre, H., and Tomasz, A. 2001. Recruitment of the *mecA* gene homologue of *Staphylococcus sciuri* into a resistance determinant and expression of the resistant phenotype in *Staphylococcus aureus*. J. Bacteriol. 183: 2417-2424.

Wu, S.W., and De Lencastre, H. 1999. Mrp - A new auxiliary gene essential for optimal expression of methicillin resistance in *Staphylococcus aureus*. Microb. Drug Res. 5: 9-18.

Wu, S.W., de Lencastre, H., and Tomasz, A. 1996. Sigma-B, a putative operon encoding alternate sigma factor of *Staphylococcus aureus* RNA polymerase: molecular cloning and DNA sequencing. J. Bacteriol. 178: 6036-6042.

Wyke, A.W., Ward, J.B., Hayes, M.V., and Curtis, C.A.M. 1981. A role *in vivo* for penicillin-binding protein-4 of *Staphylococcus aureus*. Eur. J. Biochem. 119: 389-393.

Zhang, H.Z., Hackbarth, C.J., Chansky, K.M., and Chambers, H.F. 2001. A proteolytic transmembrane signaling pathway and resistance to beta-lactams in staphylococci. Science. 291s: 1962-1965.

From: *MRSA: Current Perspectives*
Edited by: A.C. Fluit and F.-J. Schmitz

Chapter 4

MRSA Resistance Mechanisms and Surveillance Data for Non-betalactams and Non-glycopeptides

Ian Morrissey and David J. Farrell

Abstract

This chapter discusses the genetics and regulation of resistance mechanisms to the major classes of anti-staphylococcal antibiotics in methicillin-resistant *Staphylococcus aureus* (MRSA) with the exception of betalactam agents and glycopeptides, which are dealt with elsewhere. To summarise, it can be seen that MRSA from almost all regions of the world appear to be cross-resistant to many other classes of antibiotic and this emphasises the need to find new antibiotics to treat this difficult organism.

Introduction

MRSA, by definition, are resistant to methicillin and oxacillin. Unfortunately, antibiotic resistance in the majority of hospital-acquired MRSA is not confined to these anti-staphylococcal beta-lactams with multi-drug resistance being the norm. The following sections of this chapter will focus on the mechanisms of resistance to specific antibacterial classes in MRSA.

Quinolones

The quinolones are man-made antibacterials, first used clinically in the 1960s with the introduction of nalidixic acid. Around this time MRSA had only recently been described (Jevons, 1961), but compounds of the quinolone class with activity against staphylococci were not available until the introduction of fluoroquinolones, such as ciprofloxacin, some 20 years later. Up until then quinolone utility was restricted to Gram-negative bacteria, in particular the Enterobacteriaceae. Improvements to the chemical structure of early quinolones, based on the addition of a fluorine atom at the C-6 position (hence the name fluoroquinolone) and modification of the C-7 position, dramatically enhanced antibacterial spectrum and improved pharmacokinetics.

The quinolones target the essential bacterial type II DNA topoisomerases, namely DNA gyrase and topoisomerase IV (Drlica and Xhao, 1997). Both enzymes exist as tetramers, made up of 2 units each of GyrA and GyrB, in the case of DNA gyrase, or ParC and ParE for topoisomerase IV. In *Staphylococcus aureus* the ParC and ParE subunits are known as GrlA and GrlB, respectively, (Ferrero *et al.*, 1994). The type II DNA topoisomerases, are able to cut both strands of the DNA double-helix and pass one strand through this transient gap before re-sealing the DNA. With DNA gyrase, the cut DNA strand is passed through itself to propagate what is known as negative supercoiling. This is thought to be important for the release of topological stress that builds up along DNA during transcription and replication. In contrast, the DNA transfer event with topoisomerase IV involves passing one DNA molecule through a cut made in a separate DNA molecule to perform a process known as decatenation. This is vital to allow the separation of daughter chromosomes during bacterial cell division. GyrA and GrlA are the subunits responsible for the physical cutting of the DNA, whereas GyrB and GrlB provide the energy for supercoiling or decatenation by the breakdown of ATP. The precise events that lead to bacterial cell death after exposure to quinolones is unclear but it is thought that the quinolones act as poisons that trap DNA gyrase and topoisomerase IV on DNA which produces a lethal release of double-strand DNA breaks (Drlica and Xhao, 1997).

Quinolones are unique in that resistance has not emerged in a plasmid-mediated form to any great extent. Some early reports with *Shigella dysenteriae* from Kashmir and Bangladesh have since been discounted (Ambler *et al.*, 1993). Recently a more valid claim has been made for plasmid-mediated quinolone resistance in *Klebsiella pneumoniae* (Jacoby *et al.*, 2001) but this has never been implicated with other bacteria such as staphylococci. With staphylococci, clinical resistance can occur due to chromosomal mutation in the DNA gyrase genes (*gyrA* and *gyrB*) or topoisomerase IV genes (*grlA* and *grlB*). Resistance to quinolones is also associated with a chromosomally-encoded efflux pump, NorA in staphylococci (Neyfakh *et al.*, 1993).

Unlike *E.* coli, topoisomerase IV from *S. aureus* is generally more sensitive to the action of quinolones than DNA gyrase (Blanche *et al.*, 1996; Tanaka *et al.*, 1997; Gootz *et al.*, 1999; Takei *et al.*, 2001). However, there is some evidence to suggest that fluoroquinolones can be classified into 3 categories (types I, II and III) based on their ability to inhibit purified *S. aureus* DNA gyrase and topoisomerase IV and also the effect of *grlA* and *gyrA* mutations on MIC (Takei *et al.*, 2001). Most fluoroquinolones fall within the type I category (norfloxacin, enoxacin, fleroxacin, ciprofloxacin, lomefloxacin, trovafloxacin, grepafloxacin, ofloxacin and levofloxacin) because they are more influenced by topoisomerase IV inhibition than DNA gyrase inhibition. Gemifloxacin may also belong to this category (Schulte and Heisig, 2000), but data for a complete analysis is not available. Only 2 fluoroquinolones form the type II category, these being sparfloxacin and nadifloxacin, which show a greater decrease in activity against a *gyrA* mutant strain than a *grlA* mutant strain – indicating a preference for DNA gyrase (Takei *et al.*, 2001). It should be noted however, that the results of some studies do not concur with this analysis for sparfloxacin (Blanche *et al.*, 1996; Tanaka *et al.*, 1997). The type III fluoroquinolones include gatifloxacin, pazufloxacin, moxifloxacin and clinafloxacin, which have dual-targeting properties where either enzyme is inhibited at approximately the same extent and *gyrA* or *grlA* mutants strains produce only a minimal 2-fold increase in MIC. (Takei *et al.*, 2001). Although not included in the original analysis, there are other data to suggest that sitafloxacin (Tanaka *et al.*, 1997) and new non-fluorinated quinolones such as PGE9262932, PGE4175997 and PGE9509924 (Roychoudhury *et al.*, 2001) or BMS-294756 – now known as garenoxacin (Schmitz *et al.*, 2002a; Low *et al.*, 2002) are also type III quinolones.

Mutational Target Alteration

As with other bacterial species, resistance to quinolones in *S. aureus* develops in a step-wise fashion by the accumulation of mutations. Laboratory studies have shown that mutations in *grlA* occur before *gyrA* with *S. aureus* selected on ciprofloxacin (Ferrero *et al.*, 1995) and these latter mutations are silent in the absence of additional mutations in *grlA* (Ng *et al.*, 1996). This is in general agreement with ciprofloxacin being a type I fluoroquinolone. With sparfloxacin, the reverse is true with mutations in *gyrA* occurring before *grlA* (Ruiz *et al.* 2001) again in keeping with the target preference of a type II fluoroquinolone. The same classical quinolone resistance alterations in GrlA (Ser80 to Phe or Tyr and Glu84 to Lys) and GyrA (Ser84 to Leu or Lys and Glu88 to Lys or Val) have been found in laboratory mutants and clinical isolates (Drlica *et al.*, 1997; Schmitz *et al.*, 1998). Other mutations such as Asp73 to Gly in GyrA have been found clinically but are considerably less common (Wang *et al.*, 1998). Similarly, mutations in GrlB are rare, but Asp432 to Asn and Asn470 to Asp have been observed (Fournier and Hooper, 1998; Tanaka *et al.*, 1998). Mutations in *gyrB* have not been reported in *S. aureus*. The quinolones vary in their intrinsic potencies against *S. aureus* but this appears to have little effect on the development of resistance during serial passage. Despite categorisation into types I, II and III the same DNA gyrase or topoisomerase IV mutations (as discussed above) occur with differing selecting quinolones during serial passage at sub-inhibitory concentrations. However, there is some inter-strain variability, which illustrates the difficulties of extrapolating *in vitro* resistance results to predict clinical resistance (Boos *et al.*, 2001).

Active Efflux

Efflux is another method that causes quinolone resistance. *S. aureus* has a chromosomally-encoded efflux pump, known as NorA, that can efflux hydrophilic quinolones and unrelated compounds such as chloramphenicol, ethidium bromide and cetrimide (Neyfakh *et al.*, 1993). Hydrophobic quinolones such as sparfloxacin, levofloxacin, moxifloxacin or trovafloxacin are affected to a lesser extent (Muñoz-Bellido *et al.*, 1999; Aeschlimann *et al.*, 1999; Beyer *et al.*, 2000). Like other efflux pumps, such as the closely-related Bmr protein in *Bacillus subtilis* (Neyfakh, 1992), NorA depends on the transmembrane proton gradient for active efflux. The physiological role for these multi-substrate efflux pumps is to maintain a homeostatic environment within bacterial cells. The regulation of *norA* expression is complex and probably involves a two component regulatory system (ArlR-ArlS) and the binding of an undetermined 18 kDa protein to the *norA* promoter (Fournier *et al.*, 2000). Efflux-mediated resistance generally results in a lower level of resistance than that seen with topoisomerase mutations. Quinolone resistance is not due to mutation within the *norA* gene, but instead has been shown to be due to a point mutation within the regulatory region 89 base-pairs upstream of the start codon of the gene (Ng *et al.*, 1994). This mutation increases stability of *norA* mRNA by introducing an altered stem-loop structure that hinders RNAse III cleavage (Fournier *et al.*, 2001). This produces an overall reduction in accumulation of quinolones. However, other research has shown that this mutation is not the sole cause of resistance and an as yet undetermined mechanism can also result in inducible quinolone efflux (Kaatz and Seo, 1995). In laboratory plating experiments, NorA over-expression was not found as an initial mutation step (Ferrero *et al.*, 1995) but there is evidence to suggest that NorA hyper-production may enhance the emergence of secondary topoisomerase mutations leading to higher level resistance due to topoisomerase mutations (Markham and Neyfakh, 1996). As a consequence of this phenomenon, those quinolones that are poor substrates for NorA may have a reduced propensity for resistance development (Beyer *et al.*, 2000).

At the time when anti-staphylococcal quinolones were introduced, quinolone resistance in MRSA was low. Unfortunately, within a period of only a few years almost all MRSA had developed resistance to quinolones (Acar and Goldstein, 1997). This is illustrated well by a study in Hong Kong where resistance to ciprofloxacin in MRSA had reached 82% in 1993 from a relatively low level of 9% in 1988 (Scheel *et al.*, 1996). In this case, the high level of resistance was not due to any particularly dominant MRSA clone or clones, but dominant clones are a major factor in other parts of the world (Johnson *et al.*, 2001; Gomes *et al.*, 2001; De Sousa *et al.*, 2001; Simor *et al.*, 2001). Similar high levels of quinolone resistance in MRSA have been reported for other countries such as the Netherlands and Canada (De Neeling *et al.*, 1998; Simor *et al.*, 2001). Even those quinolones, such as gemifloxacin, trovafloxacin, gatifloxacin and garenoxacin, with improved activity against methicillin-susceptible *S. aureus* show reduced activity against recent isolates of MRSA (McCloskey *et al.*, 2000; Gordon *et al.*, 2002). The fluoroquinolone susceptibility of methicillin-resistant *S. aureus* compared with methicillin-susceptible *S. aureus* from recent large surveillance studies is shown in Table 1.

It was established some time ago that although initial cases of colonisation or infection with quinolone-resistant MRSA can be linked to quinolone use, the spread of MRSA is more likely due to hospital transmission of existing resistant MRSA (Smith *et al.*, 1990). Despite this there does appear to be a link between quinolone resistance and methicillin resistance. This is not due to any inherent enhanced mutation frequency in MRSA because mutation rates to ciprofloxacin resitance are the same for MSSA as they are for MRSA (Schmitz *et al.*, 2000).

Macrolides, Lincosamides, and Streptogramins

Erythromycin is a naturally occurring substance derived from *Streptomyces erythreus* with a chemical structure consisting of a 14 atom macrocyclic lactone ring with two sugar moieties (desoamine and cladinose) attached. Semisynthetic derivatives of erythromycin (such as clarithromycin, azithromycin and josamycin) differ from each other by the substitution pattern and size (14 to 16 atoms) of the lactone ring. Lincosamides are simple in chemical structure

Table 1. Selected fluoroquinolone antimicrobial surveillance susceptibility data: methicillin-susceptible *S. aureus* (MSSA) versus methicillin-resistant *S. aureus* (MRSA)

Antimicrobial	N	Organism	MIC values (mg/L)			%Susceptible
			Range	MIC$_{50}$	MIC$_{90}$	
SENTRY study 1997-1999, 30 USA hospitals, isolates from bloodstream, lung, skin/soft tissue and urine (Diekema *et al.*, 2001).						
Ciprofloxacin	4714	MSSA	-	0.25	1	94.3
Gatifloxacin			-	0.06	0.12	98.2
Ciprofloxacin	2455	MRSA	-	>2	>2	10.1
Gatifloxacin			-	4	>4	36.7
SENTRY study 1997-1999, 8 Canadian hospitals, isolates from bloodstream, lung, skin/soft tissue and urine (Diekema *et al.*, 2001).						
Ciprofloxacin	1329	MSSA	-	0.25	1	96.1
Gatifloxacin			-	0.06	0.12	99.4
Ciprofloxacin	81	MRSA	-	>2	>2	39.5
Gatifloxacin			-	4	>4	45.7
SENTRY study 1997-1999, 10 Latin American hospitals, isolates from bloodstream, lung, skin/soft tissue and urine (Diekema *et al.*, 2001).						
Ciprofloxacin	1274	MSSA	-	0.25	0.5	96.7
Gatifloxacin			-	0.06	0.12	99.7
Ciprofloxacin	682	MRSA	-	>2	>2	9.2
Gatifloxacin			-	2	4	69.5

SENTRY study 1998-1999, 17 Western Pacific hospitals, isolates from bloodstream, lung, skin/soft tissue and urine (Diekema et al., 2001).

Ciprofloxacin	770	MSSA	-	0.25	0.5	98.4
Gatifloxacin			-	0.06	0.12	99.9
Ciprofloxacin	657	MRSA	-	>2	>2	11.3
Gatifloxacin			-	2	4	65.8

SENTRY study 1997-1999, 24 European hospitals, isolates from bloodstream, lung, skin/soft tissue and urine (Diekema et al., 2001).

Ciprofloxacin	2561	MSSA	-	0.25	1	90.9
Gatifloxacin			-	0.06	0.25	97.2
Ciprofloxacin	916	MRSA	-	>2	>2	9.5
Gatifloxacin			-	2	4	68.9

PROTEKT study 1999-2000 winter season, global (69 centres in 25 countries), isolates from patients with community-acquired respiratory tract infections (http://www.protekt.org)

Ciprofloxacin	1205	MSSA	0.25 - >32	0.25	0.5	95.2
Levofloxacin			0.5 - >32	0.5	0.5	97.5
Moxifloxacin			0.03 - >4	0.03	0.06	-
Ciprofloxacin	342	MRSA	0.25 - >32	16	>32	7.0
Levofloxacin			0.5 - >32	8	32	20.5
Moxifloxacin			0.03 - >4	2	>4	-

RESIST project 1997-1998, 18 countries in Europe, Asia and Latin America, isolates from wounds (24%), blood (22%), respiratory tract infection (21%), abscesses (7%), urine (5%) (Sanches et al., 2000)

Ciprofloxacin	112	MSSA	-	-	-	78.6
Ciprofloxacin	1749	MRSA	-	-	-	10.8

consisting of an amino acid linked to an amine sugar. The lincosamide lincomycin is derived from *Streptomyces lincolnensis* and clindamycin is a chemical modification of lincomycin. Streptogramins are derived from various *Streptomyces* species and consist of two groups (A and B) of structurally unrelated compounds. Group A members are polyunsaturated macrolactones and have lactam and lactone linkages with a lactone ring whilst group B are cyclic hexadepsipeptides (Yao and Moellering Jr, 1999). Although the ketolides and the combination compound quinupristin-dalfopristin are in the MLS antibiotic class, they will be discussed in separate sections of this chapter because generalizations about MLS antibiotics do not hold true for these compounds.

Erythromycin was first introduced into clinical practice in 1952 (McGuire *et al.*, 1952). Reports of resistance to erythromycin in clinical isolates of *S. aureus* followed shortly after this from France (Chabbert, 1956), the USA (Garrod, 1957) and the UK (Jones *et al.*, 1956). Isolates resistant to erythromycin were also shown to be resistant to, or rapidly develop resistance to other macrolides, lincosamides and streptogramin type B antibiotics (MLS phenotype).

The target sites for MLS antibiotics overlap and are focussed on the 50S ribosomal subunit where they bind reversibly to the peptidyl transferase centre and inhibit RNA-dependant protein synthesis. Erythromycin A has been shown to act in the early stages of protein synthesis by blocking the production of the growing peptide chain (Andersson and Kurland, 1987). It is presumed that this mechanism is the same for other 14-membered macrolides (Vester and Douthwaite, 2001). The mechanism for 16-membered macrolides is less clear although they do bind to the same region and have been shown to inhibit peptide bond formation more directly than 14-membered macrolides (Vasquez, 1979). Macrolides also inhibit the assembly of new ribosomal 50S subunits, which ultimately leads to a decrease in functional ribosomes in the cell (Champney and Tober, 1999).

Three general categories of MLS resistance have been reported for *S. aureus* – target site modification, active efflux and inactivation. Phenotypically, several categories of resistance patterns have been described: a) "M-type", resistance to 14- and 15-membered ring macrolides alone, b) "MS-type", resistance to 14- and 15-membered macrolides and streptogramin B, c) "MLS-type", resistance to 14-, 15- and 16-membered macrolides, lincosamides and streptogramin B. MLS resistance in *S. aureus* can be expressed in either an inducible or a constitutive manner (Allen, 1977; Weisblum, 1985).

Target Modification

Lai and Weisblum (1971) demonstrated that 23S rRNA from erythromycin resistant cells contained the modified base N^6,N^6-dimethyl adenine which was absent from erythromycin sensitive cells. Subsequently, causation has been attributed to post-transcriptional methylation of A2058 (*Escherichia coli* numbering) at the peptidyl transferase centre in domain V of 23S rRNA (Barta *et al.*, 1984; Skinner *et al.*, 1983). The family of enzymes responsible for methylation of A2058 have been designated as Erm (erythromycin resistance methylase) with the corresponding genes designated as *erm*.

To date, five different methylase genes have been described in *S. aureus* – *ermA* (Murphy, 1985), *ermB* (Wu *et al.*, 1999), *ermC* (Projan *et al.*, 1987), *ermF* (Chung *et al.*, 1999) and recently, *ermY* (Matsuoka *et al.*, 2002). The most commonly found genes in clinical isolates of *S. aureus* have been *ermA* and *ermC* (Westh *et al.*, 1995; Lina *et al.*, 1999). Interestingly, *ermA* is encoded on Tn*554* which has been physically mapped to the *mec* region of an MRSA strain (Dubin *et al.*, 1991), whilst *ermC* has been found on plasmids (Thakker-Varia *et al.*, 1987). In a surveillance study in France to determine the distribution of MLS resistance mechanisms in staphylococci, Lina *et al.* (1999) found that the distribution in MRSA to be 57.6% *ermA*, 0% *ermB* and 4.9% *ermC*, and in MSSA 5.6% *ermA*, 0.7% *ermB* and 20.1% *ermC*. This data would suggest that MLS resistance in MRSA is present mainly because of the close linkage between Tn*554* and the *mec* region.

The expression of methylase genes is either constitutive or inducible. The 14-membered macrolide erythromycin A is a potent inducer whilst 15- and 16-membered macrolides, lincosamides and streptogramin B antibiotics are not (Weisblum, 1985). Phenotypically, inducible strains show resistance to erythromycin A. After induction however, cross-resistance develops to all MLS antibiotics; an important implication to consider for those responsible for antimicrobial sensitivity testing and clinical management of patients. In an inducible population, constitutively resistant mutants occur readily in the presence of MLS antibiotics with low induction potential (Weisblum, 1985). At the molecular level, gene expression in inducible *ermC* strains is controlled by translational attenuation (Dabnau, 1984; Weisblum, 1985). In the non-induced form, the ribosomal binding site and start codon for the methylase are not accessible because they are sequestered in secondary RNA structures. Inducers (such as erythromycin A) alter this secondary structure to allow translation of the methylase to proceed. Mutations in nucleotides critical to this secondary structure in the attenuator cause constitutive expression (Werckenthin *et al.*, 1999). Although more complex, the regulation of *ermA* and *ermC* genes is thought to be similar (Murphy, 1985; Sandler and Weisblum, 1988).

Active Efflux

A second resistance phenotype (MS), is encoded by the *msrA* gene, is inducible by erythromycin and features cross-resistance to other 14- and 15-membered macrolides and streptogramin B only (Ross *et al.*, 1989). *MsrA* encodes an ATP-dependant efflux pump (Ross *et al.*, 1990). Although MS resistance due active efflux encoded by the *msrA* gene is common in coagulase-negative staphylococci (Eady *et al.*, 1993), it is rare in *S. aureus*. Martineau *et al.* (2000) reported the presence of this gene in 2 out of 73 erythromycin resistant clinical isolates of *S. aureus* tested. It is possible, however, that this type of resistance could be more prevalent. A report from from Hungary found that 25.6% of macrolide-resistant strains of *S. aureus* from 27 geographical regions over a 12-year period had the MS phenotype (Janosi *et al.*, 1990). A more recent report from Hungary found that no *msrA* was detected in *S. aureus*, including 73 strains of MRSA (Milch *et al.*, 2001). Clearly more research is required to determine the prevalence of efflux mediated macrolide resistance in MSSA and MRSA.

Inactivation

Enzymes that inactivate MLS antibiotics have been described only rarely in clinical isolates of *S. aureus*. In *E. coli*, esterases encoded by *ereA* and *ereB* inactivate eryththromycin by hydrolysing the lactone ring of the macrocyclic nucleus (Ounissi and Courvalin, 1985; Arthur *et al.*, 1987). Activity similar to that encoded by the *ere* genes has been described in a strain of *S. aureus* (Wondrack *et al.*, 1996). Inactivation of macrolides by phosphotransferases has also been described in *S. aureus* (Wondrack *et al.*, 1996).

Large surveillance studies performed by different groups, in different geographical regions, reveals MLS resistance is strongly associated with oxacillin resistance (Table 2). As previously discussed, the MS phenotype and efflux genotype is rarely reported in *S. aureus*. In this case, therefore, the expected predominant phenotype for *S. aureus* would be MLS resistance. However, as shown in Table 2, clindamycin susceptibility is considerably higher than erythromycin susceptibility in all studies. The probable explanation for this is that clindamycin susceptibility is falsely elevated by inducible methylases and this is inversely related to prevalence.

Marked geographical variation in resistance profiles have been observed in MRSA. For example, in one study, erythromycin and clindamycin resistance was found to be highly prevalent in MRSA isolated in Brazil and Uruguay whilst isolates from Mexico had low levels of resistance to these antimicrobials (Sanches *et al.*, 2000). Molecular typing of large numbers of geographically and chronologically diverse MRSA strains is revealing that clonal spread of

Table 2. Selected surveillance data for erythromycin and clindamycin: MRSA compared to MSSA

Antimicrobial	N	Organism	MIC values (mg/L)			% Susceptible
			range	MIC$_{50}$	MIC$_{90}$	

SENTRY study April 1997-February 1999, 25 European hospitals, isolates from patients with bacteremia, nosocomial pneumonia, wound infections and urinary tract infections (Fluit 2001).

Erythromycin	2287	MSSA	≤0.12->8	0.5	>8	77.5
Clindamycin			≤0.12->8	0.12	0.25	93.7
Erythromycin	764	MRSA	0.12->8	>8	>8	4.8
Clindamycin			0.12->8	>8	>8	23.3

PROTEKT study 1999-2000 winter season, global (69 centres in 25 countries), isolates from patients with community-acquired respiratory tract infections (www.protekt.org).

Erythromycin	1205	MSSA	0.06->64	0.25	>64	81.9
Clindamycin			0.03->4	0.12	0.12	97.5
Erythromycin	342	MRSA	0.12->64	>64	>64	4.7
Clindamycin			0.03->4	>4	>4	16.6

RESIST project 1997-1998, 18 countries in Europe, Asia and Latin America, isolates from wounds (24%), blood (22%), respiratory tract infection (21%), abscesses (7%), urine (5%) (Sanches 2000).

Erythromycin	1749	MSSA	-	-	-	58.9
Clindamycin			-	-	-	77.7
Erythromycin	112	MRSA	-	-	-	9.1
Clindamycin			-	-	-	17.5

MRSA may be limited to a small number of closely related strains (Diekema *et al.*, 2000). Presumably the prevalence of particular clones in a geographical centre is related to exposure and local selective pressures (such as infection control and prescribing practices, etc.).

Quinupristin-Dalfopristin

Pristinamycin is a naturally occurring antibacterial agent obtained from *Streptomyces pristinaespiralis*. Although pristinamycin is a potent antibacterial agent, it has not been useful in human therapeutics due to its insolubility in water, and hence inability to be produced in a parenteral form (Pechère, 1996). Quinupristin is a semisynthetic streptogramin B antibiotic derived from pristinamycin I$_A$. Dalfopristin is a semisynthetic streptogramin A antibiotic derived from pristinamycin II$_B$. Quinupristin/Dalfopristin in a 30:70 mixture (Synercid) is water-soluble and is used extensively in the treatment of serious infections caused by multi-drug resistant, Gram-positive pathogens, including MRSA.

Individually, streptogramin A and B inhibit protein synthesis and are bacteriostatic (Ennis, 1965). When combined, however, they become bactericidal by permanently halting protein synthesis (Cocito, 1979). The peptidyl transferase centre on the 50S ribosomal subunit is the site at which amino acids are added to the growing peptide chain (elongation). Aminoacyl-tRNA (AA-tRNA) molecules attach to the donor site within the peptidyl transferase centre and an amino acid is transferred to the growing peptide chain at the acceptor site (Cocito *et al.*, 1997). Streptogramin A exerts its bacteriostatic effect by blocking attachment of AA-tRNA to both the donor and acceptor sites of the peptidyl transferase centre (Chinali *et al.*, 1984). Streptogramin B antibiotics bind to the peptidyl transferase centre and prevent the extension of the growing peptide chain and induce premature detachment of the incomplete chain (Chinali *et al.*, 1988ab).

The synergism observed between streptogramin A and B antimicrobials is a result of their different target sites within the peptidyl transferase centre and inhibition of peptide synthesis during different stages of protein synthesis, i.e. the early and late phases respectively. In addition, upon binding of streptogramin A to the ribosome, ribosomal affinity for streptogramin B increases resulting in greater inhibition of protein synthesis (Beyer and Pepper, 1998).

In staphylococci, resistance to synergistic mixtures of streptogramin A and B is always associated with resistance to streptogramin A and never with streptogramin B alone (Haroche *et al.*, 2000). Resistance to streptogramin A in *S. aureus* mainly occurs by two mechanisms, inactivation and efflux. The genes *vat*(A), *vat*(B) and *vat*(C) encode acetyltransferases which inactivate streptogramin A (Allignet *et al.*, 1993; Allignet and El Sohl, 1995; Roberts *et al.*, 1999). The *vga*(A) and *vga*(B) genes encode ATP-binding proteins most likely involved in the active efflux of streptogramin A (Allignet *et al.*, 1992; Allignet and El Sohl, 1997). A variant of *vgaA* (*vgaAv*) was found on a new transposon Tn*5406* in clinical isolates of *S. aureus* (Haroche *et al.*, 2002). Tn*5406* is common in staphylococci, is a large self-transferable plasmid, and carries two other streptogramin A resistance genes *vatB* and *vgaB* (Haroche *et al.*, 2002). Recently, mutations in the L22 riboprotein were shown to be responsible for synercid resistance in *in vitro* mutants obtained from a strain isolated from a patient undergoing synercid therapy (Malbruny *et al.*, 2002).

The prevalence of quinupristin-dalfopristin resistance is greater in MRSA than MSSA. In a large European surveillance study, quinupristin-dalfopristin resistance was found in 4.7% of MRSA isolates (n = 764) compared to 0.5% of MSSA isolates (n = 2287) (Fluit *et al.*, 2001). However, resistance prevalence is geographically diverse. In a recent study in Taiwan, 31% (n = 80) of MRSA were resistant to quinupristin-dalfopristin compared to 0% (n = 68) of MSSA (Luh *et al.*, 2000).

Oxazolidinones

The oxazolidinones are a new class of synthetic antimicrobial agents first discovered in the 1970s. It was first thought that oxazolidinones interfered with the formation of the 30 S preinitiation complex hence blocking the initiation phase of bacterial protein synthesis (Shinabarger *et al.*, 1997). This has been shown not to be so but oxazolidinones have been shown to bind preferentially to the 50 S subunit of the ribosome and inhibit the formation of N-formylmethionyl-tRNA-ribosome-mRNA ternary complex (Shinabarger *et al.*, 1997; Swaney *et al.*, 1998). The precise mechanism of action has not yet been defined. Recent data suggests that these agents inhibit bacterial protein synthesis by interfering with fMet-tRNA^{Met} binding to the ribosomal peptidyltransferase donor site (Patel *et al.*, 2001). fMet-tRNA^{Met} initiates the formation of the first peptide bond and the peptidyltransferase donor site is vacant only immediately before the formation of the first peptide bond.

So far, apart from one report of resistance in a clinical MRSA isolate (Tsiodras *et al.*, 2001), MRSA and MSSA have been shown to be fully sensitive to linezolid and a new oxazolidinones AZD2563 in a clinical isolates (Abb, 2002; Muñoz-Bellido *et al.*, 2002; Johnson *et al.*, 2002).

Ketolides

Ketolides are semisynthetic derivatives of 14-membered macrolides and differ from these compounds by the substitution of a keto group at position C3 of the macrolactone ring and a carbamate extension at position C11,12 (Bryskier, 1997). Although the basic mechanism of action of the ketolides is the same as described for the macrolides above, the two structural modifications of the ketolides result in increased activity against a range of Gram-positive pathogens, including macrolide resistant *S. aureus* (Hamilton-Miller and Shah, 1998; Nilius *et al.*, 2001). However, the ketolides are inactive against staphylococci with constitutively expressed methylase genes (Schmitz *et al.*, 2002b). *In vitro* exposure to ketolides in *S. aureus* with inducible *ermA* or *ermC* genes resulted in mutation events (tandem duplications, deletions, IS insertion) in the translational attenuator region of these genes, causing constitutive expression of the gene and elevated MICs to the ketolides, hence suggesting that the ketolides should not be used for *S. aureus* with inducible MLS$_B$ resistance (Schmitz *et al.*, 2002c, Schmitz *et al.*, 2002d).

Recent surveillance data demonstrates that significant resistance to the ketolide telithromycin has developed in MSSA and MRSA with considerable variation between geographical regions (Cantón *et al.*, 2002). In this study, although 95.3% of 1239 MSSA isolates were susceptible to telithromycin, geographical variation ranged from 79.5% in Asia to 87.9% in Europe. For 308 MRSA isolates the overall susceptibility to telithromycin was 17.9% with 97.8% of MRSA isolates in Latin America resistant to telithromycin (n=93).

Mupirocin

Mupirocin (pseudomonic acid) is a topical antibiotic derived from the fermentation of *Pseudomonas fluorescens* (Fuller *et al.*, 1971). The structure of mupirocin is unique and it contains a hydroxynonanoic moiety linked to monic acid (Chain and Mellows, 1977). Mupirocin is an analogue of isoleucine and inhibits protein synthesis by binding irreversibly to the enzyme isoleucyl t-RNA synthetase (IleS) (Hughes and Mellows, 1978).

Two forms of mupirocin resistance have been described in *S. aureus*. Low-level resistance (8 to 256 mg/L) developed after the introduction of mupirocin in the clinical setting and the mechanism has recently been shown to be a mutation in the Rossman fold of IleS (Antonio *et al.*, 2000). High-level resistance (MIC >= 512 mg/L) is less common, was found before mupirocin was introduced clinically (Rahman *et al.*, 1990), and is the result of acquisition of a new gene *mupA*, which encodes a second isoleucyl-tRNA synthetase gene (Hodgson *et al.*, 1994).

Mupirocin is mainly used as an ointment and is extremely effective in eliminating MRSA from the nasal passages (Hill, 1988) and hence reducing nosocomial infections caused by MRSA (Harbarth, 1999). Mupirocin resistance in MRSA is rare and it is unclear whether *in vitro* resistance is clinically significant for two reasons: 1) as mupirocin is administered in a topical form, the concentration of mupirocin on the mucosal surface is extremely high (20,000 mg/L) and well above the MICs found in resistant strains, and 2) there is little information concerning clinical outcomes in patients treated with mupirocin and have resistant strains (Henkel and Finlay, 1999). Clearly, further research is needed in this field before valid conclusions can be drawn regarding the significance of mupirocin resistance in MRSA.

Tetracyclines

The tetracyclines are a class of broad-spectrum bacteriostatic antibiotics produced by various streptomycetes soil bacteria. Tetracycline itself is derived from *S. aureofaciens* and oxytetracycline is from *S. rimosis*. Both of these tetracyclines were discovered in the late 1940s and are referred to as first generation tetracyclines. Later, second generation semi-synthetic analogues were developed, such as doxycycline and minocycline (Chopra and Roberts, 2001). More recently, a third generation of tetracyclines, known as glycylcyclines, has emerged. These were developed to combat the various resistance mechanisms that have plagued previous generations of tetracyclines (Testa *et al,* 1993). Tigecycline (GAR-936) is the first of these agents currently undergoing clinical trials (Johnson, 2000).

The tetracyclines inhibit bacterial protein synthesis by blocking the association of aminoacyl-tRNA with the ribosome (Schnappinger and Hillen, 1996). Tetracyclines bind to a high affinity site (probably proteins S7, S14 and S19) on the small 30S subunit of the bacterial ribosome (Buck and Cooperman, 1990). This binding to the ribosome is reversible and is thought to be the reason why tetracyclines are bacteriostatic rather than bactericidal (Chopra *et al.*, 1992). The exceptions to this are the thiatetracyclines (atypical tetracyclines) that act by a bactericidal membrane-damaging mechanism (Chopra, 1994). However, these agents are not clinically available and will not be discussed here.

Transport of tetracyclines across the cytoplasmic membrane of Gram-positive bacteria is energy dependent, relying on the ΔpH component of the proton-motive force for uptake (Schnappinger and Hillen, 1996). For Gram-negative bacteria uptake must also occur across the outer-membrane via the OmpC and OmpF porins and tetracycline resistance can occur by direct porin gene mutation or by down-regulation of expression (eg. via the *marR* region of the *mar* locus), both producing reduced tetracycline accumulation (Levy, 1992). Such mutations are not relevant for Gram-positive bacteria such as staphylococci.

The main mechanisms for resistance to tetracyclines are either active efflux or ribosome protection mechanisms. The first resistance genes were named using a sequential system of letters, i.e. *tet*(A), *tet*(B) etc., and those genes with 80% amino acid sequence homology were later re-classified accordingly (Levy *et al.,* 1989). More recently all 26 letters of the Roman alphabet have been allocated and the nomenclature for new tetracycline resistance mechanisms should now follow a numbering system (Levy *et al*, 1999). For staphylococci, four genetic elements have been attributed to tetracycline resistance, namely: *tet*(K), *tet*(L), *tet*(M) and *tet*(O) (Chopra and Roberts, 2001). The first two genes are plasmid-located efflux genes and the second two are genes involved in ribosome protection found chromosomally or on transposons.

Efflux

Tetracycline efflux genes code for membrane-associated proteins that export tetracycline from the bacterial cell. Glycylcyclines are not transported by any of these proteins, but Gram-negative bacterial proteins can efflux minocycline. However, mutations within *tet*(A) and *tet*(B) have been found to confer glycylcycline resistance *in vitro* (Chopra and Roberts, 2001). The *tet* efflux genes belong to the so-called major facilitator superfamily (MFS) and encode for membrane-bound proteins of approximately 46 k-Da in size. These have been categorised into one of six groups based on amino acid sequence identity. The staphylococcal *tet*(K) and *tet*(L) genes belong to group 2 and have predicted protein structures consisting of 14 trans-membrane helices (McMurray and Levy, 2000). These two genes confer resistance to tetracycline and chlortetracycline, but not minocycline or glycylcyclines, and are found on small plasmids, such as pT181. In these instances cross-resistance to unrelated antibiotics may occur because these plasmids can also contain other antibiotic resistance determinants (Khan and Novick, 1983). Less commonly the *tet* genes can become integrated into the chromosome or large staphylococcal plasmids (Chopra and Roberts, 2001).

For Gram-negative bacteria, the tetracycline efflux resistance mechanism relies on two genes: one encoding the efflux protein itself and the other a repressor. Both genes share overlapping operator and promoter regions that are regulated by tetracycline at the level of transcription. It is thought that a tetracycline-Mg^{2+} complex binds to the repressor protein preventing its association with the operator. This allows transcription of the efflux and repressor genes until there is insufficient tetracycline present to inhibit binding of the repressor to the operator (Chopra and Roberts, 2001). For the staphylococcal *tet*(K) and *tet*(L) genes, on the other hand, no repressor proteins have been found. In *Staphylococcus hyicus*, however, there appears to be an open reading frame for a small leader peptide upstream of the plasmid-mediated *tet*(L) gene (Schwarz *et al*, 1992). Expression of Tet(L) is thought to be controlled by translational attenuation, as seen with inducible chloramphenicol and macrolide resistance (see relevant sections in this chapter). In this model, two forms of leader peptide mRNA and Tet(L) mRNA folding are possible. Under normal circumstances, translation of the leader peptide occurs allowing two mRNA stem loop structures to occur. The second stem loop masks the ribosome binding site for *tet*(L) and prevents its translation. In the presence of low levels of tetracycline, however, ribosomes are stalled during translation of the leader peptide. This results in an alternative, more thermodynamically favourable, secondary structure where only 1 stem loop occurs. This stem loop now leaves the *tet*(L) ribosome binding site exposed and therefore translation of Tet(L) can occur (Schwarz *et al*, 1992). In contrast to this, others working with chromosomal *tet*(L) from *Bacillus subtilis* have proposed that control via the leader peptide is not translational attenuation but translational coupling between the two genes (Stasinopoulos *et al*, 1998). The subtle, but complicated, genetic differences between these mechanisms are beyond the scopeof this chapter. Furthermore, it remains to be seen what control mechanisms exist with Staphylococci and MRSA.

Table 3. Selected tetracycline surveillance susceptibility data: methicillin-susceptible *S. aureus* (MSSA) versus methicillin-resistant *S. aureus* (MRSA)

Antimicrobial	N	Organism	MIC values (mg/L)			% Susceptible
			range	MIC_{50}	MIC_{90}	
SENTRY study 1997-1999, 30 USA hospitals, isolates from bloodstream, lung, skin/soft tissue and urine (Diekema *et al.*, 2001).						
Tetracycline	4714	MSSA	-	4	4	94.2
Tetracycline	2455	MRSA	-	4	>8	83.7
SENTRY study 1997-1999, 8 Canadian hospitals, isolates from bloodstream, lung, skin/soft tissue and urine (Diekema *et al.*, 2001).						
Tetracycline	1329	MSSA	-	4	4	95.7
Tetracycline	81	MRSA	-	4	>8	82.7
SENTRY study 1997-1999, 10 Latin American hospitals, isolates from bloodstream, lung, skin/ soft tissue and urine (Diekema *et al.*, 2001).						
Tetracycline	1274	MSSA	-	≤4	8	89.2
Tetracycline	682	MRSA	-	>8	>8	35.3
SENTRY study 1998-1999, 17 Western Pacific hospitals, isolates from bloodstream, lung, skin/ soft tissue and urine (Diekema *et al.*, 2001).						
Tetracycline	770	MSSA	-	≤4	4	90.8
Tetracycline	657	MRSA	-	>8	>8	17.8
SENTRY study 1997-1999, 24 European hospitals, isolates from bloodstream, lung, skin/soft tissue and urine (Diekema *et al.*, 2001).						
Tetracycline	2561	MSSA	-	≤4	>8	88.9
Tetracycline	916	MRSA	-	>8	>8	41.4
PROTEKT study 1999-2000 winter season, global (69 centres in 25 countries), isolates from patients with community-acquired respiratory tract infections (http://www.protekt.org).						
Tetracycline	1205	MSSA	0.12 - >16	0.25	16	88.5
Tetracycline	342	MRSA	0.12 - >16	16	>32	46.5
RESIST project 1997-1998, 18 countries in Europe, Asia and Latin America, isolates from wounds (24%), blood (22%), respiratory tract infection (21%), abscesses (7%), urine (5%) (Sanches *et al.*, 2000).						
Tetracycline	112	MSSA	-	-	-	62.5
Tetracycline	1749	MRSA	-	-	-	27.7

Ribosomal Protection

Ribosomal protection proteins, in contrast to efflux proteins, act within the bacterial cytoplasm and confer tetracycline resistance by binding to the bacterial ribosome without affecting the rate or extent of protein synthesis. There are 6 classes of ribosomal protection protein, all around 70 kDa in size: TetM, TetO, TetP, TetQ, TetS and OtrA. Only Tet(M) and Tet(O) polypeptides are found in *S. aureus* and they have 75% amino acid sequence similarity (Taylor and Chau, 1996). Ribosomal protection protein N-terminal regions show close homology to those of elongation factor GTPases EF-Tu and EF-G (Taylor and Chau, 1996). Tet(M) interacts with the ribosome and confers resistance by removing bound tetracycline from the ribosome by a process reliant on the hydrolysis of GTP. This hydrolysis may provide energy for a conformational change within the ribosome (Chopra and Roberts, 2001) that directly dislodges tetracycline from the ribosome or alternatively allows amino acyl-tRNA to bind even in the presence of tetracycline (Taylor and Chau, 1996). It has been shown that Tet(M) expression and resistance in Staphylococci and other bacteria is increased in the presence of tetracycline (Nesin *et al.*, 1990; Burdett, 1991). Like the efflux proteins discussed above, expression of the ribosomal protection proteins also appears to be switched on in the presence of tetracycline.

From Table 3, it can be seen that is a strong link between MRSA phenotype and tetracycline resistance in the Western Pacific, Latin America and Europe but not so in Northern America. From a separate analysis of the Sentry study 1997-1999, where direct comparisons were made between tetracycline, doxycycline and minocycline, it would appear that most tetracycline-resistant MRSA retain susceptibility to doxycycline or minocycline (Fluit *et al.*, 2001). Analysis of the tetracycline resistance mechanisms shown with the MRSA isolates from the Sentry study (Schmitz *et al.* 2001) show that 22.5% possessed *tet*(K) only, 25.5% possessed *tet*(M) only, but 50.5% possessed both genes. A minority (1.5%) possessed *tet*(L). The acquisition of *tet*(K), *tet*(M) or a combination of both in MRSA confers resistance to tetracycline. The level of resistance can be further enhanced by pre-incubation with sub-inhibitory concentrations of tetracycline or minocycline, i.e. the resistance is inducible (Trzcinski *et al.*, 2000). The presence of *tet*(K) is not sufficient to produce doxycycline or minocycline resistance following normal susceptibility testing procedures (Bismuth *et al* 1990; Trzcinski *et al.* 2000), but doxycycline resistance can be induced in *tet*(K) isolates with tetracycline pre-incubation (Trzcinski *et al.* 2000). The majority of MRSA with both *tet*(K) and *tet*(M) have resistance to tetracycline, doxycycline and minocycline without the need for pre-incubation in the presence of a tetracycline (Bismouth *et al.* 1990, Trzcinski *et al.* 2000; Schmitz *et al* 2001). Isolates harbouring *tet*(M) alone have also been shown to be resistant to doxycycline and minocycline without pre-incubation with tetracycline (Bismouth *et al.* 1990, Schmitz *et al* 2001). However, one study has reported that minocycline resistance does not occur with the majority of *tet*(M) MRSA until induction by tetracycline or minocycline pre-incubation has taken place (Trzcinski *et al.* 2000). These differences may be due to levels of inducibility of the *tet*(M) gene due to the methodology used or the set of isolates investigated.

Rifampicin and Fusidic Acid

Rifampicin is a semisynthetic derivative of rifamycin B, which is a member of the family of antibiotics known as the rifamycins produced by *Amycolatopsis mediterranea* (Parenti and Lancini, 1997). Fusidic acid is a steroid-like antibiotic first isolated from a strain of *Fusidium coccineum* (Godtfredsen *et al.*, 1962). The mechanism by which rifampicin exerts effect is by forming a stable complex with bacterial RNA polymerase β subunit, encoded by the *rpoB* gene, and therefore preventing transcription (Aboshkiwa *et al.*, 1995). Fusidic acid interferes with elongation factor G (EF-G) to inhibit bacterial protein synthesis (Tanaka *et al.*, 1968). Because these mechanisms of action and targets are different from those of other antibiotics, cross-resistance does not occur.

Although rifampicin and fusidic acid are both extremely active against *S. aureus* and MRSA, resistance develops rapidly when either agent is used as monotherapy. Rifampicin resistance occurs when mutations in the *rpoB* gene lead to structural changes in regions of the RNA polymerase β subunit that result in reduced affinity for the antibiotic (Morrow and Harmon, 1979). Chromosomal mutations are also responsible for resistance to fusidic acid. Certain mutations in the gene *fusA* resulting in an altered EF-G result in decreased fusidic acid binding (Chopra, 1976). A second mechanism of fusidic acid resistance in *S. aureus* is caused by a exclusion of the antibiotic from the cell by a permeability barriers encoded on the plasmid pUB101 (Sinden and Chopra, 1981).

Because of the rapid development of resistance when either agent is used as monotherapy, they are always used in combination with another antimicrobial, most usually with each other (O'Neill *et al.*, 2001). Although resistance to Rifampicin in MSSA is rare (susceptibility > 95% in most geographical regions of the world), high rates of resistance have been reported in MRSA in certain geographical regions. In a recent large global study, rifampicin susceptibility in MRSA was reported to be approximately 90% or greater in the USA, Canada and the Western Pacific regions but <50% in Europe and Latin America (Table 4). Globally, resistance to Fusidic Acid in MRSA tends to be low (<5%) and sporadic (Udo and Jacob, 2000). However, higher resistance rates have been reported in specific locations: 10% in one hospital in Saudia Arabia (Zaman and Dibb, 1994), 9% in Western Australia (Torvaldsen and Riley, 1996), and 11% in Auckland New Zealand (Lang *et al.*, 1992).

Table 4. Selected surveillance data for rifampicin: MRSA compared to MSSA. SENTRY study April 1997-February 1999, isolates from patients with bacteremia, nosocomial pneumonia, wound infections and urinary tract infections (Diekema *et al.*, 2001)

N	Organism	MIC values (mg/L)			%Susceptible
		Range	MIC$_{50}$	MIC$_{90}$	
Europe					
2561	MSSA	≤0.25-≥2	0.25	0.25	97.3
764	MRSA	≤0.25-≥2	2	>2	46.7
USA					
4714	MSSA	≤0.25-≥2	0.25	0.25	99.2
2455	MRSA	≤0.25-≥2	≤0.25	1	90.1
Canada					
1329	MSSA	≤0.25-≥2	≤0.25	0.25	99.3
81	MRSA	≤0.25-≥2	0.25	0.5	92.6
Latin America					
1274	MSSA	≤0.25-≥2	≤0.25	≤0.25	97.4
682	MRSA	≤0.25-≥2	2	>2	40.6
Western Pacific					
770	MSSA	≤0.25-≥2	≤0.25	≤0.25	99.1
657	MRSA	≤0.25-≥2	≤0.25	>2	88.6

Aminoglycosides

The aminoglycosides were first discovered in 1944 and are a diverse group of potent broad-spectrum bactericidal antibiotics with a chemical structure made up of aminated sugars joined to a dibasic cyclitol (this commonly being 2-deoxystreptamine). The aminoglycosides inhibit protein synthesis by binding to the 16S rRNA component of the ribosomal 30S subunit. Formation of the initiation complex between the ribosomal subunits, mRNA and tRNA is not inhibited, but elongation of the nascent peptide is disrupted to cause protein misreading. It is thought that some of these aberrant proteins are inserted into the cell membrane, which alters permeability and stimulates aminoglycoside transport into the bacterial cell (Busse *et al.*, 1992).

The various classes of aminoglycoside bind to different sites of the rRNA based on specific regions of the antibiotic and rRNA that compliment each other. One might anticipate, therefore, that bacterial resistance to aminoglycosides would be likely to occur via an alteration of the 16S rRNA either enzymatically or by chromosomal mutation (analogous to 23S rRNA methylases and/or mutations conferring macrolide resistance). Although

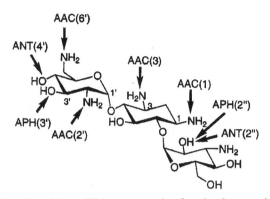

Figure 1. Aminoglycoside-modifying enzyme sites found on kanamycin B (from Kotra *et al.* 2000).

methylation of 16S rRNA is a mechanism employed by those organisms that naturally produce aminoglycosides to circumvent aminoglycoside action (Thompson *et al.*, 1985), clinical resistance occurs almost entirely by enzymatic modification of aminoglycoside molecules.

The enzymes that cause resistance to this class of antibiotics are *N*-acetyltransferases (AAC), *O*-nucleotidyltransferases (ANT), and *O*-phosphotransferases (APH) which all modify specific amino or hydroxyl groups. An example of the modification sites for kanamycin B is shown in Figure 1. The naming of these enzymes is based on the site of modification eg. AAC(6') with a further designation eg. AAC(6')-I to differentiate distinct proteins with the same modifying activity (Vanhoof *et al.*, 1998). These subtypes can vary by only a few and sometimes a single amino acid to produce variable substrate specificities (Mingeot-Leclercq *et al.*, 1999). There are four AACs, four ANTs and seven APHs. In addition there is one enzyme, AAC(6')-APH(2"), that is bifunctional. The structural modifications caused by these enzymes severely reduce the ability of aminoglycosides to bind to the target RNA (Kotra *et al.*, 2000).

In staphylococci, resistance to gentamicin, tobramycin and kanamycin is mediated by AAC(6')-APH(2") encoded by the *aac(6')-Ie+aph(2")* gene. This gene is commonly located on the transposon Tn*4001* and is widely distributed both chromosomally and on many plasmid types (Ubukata *et al.*, 1984). Other aminoglycoside modification resistance genes in staphylococci include *ant(4')-Ia)* which encodes ANT(4')-I and confers resistance to neomycin, kanamycin, tobramycin and amikacin. This gene is associated with small and large plasmids and also the *mec* region of the chromosome of some isolates (Stewart *et al.*, 1994). The *aph(3')-IIIa* gene, which encodes APH(3')-III, is also important, conferring resistance to neomycin and kanamycin. This gene is associated with transposon Tn*5405* (Derbise *et al.*, 1996).

Table 5. Selected gentamicin surveillance susceptibility data: methicillin-susceptible *S. aureus* (MSSA) versus methicillin-resistant *S. aureus* (MRSA)

Antimicrobial	N	Organism	MIC values (mg/L)			%Susceptible
			range	MIC_{50}	MIC_{90}	
SENTRY study 1997-1999, 30 USA hospitals, isolates from bloodstream, lung, skin/soft tissue and urine (Diekema *et al.,* 2001).						
Gentamicin	4714	MSSA	-	0.5	1	98.4
Gentamicin	2455	MRSA	-	1	>8	63.1
SENTRY study 1997-1999, 8 Canadian hospitals, isolates from bloodstream, lung, skin/soft tissue and urine (Diekema *et al.,* 2001).						
Gentamicin	1329	MSSA	-	0.5	1	98.5
Gentamicin	81	MRSA	-	0.5	>8	74.1
SENTRY study 1997-1999, 10 Latin American hospitals, isolates from bloodstream, lung, skin/soft tissue and urine (Diekema *et al.,* 2001).						
Gentamicin	1274	MSSA	-	0.5	1	95.4
Gentamicin	682	MRSA	-	>8	>8	7.0
SENTRY study 1998-1999, 17 Western Pacific hospitals, isolates from bloodstream, lung, skin/soft tissue and urine (Diekema *et al.,* 2001).						
Gentamicin	770	MSSA	-	0.5	1	98.4
Gentamicin	657	MRSA	-	>8	>8	25.3
SENTRY study 1997-1999, 24 European hospitals, isolates from bloodstream, lung, skin/soft tissue and urine (Diekema *et al.,* 2001).						
Gentamicin	2561	MSSA	-	0.5	1	95.0
Gentamicin	916	MRSA	-	>8	>8	26.2
RESIST project 1997-1998, 18 countries in Europe, Asia and Latin America, isolates from wounds (24%), blood (22%), respiratory tract infection (21%), abscesses (7%), urine (5%) (Sanches *et al.,* 2000).						
Gentamicin	112	MSSA	-	-	-	
Gentamicin	1749	MRSA	-	-	-	

In an almost identical way to that seen with tetracyclines there is a link between MRSA and resistance to aminoglycosides in Latin America, Europe and the Western Pacific but not North America (Table 5).

In the Schmitz *et al.* (1999) study of staphylococci circulating in Europe in 1997, the most common aminoglycoside resistance gene was found to be *aac(6')-Ie+aph(2")* with 76% of MRSA possessing this gene, followed by *ant(4')-Ia* (53%). Only 7% of MRSA isolates possessed *aph(3')-IIIa*. This contrasts with a similar European study carried out 10 years previously, where prevalence of *ant(4')-Ia* in all staphylococci was low at 10% and *aph(3')-III* was found in only 3 isolates (1.1%) (European Study Group on Antibiotic Resistance, 1987). Interestingly, in a recent study of Japanese MRSA, the *ant(4')-I* gene was the most common aminoglycoside resistance gene at 84.5%, followed by *aac(6')-aph(2")* at 61.7% and *aph(3')-III* at 8.9% (Ida *et al.*, 2001). Over 50% of the MRSA investigated possessed more than 1 resistance gene; most commonly being *ant(4')-I* and *aac(6')-(2")* at 48.0% (Ida *et al.*, 2001). Tracking changes in patterns of resistance mechanisms either geographically or over time are important because the various resistance genes do not show complete cross-resistance to aminoglycosides. For example, the presence of *aac(6')-aph(2")* will cause resistance to gentamicin, tobramycin and kanamycin but not to streptomycin, lividomycin or in most cases arbekacin. For *ant(4')-I*, on the other hand, only tobramycin and kanamycin are affected and for *aph(3')-III* only lividomycin, streptomycin and kanamycin are affected by the gene product (Ida *et al.*, 2001). This may explain the growing phenomenon of MRSA with more than one aminoglycoside-modifying gene. In addition to the acquisition of more than 1 resistance gene, mutation of known modifying enzymes can also change aminoglycoside specificity. For example, arbekacin resistance in a Japanese MRSA isolate has been attributed to a mutation in *aac(6')-aph(2")* producing a novel aminoglycoside-modifying enzyme with the ability to undertake 4'''-N-acetylation (Fujimura *et al.*, 2000).

Co-Trimoxazole

Co-trimoxazole is the commercially-available combination of the sulphonamide sulphamethoxazole and trimethoprim. The sulphonamides have been in clinical use since the 1930s and trimethoprim since the 1960s. The rationale for the ready-made availability of both compounds together is based on observed synergistic interactions *in vitro* (Bushby and Hitchings, 1968), which are not necessarily corroborated in clinical use (O'Brien *et al.*, 1982). Both agents have broad-spectrum antimicrobial activity and act to inhibit the folic acid biosynthesis pathway. Sulphamethoxazole and other sulphonamides act by mimicking *p*-aminobenzoic acid (PABA), the natural substrate for the bacterial enzyme dihydropteroate synthetase (DHPS) involved in the ultimate production of folic acid. In a later step in the pathway, trimethoprim inhibits dihydrohydrofolate reductase (DHFR), which is involved in the formation of tetrahydrofolate from dihydrofolate, due to the fact it is an analogue of dihydrofolate.

Despite the fact that sulphonamides and trimethoprim are synthetic agents to which bacteria had not been exposed to prior to clinical use, bacterial resistance to sulphonamides and trimethoprim is caused by numerous mechanisms including permeability/efflux changes, up-regulation of target enzymes, mutational changes in the target enzymes and the acquisition of additional drug-resistant target enzymes. There appears to be a larger number of resistance mechanisms with Gram-negative bacteria (based on a large number of DHFR genes (*dfr*)) compared with Gram-positive bacteria (Huovinen, 2001).

For staphylococci, the major mechanism of trimethoprim resistance is the expression of a type S1 trimethoprim-resistant DHFR encoded by the *dfrA* gene located on transposon T*n4003*. This transposon seems to have been disseminated world wide via various plasmids (Burdeska *et al.*, 1990) but can also be located chromosomally (Dale *et al.*, 1995). Trimethoprim resistance conferred by Tn*4003* can be either low or high level based on the extent of *drfA* transcription. Deletions adjacent to the insertion sequence IS*257* found on Tn*4003* are thought to reduce the level of transcription and produce low-level resistance. It has been postulated that this type of resistance is advantageous because it involves reduced 'metabolic overhead' compared to high-level resistance but still produces a sufficient level of trimethoprim resistance to be clinically beneficial (Leelaporn *et al.*, 1994). The type S1 DHFR has 3 amino acid changes compared to wild-type DHFR, but trimethoprim resistance is caused by a just one of these mutations: the Phe-98-Tyr alteration (Dale *et al.*, 1997). A type S2 DHFR (encoded by *dfrD*) has also been observed in *S. haemolyticus* and *Listeria monocytogenes* (Charpentier and Courvalin, 1997) but there have not been any reports of this gene in *S. aureus*.

Resistance to sulphonamides in *S. aureus* occurs due to mutations within the *dhps(folP)* gene. In contrast to trimethoprim resistance, the resistance to sulphonamides does not appear to be plasmid-mediated. Furthermore, numerous mutations affecting as many as 14 residues could be involved in sulphonamide resistance in *S. aureus* (Hampele *et al.*, 1997). It may be possible that some of these mutations are compensatory mutations that allow DHPS to function normally but retain sulphonamide resistance (Sköld, 2000).

As with most antibacterials, resistance to trimethoprim-sulphamethoxazole is associated with MRSA. For example, resistance to trimethoprim-sulphamethoxazole in MSSA from the Sentry Antimicrobial Program, 1997-1999 ranged from 1.2% to 2.6% world-wide whereas for MRSA, trimethoprim-sulphamethoxazole resistance ranged from 16% to 65.4% (Diekema *et al.*, 2001).

Conclusion

The mechanisms of antibacterial resitance shown by MRSA are diverse and include all known types of resistance such as target alteration, efflux and permeability changes. In addition a number of these mechanisms can be expressed in single isolates often contained within the same transposon or plasmid. This gives MRSA a huge selective advantage in the hospital setting and allows the spread of multi-drug resistant clones from ward to ward. Although strict hospital isolation and infection control procedures have helped in the battle to control MRSA, it is quite clear that we still require new antibacterial agents to treat infections caused by this problem bacterium.

References

Abb, J. 2002. *In vitro* activity of linezolid, quinupristin-dalfopristin, vancomycin, teicoplanin, moxifloxacin and mupirocin against methicillin-resistant *Staphylococcus aureus*: comparative evaluation by the E test and broth microdilution method. Diag. Microbiol. Infect. Dis. 43: 319-321.

Aboshkiwa, M., Rowland, G., and Coleman, G. 1995. Nucleotide sequence of the *Staphylococcus aureus* RNA polymerase *rpoB* gene and comparison of its predicted amino acid sequence with those of other bacteria. Biochem. Biophys. Acta. 1262: 73-78.

Acar, J.F., and Goldstein, F.W. 1997. Trends in bacterial resistance to fluoroquinolones. Clin. Infect. Dis. 24 (Suppl. 1): S67-73.

Aeschlimann, J.R., Dresser, L.D., Kaatz, G.W., and Rybak, M.J. 1999. Effects of NorA inhibitors on *in vitro* antibacterial activities and postantibiotic effects of levofloxacin, ciprofloxacin, and norfloxacin in genetically related strains of *Staphylococcus aureus*. Antimicrob. Agents Chemother. 43: 335-340.

Allen, N.E. 1977. Macrolide resistance in *Staphylococcus aureus*: inducers of macrolide resistance. Antimicrob. Agents Chemother. 11: 669-674.

Allignet, J., Loncle, V., and El Sohl, N. 1992. Sequence of a staphylococcal plasmid gene, *vga*, encoding a putative ATP-binding protein involved in resistance to virginiamycin A-like antibiotics. Gene 117: 45-51.

Allignet, J., Loncle, V., Simenel, C., Delepierre, M., and El Sohl, N. 1993. Sequence of staphylococcal gene, *vat*, encoding an acetyltransferase inactivating the A-type compounds of virginiamycin-like antibiotics. Gene 130: 91-98.

Allignet, J., and El Sohl, N. 1997. Characterization of a new staphylococcal gene *vgaB*, encoding a positive ABC transporter conferring resistance to streptogramin A and related compounds. Gene 202: 133-138.

Allignet, J., and El Sohl, N. 1995. Diversity among the Gram-positive acetyltransferases inactivating streptogramin A and structurally related compounds and characterization of a new staphylococcal determinant, *vatB*. Antimicrob. Agents Chemother. 39: 2027-2036.

Ambler, J.E., Drabu, Y.J., Blakemore, P.H., and Pinney, R.J. 1993. Mutator plasmid in a nalidixic acid-resistant strain of *Shigella dysenteriae* type 1. J. Antimicrob. Chemother. 31: 831-839.

Andersson, S., and Kurland, C.G. 1987. Elongating ribosomes *in vivo* are refractory to erythromycin. Biochim. 69: 901-904.

Antonio, M., McFerran, N., and Pallen, M.J. 2000. Mutations affecting the Rossman fold of isoleucyl-tRNA synthetase are correlated with low-level mupirocin resistance in *Staphylococcus aureus*. Antimicrob. Agents Chemother. 46: 438-442.

Arthur, M., Andremont, A., and Courvalin, P. 1987. Distribution of erythromycin esterase and rRNA methylase genes in members of the family Enterobacteriaceae highly resistant to erythromycin. Antimicrob. Agents Chemother. 31: 404-409.

Barta, A., Steiner, G., Brosius, J., Noller, H.F., and Kuechler, E. 1984. Identification of a site on 23S ribosomal RNA located at the peptidyl transferase center. Proc. Nat. Acad. Sci. USA 81: 3607-3611.

Beyer, D., and Pepper, K. 1998. The streptogramin antibiotics: update on their mechanism of action. Exp. Opin. Invest. Drugs. 7: 591-599.

Beyer, R., Pestova, E., Millichap, J.J., Stosor, V., Noskin, G.A., and Peterson, L.R. 2000. A convenient assay for estimating the possible involvement of efflux of fluoroquinolones by *Streptococcus pneumonaie* and *Staphylococcus aureus*: evidence for diminished moxifloxacin, sparfloxacin and trovafloxacin efflux. Antimicrob. Agents Chemother. 44: 798-801.

Bismuth, R., Zilhao, R., Sakamoto, H., Guesdon, J.L., and Courvalin, P. 1990. Gene heterogeneity for tetracycline resistance in *Staphylococcus* spp. Antimicrob Agents Chemother 34: 1611-1614.

Blanche, F., Cameron, B., Bernard, F-X., Maton, L., Manse, B., Ferrero, L., Ratet, N., Lecoq, C., Goniot, A., Bisch, D., and Crouzet, J. 1996. Differential behaviors of *Staphylococus aureus* and *Escherichia coli* type II topoisomerases. Antimicrob. Agents Chemother. 40: 2714-2720.

Boos, M., Mayer, S., Fischer, A., Köhrer, K., Scheuring, S., Heisig, P., Verhoef, J., Fluit, A.C., and F.-J. Schmitz. 2001. *In vitro* development of resistance to six quinolones in *Streptococcus pneumoniae*, *Streptococcus pyogenes*, and *Staphylococcus aureus*. Antimicrob. Agents Chemother. 45: 938-942.

Bryskier, A. 1997. Ketolides : new semisynthetic 14-membered macrolides. In: Expanding Applications for New Macrolides, Azalides and Streptogramins. S. H. Zinner, L. S. Young, J. F. Acar, and H. C. Neu, eds. Marcel Dekker, New York. p. 39-53.

Buck, M. A., and Cooperman B. S. 1990. Single protein omission reconstitution studies of tetracycline binding to the 30S subunit of *Escherichia coli* ribosomes. Biochem. 29: 5374–5379.

Burdeska, A., Ott, M., and Then, R.L. 1990. Identical genes for trimethoprim-resistant dihydrofolate reductase from *Staphylococcus aureus* in Australia and central Europe. FEBS Lett. 266: 159-162.

Burdett, V. 1991. Purification and characterization of Tet(M), a protein that renders ribosomes resistant to tetracycline. J. Biol. Chem 266: 2872-2877.

Bushby, S.R.M., and Hitchings, G.H. 1968. Trimethoprim, a sulphonamide potentiator. Br. J. Pharmacol. Chemother. 33: 72-90.

Busse, H.J., Wostmann, C., and Bakker, E.P. 1992. The bactericidal action of streptomycin: membrane permeabilisation caused by the insertion of mistranslated proteins into the cytoplasmic membrane of *Escherichia coli* and subsequent caghing of the antibiotic inside the cells: degredation of these proteins. J. Gen. Microbiol. 138: 551-561.

Cantón, R., Loza, E., Morosini, M.I., and Baquero, F. 2002. Antimicrobial resistance amongst isolates of *Streptococcus pyogenes* and *Staphylococcus aureus* in the PROTEKT antimicrobial surveillance programme during 1999-2000. J. Antimicrob. Chemother. 50 (Suppl. S1): 9-24.

Chabbert, Y. 1956. Antagonisme *in vitro* entre l'erythromycine et la spiroamycine. Ann. Instit. Past. (Paris) 90 : 787-790.

Chain, E.B., and Mellows, G. 1977. The structure of psudomonic acid A, a novel antibiotic produced by *Pseudomonas fluorescens*. J. Chem. Soc.: 294-309.

Champney, W.S., and Tober, C.L. 1999. Superiority of 11,12 carbonate macrolides antibiotics as inhibitors of translation and 50S ribosomal subunit formation in *Staphylococcus aureus*. Curr. Microbiol. 38: 342-348.

Charpentier, E., and Courvalin, P. 1997. Emergence of the trimethoprim gene *dfrD* in *Listeria monocytogenes* BM4293. Antimicrob. Agents Chemother. 41: 1134-1136.

Chinali, G., Moreau, P., and Cocito, C. 1984. The action of virginiamycin M on the acceptor, donor and catalytic sites of peptidyltransferase. J. Biol. Chem. 259: 9563-9569.

Chinali, G., Nyssen, E., Di Giambattista, M., and Cocito, C. 1988a. Inhibition of polypeptide synthesis in cell-free systems by virginiamycin S and erythromycin. Evidence for a common mode of action of type B synergimycins and 14-membered macrolides. Biochim. Biophys. Acta 949: 71-78.

Chinali, G., Nyssen, E., Di Giambattista, M., and Cocito, C. 1988b. Action of erythromycin and virginiamycin S on polypeptides synthesis in cell-free systems. Biochim. Biophys. Acta 951: 42-52.

Chopra, I. 1976. Mechanisms of resistance to fusidic acid in *Staphylococcus aureus*. J. Gen. Microbiol. 96: 229–238

Chopra, I. 1994. Tetracycline analogs whose primary target is not the bacterial ribosome. Antimicrob. Agents Chemother. 38: 637-640.

Chopra, I., and Roberts, M. 2001. Tetracycline antibiotics: mode of action, applications, molecular biology, and epidemiology of bacterial resistance. Microb. Mol. Rev. 65: 232-260.

Chopra, I., Hawkey, P.M., and Hinton, M. 1992. Tetracyclines, molecular and clinical aspects. J. Antimicrob. Chemother. 29: 245-277.

Chung, W.O., Werckenthin, C., Schwarz, S., and Roberts, M.C. 1999. Host range of the *ermF* rRNA methylase gene in bacteria of human and animal origin. J. Antimicrob. Chemother. 43: 5-14.

Cocito, C. 1979. Antibiotics of the virginiamycin family, inhibitors which contain synergistic components. Microbiol. Rev. 43: 145-198.

Cocito, C., Di Giambattista, M., Nyssen, G., and Vannufel, P. 1997. Inhibition of protein synthesis by streptogramins and related antibiotics. J. Antimicrob. Chemother. 39 (Suppl. A): 7-13.

Dabnau, D. 1984. Translational attenuation: the regulation of bacterial resistance to the macrolide-lincosamide-streptogramin B antibiotics. Crit. Rev. Biochem. 16: 103-132.

Dale, G.E., Broger, C., Hartman, P.G., Langen, H., Page, M.G., Then, R.L., and Stüber, D. 1995. Characterisation of the gene for chromosomal dihydrofolate reductase (DHFR) of *Staphylococcus epidermidis* ATCC 14990: the origin of the trimethoprim-resistant S1 DHFR from *Staphylococcus aureus*? J. Bacteriol. 177: 2965-2970.

Dale, G.E., Broger, C., D'Arcy, A., Hartman, P.G., DeHoogt, R., Jolidon, S., Kompis, I., Labhardt, A.M., Langen, H., Locher, H., Page, M.G., Stuber, D., Then, R.L., Wipf, B., and Oefner, C. 1997. A single amino acid substitution in *Staphylococcus aureus* dihydrofolate reductase determines trimethoprim resistance. 266: 23-30.

Derbise, A., Dyke, K.G.H., and El Solh, N. 1996. Characterisation of a *Staphylococcus aureus* transposon Tn*5405*, located within Tn*5404* carrying aminoglycoside resistance genes, *aphA-3* and *aadE*. Plasmid 35: 74-88.

Diekma, D.J., Pfaller, M.A, Turnidge, J., Verhoef, J., Bell, J., Fluit, A.C., Doern, G.V., Jones, R., and the SENTRY participants group. 2000. Genetic relatedness of multidrug-resistant, methicillin (oxacillin)-resistant *Staphylococcus aureus* bloodstream isolates from SENTRY antimicrobial resistance surveillance centres worldwide, 1998. Microb. Drug Res. 6: 213-221.

Diekema, D.J., Pfaller, M.A., Schmitz, F.J., Smayevsky, J., Bell, J., Jones, R.N., Beach, M., and the SENTRY Participants Group. 2001. Survey of infections due to *Staphylococcus* species: frequency of occurrence and antimicrobial susceptibility of isolates collected in the United States, Canada, Latin America, Europe, and the Western Pacific Region for the SENTRY Antimicrobial Program, 1997-1999. Clin. Infect. Dis. 32 (Suppl. 2): S114-132.

Drlica, K., and Zhao, X. 1997. DNA gyrase, topoisomerase IV, and the 4-quinolones. Microbiol. Mol. Biol. Rev. 61: 377-392.

Dubin, D.T., Matthews, P.R., Chikramane, S.G., and Stewart PR. 1991. Physical mapping of the *mec* region of an American methicillin-resistant *Staphylococcus aureus* strain. Antimicrob. Agents Chemother. 35: 1661-1665.

Eady, E.A., Ross J.I., Tipper, J.L., Walters, C.E., Cove, J.H., and Noble, W.C. 1993. Distribution of genes encoding erythromycin ribosomal methylases and an erythromycin efflux pump in epidemiologically distinct groups of staphylococci. J. Antimicrob. Chemother. 31: 211-217.

Ennis, H.L. 1965. Inhibition of protein synthesis by polypeptide antibiotics. I. Inhibition of intact bacteria. J. Bacteriol. 90: 1102-1108.

European Study Group on Antibiotic Resistance. 1987. *In-vitro* susceptibility to aminoglycoside antibiotics in blood and urine isolates consecutively collected in twenty-nine European laboratories. Eur. J. Clin. Microbiol. 6: 378-385.

Ferrero, L., Cameron, B., Manse, B., Lagneaux, D., Crouzet, J., Famechon, A., and Blanche, F. 1994. Cloning and primary structure of *Staphylococcus aureus* DNA topoisomerase IV: a primary target of fluoroquinolones. Mol. Microbiol. 13: 641-653.

Ferrero, L., Cameron, B., and Crouzet, J. 1995. Analysis of *gyrA* and *grlA* mutations in stepwise-selected ciprofloxacin-resistant mutants of *Staphylococcus aureus*. Antimicrob. Agents Chemother. 39: 1554-1558.

Fluit, A.C., Wielders, C.L.C., Verhoef, J., and Schmitz, F.-J. 2001. Epidemiology and susceptibility of 3,051 *Staphylococcus aureus* isolates from 25 university hospitals participating in the European SENTRY study. J. Clin. Microbiol. 39: 3727-3732.

Fournier, B., and Hooper, D.C. 1998. Mutations in topoisomerase IV and DNA gyrase of *Staphylococus aureus*: novel pleiotropic effects on quinolone and coumarin activity. Antimicrob. Agents Chemother. 42: 121-128.

Fournier, B., Aras, R., and Hooper, D.C. 2000. Expression of the multidrug resistance transporter NorA from *Staphylococcus aureus* is modified by a two-component regulatory system. J. Bacteriol. 182: 664-671.

Fournier, B., Truong-Bolduc, Q.C., Zhang, X., and Hooper, D.C. 2001. A mutation in the 5' untranslated region increases stability of *norA* mRNA, encoding a multidrug resistance transporter of *Staphylococcus aureus*. J. Bacteriol. 183: 3267-2371.

Fujimura, S., Tokue, Y., Takahashi, H., Kobayashi, T., Gomi, K., Abe, T., Nukiwa, T., and Watanabe, A. 2000. Novel arbekacin- and amikacin-modifying enzyme of methicillin-resistant *Staphylococcus aureus*. FEMS Microbiol. Lett. 190: 299-303.

Fuller, A.T., Mellows, G., Woolford, M., Banks, G.T., Barrow, K.D., and Chain, E.B. 1971. Pseudomonic acid: an antibiotic produced by *Pseudomonas fluorescens*. Nature 234: 416-417.

Garrod, L.P. 1957. The erythromycin group of antibiotics. Br. Med. J. II: 57-63.

Godtfredsen, W., Roholt, K., and Tybring, L. 1962. Fucidin. A new orally active antibiotic. Lancet I: 928-931.

Gomes, A.R., Sanches, I.S., De Sousa, M.A., Castañeda, E., and De Lencastre, H. 2001. Molecular epidemiology of methicillin-resistant *Staphylococcus* in Colombian hospitals: dominance of a single unique multidrug-resistant clone. Microb. Drug Res. 7: 23-32.

Gootz, T.D., Zaniewski, R.P., Haskell, S.L., Kaczmarek, F.S., and Maurice, A.E. 1999. Activities of trovafloxacin compared with those of other fluoroquinolones against purified topoisomerases and *gyrA* and *grlA* mutants of *Staphylococcus aureus*. Antimicrob. Agents Chemother. 43: 1845-1855.

Gordon, K.A., Pfaller, M.A., Jones, R.N., and the Sentry participants group. 2002. BMS284756 (formerly T-3811, a des-fluoroquinolone) potency and spectrum testsed against over 10,000 bacterial bloodstream infection isolates from the SENTRY antimicrobial surveillance programme. J. Antimicrob. Chemother. 49: 851-855.

Hamilton-Miller, J.M.T., and Shah, S. 1998. Comparative *in-vitro* activity of ketolide HMR 3647 and four macrolides against Gram-positive cocci of known erythromycin susceptibility status. J. Antimicrob. Chemother. 41: 649-653.

Hampele, I.C., D'Arcy, A., Dale, G.E., Kostrewa, D., Nielson, J., Oefner, C., Page, M.G., Schonfeld, H.J., Stuber, D., and Then, R.L. 1997. Structure and function of the dihydropteroate synthase from *Staphylococcus aureus*. J. Mol. Biol. 268: 21-30.

Harbarth, S., Dharan, S., Liassine, N., Herrault, P., Auckenthaler, R, and Pittet, D. 1999. Randomized, placebo-controlled, double-blind trial to evaluate the efficacy of mupirocin for eradicating carriage of methicillin-resistant *Staphylococcus aureus*. Antimicrob Agents Chemother. 43:1412-1416.

Haroche, J., Allignet, J., Buchrieser, C., and El Sohl, N. 2000. Characterization of a variant of *vga*(A) conferring resistance to streptogramin A and related compounds. Antimicrob. Agents Chemother. 44: 2271-2275.

Haroche, J., Allignet, J., and El Sohl, N. 2002. Tn*5406*, a new staphylococcal transposon conferring resistance to streptogramin A and related compounds including dalfopristin. Antimicrob. Agents Chemother. 46: 2337-2343.

Henkel, T., and Finlay, J. 1999. Emergence of resistance during mupirocin treatment: is it a problem in clinical practice? J. Chemother. 11: 331-337.

Hill, R.L., Duckworth, G.J., Casewell, M.W. 1988. Elimination of nasal carriage of methicillin-resistant *Staphylococcus aureus* with mupirocin during a hospital outbreak. J. Antimicrob Chemother. 22: 377-384.

Hodgson, J.E., Curnock, S.P., Dyke, K.G., Morris, R., Sylvester, D.R., and Gross, M.S. 1994. Molecular characterization of the gene encoding high-level mupirocin resistance in *Staphylococcus aureus* J2870. Antimicrob. Agents Chemother. 38: 1205-1208.

Hughes, J., and Mellows, G. 1978. On the mode of action of pseudomonic acid: Inhibition of protein synthesis in *Staphylococcus aureus*. J. Antibiot. 31: 330-335.

Huovinen, P. 2001. Resistance to trimethoprim-sulphamethoxazole. Clin. Infect. Dis. 32: 1608-1614.

Ida, T., Okamoto, R., Shimauchi, C., Okubo, T., Kuga, A., and Inoue, M. 2001. Idnetification of aminoglycoside-modifying enzymes by susceptibility testing: epidemiology of methicillin-resistant *Staphylococcus aureus* in Japan. J. Clin. Micro. 39: 3115-3121.

Jacoby, G., Chow, N., and Waites, K. 2001. Prevalence of plasmid-mediated quinolone resistance. In: Abstracts 41st Interscience Conference on Antimicrobial Agents and Chemotherapy. American Society for Microbiology, Washington. Abst. C2-2120, p. 145.

Janosi, L., Nakajima, Y., and Hashimoto, H. 1990. Characterization of plasmids that confer inducible resistance to 14-membered macrolides and streptogramin B antibiotics in *Staphylococcus aureus*. Microbiol. Immunol. 34: 723-735.

Jevons, M.P. 1961. 'Celbenin'-resistant staphylococci. Br. Med. J. 1: 124-125.

Johnson, A.P., Warner, M., and Livermore, D.M. 2002. *In vitro* activity of a novel oxazolidinones, AZD2563, against randomly selected and multiresistant Gram-positive cocci. J. Antimicrob. Chemother. 50: 89-93.

Johnson, A.P. 2000. GAR-936. Curr. Opin. Anti-infect. Investig. Drugs 2: 164-170.

Johnson, A.P., Aucken, H.M., Cavendish, S., Ganner, M., Wale, M.C.J., Warner, M., Livermore, D.M., Cookson, B.D., and the UK EARSS participants. 2001. Dominance of EMRSA-15 and −16 among MRSA causing nosncomial bacteraemia in the UK: analysis of isolates from the European Resistance Surveillance System (EARSS). J. Antimicrob. Chemother. 48: 143-144.

Jones, W.F. Jr, Nichols, R.L., and Finland, M. 1956. Development of resistance and cross- resistance *in vitro* to erythromycin, carbomycin, oleandomycin, and streptogramin. Proc.Soc.Exp. Biol. Med. 93: 388-393.

Kaatz, G.W., and Seo, S.M. 1995. Inducible NorA-mediated multidrug resistance in *Staphylococcus aureus*. Antimicrob. Agents Chemother. 39: 2650-2655.

Khan, S.A., and Novick, R.P. 1983. Complete nucleotide sequence of pT181, a tetracycline resistance plasmid from *Staphylococcus aureus*. Plasmid 30: 163-166.

Kotra, L.P., Haddad, J., and Mobashery, S. 2000. Aminoglycosides: perspectives on mechanisms of action and resistance and strategies to counter resistance. Antimicrob. Agents Chemother. 44: 3249-3256.

Lai, C.J., and Weisblum, B. 1971. Altered methylation of ribosomal RNA in an erythromycin-resistant strain of *Staphylococcus aureus*. Proc. Nat. Acad. Sci. USA 68: 856-860.

Lang, S., Raymond, N., and Brett, M. 1992. Mupirocin-resistant *S. aureus* in Auckland. NZ. Med. J. 105: 438.

Leelaporn, A., Firth, N., Byrne, M.E., Roper, E., and Skurray, R.A. 1994. Possible role of insertion sequence IS257 in dissemination and expression of high- and low-level trimethoprim resistance in staphylococci. Antimicrob. Agents Chemother. 38: 2238-2244.

Levy, S.B. 1992. Active efflux mechanisms for antimicrobial resistance. Antimicrob. Agents Chemother. 36: 695-703.

Levy, S.B., McMurry, L.M., Burdett, V., Courvalin, P., Hillen, W., Roberts, M.C., and Taylor, D.E. 1989. Nomenclature for tetracycline resistance determinants. Antimicrob. Agents Chemother. 33: 1373–1374.

Levy, S.B., McMurray, L.M., Barbosa, T.M., Burdett, V., Courvalin, P., Hillen, W., Roberts, M.C., Rood, J.I., and Taylor, D.E. 1999. Nomenclature for new tetracycline resistance determinants. Antimicrob. Agents Chemother. 43: 1523-1524.

Lina, G., Quaglia, A., Reverdy, M.-E., Leclercq, R., Vandenesch, F., and Etienne, J. 1999. Distribution of genes encoding resistance to macrolides, lincosamides, and streptogramins among staphylococci. Antimicrob. Agents Chemother. 43: 1062-1066.

Low, D.E., Muller, M., Duncan, C.L., Willey, B.M., de Azavedo, J.C., McGeer, A., Krieswirth, B.N., Pong-Porter, S., and Bast, D.J. 2002. Activity of BMS-284756, a novel des-fluoro(6)quinolone, against *Staphylococcus aureus*, including contributions of mutations to quinolone resistance. Antimicrob. Agents Chemother. 46: 1119-1121.

Luh, K.-T., Hsueh, P.-R., Teng, L.-J., Pan, H.-J., Chen, Y.-C., Lu, J.-J., Wu, J.-J., and Ho, S.-W. 2000. Quinupristin-dalfopristin resistance among Gram-positive bacteria in Taiwan. Antimicrob. Agents Chemother. 44: 3374-80.

Malbruny, B., Canu, A., Bozdogan, B., Fantin, B., Zarrouk, V., Dutka-Malen, S., Feger, C., and Leclercq, R. 2002. Resistance to quinupristin-dalfopristin due to mutation of L22 ribosomal protein in *Staphylococcus aureus*. Antimicrob. Agents Chemother. 46: 2200-2207.

Markham, P.N., and Neyfakh, A.A. 1996. Inhibition of the multidrug transporter NorA prevents emergence of norfloxacin resistance in *Staphylococcus aureus*. Antimicrob. Agents Chemother. 40: 2673-2674.

Martineau, F., Picard, F.J., Lansac, N., Ménard, C., Roy, P.H., Ouellette, M., and Bergeron, M.G. 2000. Correlation between the resistance genotype determined by multiplex PCR assays and the antibiotic susceptibility patterns of *Staphylococcus aureus* and *Staphylococcus epidermidis*. Antimicrob. Agents Chemother. 44: 231-238.

Matsuoka, M., Inoue, M., Nakajima, Y., and Endo, Y. 2002. New *erm* gene in *Staphylococcus aureus* clinical isolates. Antimicrob. Agents Chemother. 46: 211-215.

McCloskey, L., Moore, T., Niconovich, N., Donald, B., Broskey, J., Jakielaszek, C., Rittenhouse, S., and Coleman, K. 2000. *In vitro* activity of gemifloxacin against a broad range of recent clinical isolates from the USA. J. Antimicrob. Chemother. 45 (Suppl. S1): 13-21.

McGuire, J.M., Bunch R.L., Anderson, R.C., Boaz, H.E., Flynn, E.H., Powell, H.M., and Smith, H.W. 1952. 'Ilotycin', a new antibiotic. Antibiot. Chemother. 2: 281-283.

McMurray, L.M., and Levy, S.B. 2000. Tetracycline resistance in Gram-positive bacteria. In: Gram Positive Pathogens. Fischetti, R.P., Novick, R.P., Ferreti, J.J., Portnoy, D.A., and Rood, J.I., eds. American Society for Microbiology, Washington DC, USA. p. 660-677.

Milch, H., Paszti, J., Erdosi, T., and Hetzmann, M. 2001. Phenotypic and genotypic properties of methicillin resistant *Staphylococcus aureus* strains isolated in Hungary, 1997-2000. Acta Microbiol. Immunol. Hung. 48: 457-77.

Mingeot-Leclercq, M.-P., Glupczynski, Y., and Tulkens, P.M. 1999. Aminoglycosides: activity and resistance. Antimicrob. Agents Chemother. 43: 727-737.

Morrow, T.O., and Harmon, S.A. 1979. Genetic analysis of *Staphylococcus aureus* RNA polymerase mutants. J. Bacteriol. 137: 374-383.

Muñoz-Bellido, J.L., Manzanares, M.A.A., Andrés, J.A.M., Zufiaurre, M.N.G., Ortiz, G., Hernández, M.S., and García-Rodríguez, J.A. 1999. Efflux pump-mediated quinolones resistance in *Staphylococcus aureus* strains wild type for *gyrA, gyrB, grlA,* and *norA*. Antimicrob. Agents Chemother. 43: 354-356.

Muñoz-Bellido, J.L., Zufiaurre, M.N.G., Hernandez, F.J.S., Guirao, G.Y., Hernández, M. S., and García-Rodríguez, J.A. 2002. *In vitro* activity of linezolid, synercid and telithromycin against genetically defined high level fluoroquinolone-resistant *Staphylococcus aureus*. Int. J. Antimicrob. Agents 20: 61-64.

Murphy, E. 1985. Nucleotide sequence of *ermA*, a macrolide-lincosamide-streptogramin B determinant in *Staphylococcus aureus*. J. Bacteriol. 162: 633-640.

De Neeling, A.J., van Leeuwen, W.J., Schouls, L.M., Schot, C.S., van Veen-Rutgers, A., Beunders, A.J., Buiting, A.G.M., Hol, C., Ligtvoet, E.E.J., Petit, P.L., Sabbe, L.J.M., van Griethuysen, A.J.A., and van Embden, J.D.A. 1998. Resistance of staphylococci in The Netherlands: surveillance by an electronic network during 1989-1995. J. Antimicrob. Chemother. 41: 93-101.

Nesin, M., Svec, P., Lupski, J.R., Godson, G.N., Krieswirth, B., Kornblum, J., and Projan, S.J. 1990. Cloning and nucleotide sequence of a chromosomally-encoded tetracycline resistance determinant, *tetA*(M), from a pathogenic, methicillin-resistant strain of *Staphylococcus aureus*. Antimicrob. Agents Chemother. 34: 2273-2276.

Neyfakh, A.A. 1992. The multidrug efflux transporter of *Bacillus subtilis* is a structural and functional homolog of the staphylococcus NorA protein. Antimicrob. Agents Chemother. 36: 484-485.

Neyfakh, A.A., Borsch, C.M., and Kaatz, G.W. 1993. Fluoroquinolone efflux protein NorA of *Staphylococcus aureus* is a multidrug efflux transporter. Antimicrob. Agents Chemother. 37: 128-129.

Ng, E., Trucksis, Y.M., and Hooper, D.C. 1994. Quinolone resistance mediated by *norA*: physiologic characterisation and relationship to *flqB*, a quinolone resistance locus on the *Staphylococcus aureus* chromosome. Antimicrob. Agents Chemother. 38: 1345-1355.

Ng, E., Trucksis, Y.M., and Hooper, D.C. 1996. Quinolone resistance mutations in topoisomerase IV: relationship to the *flqA* locus and genetic evidence that topoisomerase IV is the primary target and DNA gyrase is the secondary target of fluoroquinolones in *Staphylococcus aureus*. Antimicrob. Agents Chemother. 40: 1881-1888.

Nilius, A.M., Bui, M.H., Almer, L., Hensey-Rudloff, D., Beyer, J., May, Z., Or Y,S, and Flamm, R.K. 2001. Comparative *in-vitro* activity of ABT-773, a novel antibacterial ketolide. Antimicrob. Agents Chemother. 45: 2163-2168.

O'Brien, T.F., Acar, J.F., Altmann, G., Blackburn, B.O., Chao, L., Courtieu, A.-L., Evans, D.A., Guzman, M., Holmes, M., Jacobs, M.R., Kent, R.L., Norton, R.A., Koornhof, H.J., Mdediros, A.A., Pasculle, A.W., Surgalla, M.J., and Williams, J.D. 1982. Session II. Laboratory surveillance of synergy between and resistance to trimethoprim and sulfonamides. Rev. Infect. Dis. 4: 351-357.

O'Neill, A.J., Cove, J.H., and Chopra, I. 2001. Mutation frequencies for resistance to fusidic acid and rifampicin in *Staphylococcus aureus*. J. Antimicrob. Chemother. 47: 647-650.

Ounissi, H., and Courvalin, P. 1985. Nucleotide sequence of the gene *ereA* encoding the erythromycin esterase in *Escherichia coli*. Gene 35: 271-278.

Parenti, F., and Lancini, G. 1997. Rifamycins. In: Antibiotic and Chemotherapy: Anti-infective Agents and Their Use in Therapy 7th Edition. F. O'Grady, H.P. Lambert, R.G. Finch, and D. Greenwood, eds. Churchill Livingstone, London. p. 453-459.

Patel U, Yan YP, Hobbs FW Jr, Kaczmarczyk J, Slee AM, Pompliano DL, Kurilla MG, Bobkova EV. 2001. Oxazolidinones mechanism of action: inhibition of the first peptide bond formation. J. Biol. Chem. 276: 37199-37205.

Pechère J-C. 1996. Streptogramins: a unique class of antibiotics. Drugs 51 (Suppl. 1): 13-19.

Projan, S.J., Monod, M., Narayanan, C.S., and Dubnau, D. 1987. Replication properties of pIM13, a naturally occurring plasmid found in *Bacillus subtilis*, and of its close relative pE5, a plasmid native to *Staphylococcus aureus*. J. Bacteriol. 169: 5131-5139.

Rahman, M., Connolly, S., Noble, W.C., Cookson, B., and Phillips, I. 1990. Diversity of staphylococci exhibiting high-level resistance to mupirocin. J. Med. Microbiol. 33: 97-100.

Roberts, M.C., Sutcliffe, J., Courvalin, P., Bogo, J. L., Rood, J., and Seppala, H. 1999. Nomenclature for macrolides and macrolides-lincosamide-streptogramin B resistance determinants. Antimicrob. Agents Chemother. 43: 2823-2830.

Ross, J.I., Farrell, A.M., Eady, E.A., Cove, J.H., and Cunliffe W.J. 1989. Characterization and molecular cloning of the novel macrolide-streptogramin B resistance determinant from *Staphylococcus epidermidis*. J. Antimicrob. Chemother. 24: 851-862.

Ross, J.I., Eady, E.A, Cove, J.H., Cunliffe, W.J., Baumberg, S., and Wootton, J.C. 1990. Inducible erythromycin resistance in staphylococci is encoded by a member of the ATP-binding transport super-gene family. Mol. Microbiol. 4: 1207-1214.

Roychoudhury, S., Catrenich, C.E., McIntosh, E.J., McKeever, H.D., Makin, K.M., Koenigs, P.M., and Ledoussal, B. 2001. Quinolone resistance in staphylococci: activities of new nonfluorinated quinolones against molecular targets in whols cells and clinical isolates. Antimicrob. Agents Chemother. 45: 1115-1120.

Ruiz, J., Sierra, J.M., De Anta, M.T., and Vila, J. 2001. Characterisation of sparfloxacin-resistant mutants of *Staphylococcus aureus* obtained *in vitro*. Int. J. Antimicrob. Agents 18: 107-112.

Sanches, I.S., Mato, R., De Lancastre, H., Tomasz, A., Cem/Net Collaborators and International Collaborators. 2000. Patterns of multidrug resistance among methicillin-resistant hospital isolates of coagulase-positive and coagulase-negative staphylococci collected in the international multicenter study RESIST in 1997 and 1998. Microb. Drug Res. 6: 199-211.

Sandler, P., and Weisblum, B. 1988. Erythromycin-induced stabilization of *ermA* messenger RNA in *Staphylococcus aureus* and *Bacillus subtilis*. J. Mol. Biol. 203: 905-915.

Scheel, O., Lyon, D.J., Rosdahl, V.T., Adeyemi-Doro, F.A., Ling, T.K., and Cheng, A.F. 1996. *In-vitro* susceptibility of isolates of methicillin-resistant *Staphylococcus aureus* 1988-1993. J. Antimicrob. Chemother. 37: 243-251.

Schmitz, F.-J., Hofmann, B., Hansen, B., Scheuring, S., Luckefahr, M., Klootwijk, M., Verhoef, J., Fluit, A., Heinz, H.P., Kohrer, K., and Jones, M.E. 1998. Relationship between ciprofloxacin, ofloxacin, levofloxacin, sparfloxacin and moxifloxacin (BAY 12-8039) MICs and mutations in *grlA*, *grlB*, *gyrA* and *gyrB* in 116 unrelated clinical isolates of *Staphylococcus aureus*. J. Antimicrob. Chemother. 41: 481-484.

Schmitz, F.-J., Boos, M., Mayer, S., Jagusch, H., and Fluit, A.C. 2002a. Increased *in vitro* activity of the novel des-fluoro(6) quinolone BMS-284756 against genetically defined clinical isolates of *Staphylococcus aureus*. J. Antimicrob. Chemother. 49: 283-287.

Schmitz, F.-J., Petridou, J., Milatovic, D., Verhoef, J., Fluit, A.C., and Schwartz, S. 2002b. *In-vitro* activity of new ketolides against macrolides-susceptible and macrolides-resistant *Staphylococcus aureus* isolates with defined resistance gene status. J. Antimicrob. Chemother. 49: 580-582.

Schmitz, F.-J., Petridou, J., Jagusch, H., Astfalk, N., Scheuring, S., and Schwartz, S. 2002c. Molecular characterization of ketolide-resistant *ermA*-carrying *Staphylococcus aureus* isolates selected *in vitro* by telithromycin, ABT-773 and clindamycin. J. Antimicrob. Chemother. 49: 611-617.

Schmitz, F.-J., Petridou, J., Astfalk, N., Köhrer, K., Scheuring, S., and Schwartz, S. 2002d. Molecular analysis of constitutively expressed *ermC* genes selected *in vitro* by incubation in the presence of the noninducers quinupristin, telithromycin, or ABT-773. Microb. Drug Res. 8: 171-177.

Schmitz, F.-J., Fluit, A.C., Gondolf, M., Beyrau, R., Lindenlauf, E., Verhoef,

J., Heinz, H.-P., and Jones, M.E. 1999. The prevalence of aminoglycoside resistance and corresponding resistance genes in clinical isolates of staphylococci from 19 European hospitals. J. Antimicrob. Chemother. 43: 253-259.

Schmitz, F.-J., Fluit, A.C., Hafner, D., Beeck, A., Perdikouli, M., Boos, M., Scheuring, S., Verhoef, J., Köhrer, K., and Von Eiff, C. 2000. Development of resistance to ciprofloxacin, rifampin, and mupirocin in methicillin-susceptible and -resistant *Staphylococcus aureus* isolates. Antimicrob. Agents Chemother. 44: 3229-3231.

Schmitz, F.-J., Krey, A., Sadurski, R., Verhoef, J., Milatovic, D., and Fluit, A.C. 2001. Resistance to tetracycline and distribution of tetracycline resistance genes in European *Staphylococcus aureus* isolates. 47: 239-240.

Schnappinger, D., and Hillen, W. 1996. Tetracyclines: antibiotic action, uptake, and resistance mechanisms. Arch. Microbiol. 165: 359-369.

Schulte, A., and Heisig, P. 2000. *In vitro* activity of gemifloxacin and five other fluoroquinolones against defined isogenic mutants of *Escherichia coli*, *Pseudomonas aeruginosa* and *Staphylococcus aureus*. J. Antimicrob. Chemother. 46: 1037-1046.

Schwarz, S., Cardoso, M., Wegener, H.C. 1992. Nucleotide sequence and phylogeny of the *tet*(L) tetracycline resistance determinant encoded by plasmid pSTE1 from *Staphylococcus hyicus*. Antimicrob. Agents Chemother. 36: 580-588.

Shinabarger, D.L., Marotti, K.R., Murray, R.W., Lin, A.H., Melchoir, E.P., Swaney, S.M., Dunyak, D.S., Demyan, W.F., and Buysse, J.M. 1997. Mechanism of action of oxazolidinones: effects of linezolid and eperzolid on translation reactions. Antimicrob. Agents Chemother. 41: 2132-2136.

Sköld, O. 2000. Sulfonamide resistance: mechanisms and trends. Drug Res. Update 3: 155-160.

Simor, A.E., Ofner-Agostini, M., Bryce, E., Green, K., McGeer, A., Mulvey, M., Paton, S., and the Canadian Nosocomial Infection Surveillance Program, Health Canada. 2001. The evolution of methicillin-resistant *Staphylococcus aureus* in Canadian hospitals: 5 years of national surveillance. Can. Med. Assoc. J. 165: 21-26.

Sinden, D. & Chopra, I. 1981. Fusidic acid resistance in *Staphylococcus aureus*. Zentralblatt fuer Bakteriologie, Mikrobiologie und Hygiene 251 (Suppl. 10): 571–574.

Skinner, R., Cundliffe, E., and Schmidt, F.-J. 1983. Site of action of a ribosomal RNA methylase responsible for resistance to erythromycin and other antibiotics. J. Biol. Chem. 258: 12702-12706.

Smith, S.M., Eng, R.H., Bais, P., Fan-Havard, P and Tecson-Tumang, F. 1990. Epidemiology of ciprofloxacin resistance among patients with methicillin-resistant *Staphylococcus aureus*. J. Antimicrob. Chemother. 26: 567-572.

De Sousa, M.A., Miragaia, M., Sanches, I.S., Ávila, S., Adamson, I.,

Casagrande, S.T., Brandileone, M.C.C., Palacio, R., Dell'Acqua, L., Hortal, M., Camou, T., Rossi, A., Velazquez-Meza, M.E., Echaniz-Aviles, G., Soloranzo-Santos, F., Heitmann, I., and De Lencastre, H. 2001. Three-year assessment of methicillin-resistant *Staphylococcus aureus* clones in Latin America from 1996 to 1998. J. Clin. Microbiol. 39: 2197-2205.

Stasinopoulos, S.J., Farr, G.A., and Bechhofer, D.H. 1998. *Bacillus subtilis tetA*(L) gene expression: evidence for regulation by translational reinitiation. Mol. Microbiol. 30: 923-932.

Stewart, P.R., Dubin, D.T., Chikramane, S.G., Inglis, B., Matthews, P.R., and Poston, S.M. 1994. IS257 and small plasmid insertions in the *mec* region of the chromosome of *Staphylococcus aureus*. Plasmid 31: 12-20.

Swaney, S.M., Aoki, H., Ganoza, M.C., and Shinabarger, D.L. 1998. The oxazolidinone linezolid inhibits initiation of protein synthesis in bacteria. Antimicrob. Agents Chemother. 42: 3251-3255.

Takei, M., Fukuda, H., Kishii, R., and Hosaka, M. 2001. Target preference of 15 quinolones against *Staphylococcus aureus*, based on antibacterial activities and target inhibition. Antimicrob. Agents Chemother. 45: 3544-3547.

Tanaka, N., Kinoshita, T., and Masukawa, H. 1968. Mechanism of protein synthesis inhibition by fusidic acid and related antibiotics. Biochem. Biophys. Res. Comm. 30: 278-283.

Tanaka, M., Onodera, Y., Uchida, Y., Sato, K., and Hayakawa, I. 1997. Inhibitory activities of quinolones against DNA gyrase and topoisomerase IV purified from *Staphylococcus aureus*. Antimicrob. Agents Chemother. 41: 2362-2366.

Tanaka, M., Onodera, Y., Uchida, Y., and Sato, K. 1998. Quinolone resistance mutations in the GrlB protein of *Staphylococcus aureus*. 42: 3044-3046.

Taylor, D.E., and Chau, A. 1996. Tetracycline resistance mediated by ribosomal protection. Antimicrob. Agents Chemother. 40: 1-5.

Testa, R.T., Petersen, P.J., Jacobus, N.L., Sum, P.-E., Lee, V.J., and Tally, F.P. 1993. *In vitro* and *in vivo* antibacterial activities of the glycylcyclines, a new class of semisynthetic tetracyclines. Antimicrob. Agents Chemother. 37: 2270-2277.

Thakker-Varia, S., Jenssen, W.D., Moon-McDermott, L., Weinstein, M.P., and Dubin, D.T. 1987. Molecular epidemiology of macrolide-lincosamide-streptogramin B resistance in *Staphylococcus aureus* and coagulase-negative staphylococci. Antimicrob. Agents Chemother. 31: 735-743.

Thompson, J., Skeggs, P.A., and Cundliffe, E. 1985. Methylation of 16S ribosomal RNA and resistance to the aminoglycoside antibiotics gentamicin and kanamycin determined by DNA from the genatmicin-producer, *Micromonospora purpurea*. Mol. Gen. Genet. 201: 168-173.

Torvaldsen, S, and Riley, T. 1996. Emerging sodium fusidate resistance in Western Australian methicillin-resistant *Staphylococcus aureus*. Comm. Dis. Int. 20 : 492-494.

Trzcinski, K., Cooper, B.S., Hryniewski, W., and Dowson, C.G. 2000.

Expression of resistance to tetracyclines in strains of methicillin-resistant *Staphylococcus aureus*. J. Antimicrob. Chemother. 45: 763-770.

Tsiodras, S., Gold, H.S., Sakoulas, G., Eliopoulos, G.M., Wennersten, C., Venkataraman, L., Moellering, R.C., and Ferraro, M.J. 2001. Linezolid resistance in a clinical isolate of *Staphylococcus aureus*. Lancet 358: 207-208.

Ubukata, K., Yamashita, N., Gotoh, A., and Konno, M. 1984. Purification and characterisation of aminoglycoside-modifying enzymes from *Staphylococcus aureus* and *Staphylococcus epidermidis*. Antimicrob. Agents Chemother. 25: 754-759.

Udo, E.E., and Jacob, L.E. 2000. Characterisation of methicillin-resistant *Staphylococcus aureus* from Kuwait hospitals with high-level fusidic acid resistance. J. Med. Microbiol. 49: 419-426.

Vanhoof, R., Hannecart-Pokorni, E., and Content, J. 1998, Nomenclature of genes encoding aminoglycoside-modifying enzymes. Antimicrob. Agents Chemother. 42: 483.

Vasquez, D. 1979. Inhibitors of protein synthesis. Springer-Verlag, Berlin, Germany.

Vester, B., and Douthwaite, S. 2001. Macrolide resistance conferred by base substitutions in 23S rRNA. Antimicrob. Agents Chemother. 45: 1-12.

Wang, T., Tanaka, M., and Sato, K. 1998. Detection of *grlA* and *gyrA* mutations in 344 *Staphylococus aureus* strains. Antimicrob. Agents Chemother. 42: 236-240.

Weisblum, B. 1985. Inducible resistance to macrolides, lincosamides and streptogramin type B antibiotics: the resistance phenotype, its biological diversity, and structural elements that regulate expression – a review. J. Antimicrob.Chemother. 16 (Suppl A): 63-90.

Werckenthin, C., Schwarz, S., and Westh, H. 1999. Structural alterations in the translational attenuator of constitutively expressed *ermC* genes. Antimicrob. Agents Chemother. 43: 1681-5.

Westh, H., Hougaard, D.M., Vuust, J., and Rosdahl, V.T. 1995. Prevalence of *erm* gene classes in erythromycin-resistant *Staphylococcus aureus* strains isolated between 1959 and 1988. Antimicrob. Agents Chemother. 39: 369-373.

Wondrack, L., Massa, M., Yang, B.V., and Sutcliffe, J. 1996. Clinical strains of *Staphylococcus aureus* inactivates and causes efflux of macrolides. Antimicrob. Agents Chemother. 40: 992-998.

Wu, S., de Lancastre, H., and Tomasz, A. 1999. The *Staphylococcus aureus* transposon Tn*551*: complete nucleotide sequence and transcriptional analysis of expression of the erythromycin resistance gene. Microb. Drug Res. 5: 1-7.

Yao, J.D.C., and Moellering, R.C. Jr. 1999. Antibacterial agents. In: Manual of Clincal Microbiology 7th Edition. P.R. Murray, E. J. Baron, M. A. Pfaller, F. C. Tenover and R. H. Yolken,eds. American Society for Microbiology, Washington D.C.

Zaman, R., and Dibb, W.L. 1994. Methicillin-resistant *Staphylococcus aureus*

(MRSA) isolated in Saudia Arabia: epidemiology and antimicrobial resistance patterns. J. Hosp. Infect. 26: 297-300.

From: *MRSA: Current Perspectives*
Edited by: A.C. Fluit and F.-J. Schmitz

Chapter 5

Molecular Approaches for the Epidemiological Characterization of *Staphylococcus aureus* Strains

Willem B. van Leeuwen

Abstract

Staphylococcus aureus has remained a significant cause of nosocomial morbidity and mortality. For almost four decades the increasing prevalence of methicillin-resistant *S. aureus* (MRSA) has posed an additional major clinical threat worldwide. During this period, multiple DNA-based methods have been introduced to genetically type *S. aureus* strains, but not a single technique appeared to be universally applicable. Most of the current image-based approaches generate complex banding patterns and lack generally accepted interpretation criteria. The need for straightforward and reproducible techniques generating simple output that can be used for computerized data-management, still is an important research topic.

This review summarizes the chronology of development and the qualities of the various laboratory techniques and includes the description of the generation, application and validation of novel DNA-based strategies rendering a binary outcome, such as multilocus sequence typing (MLST), binary typing (BT) or Micro Array technologies. The value of these novel typing procedures, with respect to simplicity and performance in the elucidation of complex biological phenomena such as epidemicity, relevance to pathogenesis of both human and bovine strains, antimicrobial resistance and genomic evolution of *S. aureus* strains, will be discussed.

Introduction

Over the past century microbiologists have searched for increasingly rapid and efficient means for the differentiation of *S. aureus* strains. The discrimination between strains of *S. aureus* has basically relied on phenotypic characteristics including phage typing (Blair and Williams, 1961, Tambic-Andrasevic *et al.*, 1999). Consecutive phases in the development of molecular biology approaches over the past decades have opened new avenues for the exploitation of the genetic diversity within *S. aureus* as a species (Goering, 2000). Molecular techniques were applied to elucidate the basic mechanisms of staphylococcal pathogenicity and to track the spread of *S. aureus* clones, especially the methicillin-resistant ones. A clone is defined as the progeny of a common ancestor and the result of a direct chain of replication of that ancestor. The discrimination between isolates of a bacterial species is central to many aspects of clinical microbiology (Tenover, *et al.*, 1997). Genetic typing systems with different levels of resolution, but applied on the same set of strains, are likely to reveal different degrees of genetic variability. Diverse evolutionary processes are responsible for this genetic polymorphism (*see the* "Genotyping of *S. aureus* Strains" *section*). Various forms of intraspecies variation are used for molecular typing and the appropriateness of a given typing method depends on the epidemiological questions that are posed and the population structure (for a definition, *see the* "Genotyping of *S. aureus* Strains" *section)* of the species under study.

Fundamentals of Bacterial Genomic Variability

The basic assumption in any microbiological typing system is that epidemiologically related strains are relatively recent descendants of a single precursor. Consequently, these descendants share characteristics that differ from those of epidemiologically unrelated strains. The utility of such characteristics for typing is related to their stability within a strain and general diversity at the species level. Genetic diversity arises by various mutational processes. These processes include the accumulation of spontaneous point mutations, diverse types of genetic rearrangements, and acquisition and loss of chromosomal and extra-chromosomal DNA sequences. The *Staphylococcus* genome is composed of multiple genes that seem to have acquired by horizontal transmission between genes. These mobile genetic elements include bacteriophages, plasmids, and multiple gene islands. Most of these genes refer to antibiotic resistance determinants or virulence factors (Kuroda *et al.*, 2001). Polymorphism in the staphylococcal cassette chromosome *mec* (SSC*mec*) forms an excellent example of recombination. SSC*mec* has a mosaic structure. This antibiotic resistance island primarily harbors the methicillin resistance determinant *mecA*. Moreover, various combinations of mobile elements, such as transposons, insertion- and plasmid sequences, encoding antibiotic resistance, are found integrated in the cassette (Oliveira, *et al.* 2000; Hiramatsu *et al.*, 2001; Kuroda *et al.*, 2001). SSC*mec* could be actively transmitted among the *Staphylococcus* species *in vivo* (Wielders, *et al.*, 2001). Variability among SSC*mec* types may originate form horizontal gene exchange with different *Staphylococcus* species and other forms of molecular evolution that depend on a large array of environmental influences.

Information on factors that are involved in the generation of genetic variation and in the modulation of the frequency of genetic variation is scarce (Arber, 1999). However, genetic variation is described in many papers. It is a challenge for future studies to investigate the more general nature of unexpected gene activities with respect to biological evolution, which ensures biodiversity and represents a guarantee for maintenance and development of certain microbial species over a prolonged period of time. In contrast to mutation, selection factors acting on bacterial populations limit the level of diversity. The factors, that enable bacteria to cause infection, generally are not uniformly distributed within a species. The organisms associated with infections often represent a smaller subset of the many strains or genotypes that constitute a species. Consequently, these pathogenic strains exhibit relatively little genetic diversity. Antibiotic resistance is another selection factor. Kreiswirth *et al.* (1995) suggested that MRSA strains are derived from a single ancestor and, in this way, represent a restricted subset of lineages in comparison with the diversity among the methicillin-susceptible *Staphylococcus aureus* strains. In conclusion, most clinically relevant strains, sharing virulence factors or resistance determinants, are often difficult to differentiate with a single typing system as a result of selection.

Purpose of Epidemiological Typing

Epidemiological studies, such as for instance infectious disease outbreak investigations, aim to define relationships between strains, which are isolated in a restricted area (hospital unit) within a short period of time (days, weeks). Other studies, e.g. epidemiological surveillance of infectious diseases or the analysis of the population genetics, address the relationship between strains recovered during extended periods of time (years, decades) and over a broader geographical range (nation-, worldwide).

Outbreak Investigation

An outbreak is defined as a local and temporal increase in the frequency of colonization and/or infection by a given microorganism. For example, hospital infection control is alerted in the case of a conspicuous increase in the rate of isolation of a specific pathogen (possibly exhibiting a unusual antibiogram) or a cluster of infections in a hospital ward. In these situations, answers to questions of strain relatedness may be elucidated by comparison of typing data (Kluytmans *et al.*, 1995; Andersen *et al.*, 2002; Nakano *et al.*, 2002). Comparative typing is applied to facilitate the development of outbreak control strategies, and address questions regarding the extent of epidemic spread of microbial clones, the number of clones involved in transmission and infection, the monitoring of reservoirs of epidemic clones or for the evaluation of the efficacy of control measures.

Pathogenesis

Genetic variation in *S. aureus* is very extensive. More than 20% of the genome comprises of apparently dispensable genetic material (Fitzgerald *et al.*, 2001; Kuroda *et al.*, 2001). Eighteen large strain-specific DNA regions have been defined and the majority of these elements encode putative virulence factors (Fitzgerald *et al.*, 2001). Typing techniques that are able to detect distinct epidemiological markers associated with pathogenic isolates within a species, provide insight into the mechanism of pathogenesis. This typing strategy was demonstrated in a study in which the authors showed that the presence of a gene coding for Panton-Valentine leukocidine in *S. aureus* strains was related to the severity of necrotising pneumonia in young immunocompetent patients (Gillet *et al.*, 2002).

Surveillance of Infectious Diseases

Surveillance is defined as the ongoing systematic collection analysis, and interpretation of data regarding the occurrence of infectious diseases or agents. The reporting of the results of such analyses to those individuals and authorities who need to know is the ultimate goal, which in turn can be translated into preventive or therapeutic measures. Typing methods provide essential information in the epidemiological surveillance of infectious diseases. For example, in a worldwide multicenter antibiotic resistance surveillance study, pulsed-field gel electrophoresis was used for the analysis of the genetic relatedness of multidrug-resistant *S. aureus* bloodstream isolates. They demonstrated that many MRSA strains were grouped geographically and, moreover, a limited number of closely related strains were spread nation- or even worldwide (Diekema *et al.*, 2000). The selection of stable, discriminatory and definitive epidemiological markers for a typing system in surveillance studies is essential. These typing systems have a standardized scoring method, using a uniform nomenclature of types and produce high throughput results. Methods, which comply with the abovementioned conditions are called; library typing systems. Data obtained from library typing techniques can be compared over time and place for diverse bacterial species and are, for that reason, essential components of surveillance studies.

Population Structure

Genetic diversity within a population is generated by the accumulation of mutations. The significance of these mutations for the behavior of certain genotypes in a bacterial population (epidemicity, virulence, antibiotic resistance, etc.) can be elucidated by typing systems. In order to determine the population structure within a species, specific high-throughput typing systems are needed. Such typing data are crucial in defining the level of similarity within a given species. Data from multilocus enzyme electrophoresis (Selander *et al.*, 1986, 1987; Kapur *et al.*, 1995) and recently from multilocus sequence typing (Enright and Spratt, 1998; Maiden *et al.*, 1998; Spratt, 1999; Enright, and Spratt, 1999; Enright *et al.*, 2000), designed to detect associations between several conserved genes at different sites (loci) on the chromosome, have revealed that among bacteria different types of population structures exist. Some species were found to have a clonal structure (*Escherichia coli*, *Salmonella* spp.). This implies that propagation occurs through simple genome copying during cell division and that the intermolecular recombination is infrequent or even absent. The clonality of a population can be deduced from the global distribution of certain constant genotypes. At the other extreme, some species exhibit a panmictic population structure (e.g. *Neisseria gonorrhoeae),* and excessive recombination may occur between isolates. An intermediate type of population structure, the so-called epidemic structure was also found (e.g. *N. meningitidis*) (Maynard Smith *et al.*, 1993; Maynard Smith, 2000). This population structure is essentially panmitic. However, occasionally a highly successful clone arises and increases rapidly in frequency to produce an outbreak.

Criteria for the Evaluation of Typing Systems

Several parameters should be considered when evaluating typing systems (Maslow *et al.*, 1993; Arbeit, 1997, Struelens and the members of the European Study Group on Epidemiological Markers (ESGEM) of the European Society for Clinical Microbiology and Infectious Diseases (ESCMID), 1996). The performance criteria include the typeability, reproducibility, stability, and discriminatory power of a typing system. Typeability refers to the ability of a system to obtain a positive result for each isolate analyzed and is influenced by both technical and biological factors. The technical reproducibility is the ability to assign the same type to a strain tested on independent occasions. The biological reproducibility or stability of epidemiological markers is the ability of a typing system to recognize clonal relatedness of strains derived from a common ancestor. Phenotypic or genomic variation may occur during storage or replication of strains in the laboratory (*in vitro* stability). Clonal expansion of a strain over a long period of time or during geographically widespread outbreaks (*in vivo* stability) can also be result in various degrees of genetic variation. The discriminatory power refers to the average probability that a typing system will assign different types to two unrelated strains. Ideally, each unrelated strain is identified as unique. In practice a method is statistically useful when the most commonly detected type represents less than 5% of random unrelated strains (Hunter and Gaston, 1988; Hunter, 1990).

The applicability of typing techniques is affected by the efficiency of the method (Struelens and the members of the European Study Group on Epidemiological Markers (ESGEM) of the European Society for Clinical Microbiology and Infectious Diseases (ESCMID), 1996). These criteria include the presentation of the epidemiological question (e.g. local outbreak, surveillance studies), data processing (analysis, storage), or the financial and technical resources available. Flexibility, rapidity, accessibility and ease of use may be considered as convenience criteria. Flexibility reflects the typeability of a wide range of species with minimal modification of the system. The rapidity of typing techniques ranges from one day to several weeks. Accessibility is based on the cost aspects (reagents, equipment) and the level of technical skill required for the performance of a distinct method. The ease of use is reflected by the simplicity, the workload and the ease of interpretation of the results. For evaluation of the performance criteria mentioned above, special attention should be paid to the appropriate selection of a bacterial test population. This collection should contain well-characterized strains, both epidemiologically unique and clonally related ones (Maslow *et al.*, 1993;

Struelens, and the members of the European Study Group on Epidemiological Markers (ESGEM) of the European Society for Clinical Microbiology and Infectious Diseases (ESCMID) 1996; Arbeit, 1997).

Classification of Typing Methods

A convenient basis for classifying typing systems is to recognize them as phenotypic techniques, those that detect characteristics expressed by microorganisms, and genotypic techniques, those that involve direct nucleic acid-based analysis of chromosomal or extra-chromosomal genetic elements.

Phenotypic Techniques for *S. aureus*

Historically, the identification and characterization of MRSA strains has been achieved by phenotypic analyses and for many decades, have served as the basis for epidemiological analyses. Phenotypic methods are those that characterize products of gene expression in order to identify to the species level or to differentiate strains. Properties such as biochemical profiles, susceptibility to bacteriophages, antigens present on the cell's surface, whole protein analysis (Clink and Pennington, 1987; Gaston *et al.*, 1988; Costas *et al.*, 1989) and antimicrobial susceptibility patterns were used as epidemiological targets. All are examples of phenotypic properties that can be determined in the microbiology laboratory. Because they involve gene expression, these properties all have a tendency to vary, based on environmental influences. For this reason, phenotyping assays are often limited in reproducibility or reliability. Moreover, these systems lack typeability, discriminatory power and, consequently, are not the most adequate approaches for bacterial comparison.

Phagetyping

For decades *Staphylococcus* strains have been characterized by their differential susceptibility to bacteriophage infection. This approach was introduced in 1961 by Blair and Williams (1961) and was based on the observation that most *S. aureus* isolates carry bacteriophages that lysed some, but not all, epidemiologically unrelated strains. The response of isolates representing the same strain, to a specific bacteriophage was identical. A panel of diverse bacteriophages could be used to differentiate distinct *S. aureus* strains. Phagetyping has been particularly successfully as a centralized typing method for nationwide surveillance studies in several European countries (Rosdahl *et al.*, 1994; Voss *et al.*, 1994). Despite the popularity of phagetyping, to date many reference centers switched to molecular typing techniques. Various factors, such as the enormous effort to maintain high quality phage sets, the technical skill, poor typeability among MRSA strains (Tambic *et al.*, 1997; Murchan *et al.*, 2000) and the limitation in reproducibility (Tambic-Andrasevic *et al.*, 1999; O'Neill *et al.*, 2001) contributed to that decision.

Antimicrobial Susceptibility Analysis

The susceptibility of MRSA for a panel of antimicrobial agents is routinely tested in the clinical microbiological laboratory. Comparison of the antibiotic susceptibility patterns will show association among outbreak MRSA strains and non-related strains will be distinguished to a certain level (Mulligan and Arbeit, 1991). The assay is highly quality-controlled, easy to perform, relatively inexpensive and the method can be improved by quantitative analysis of the disk-susceptibility zone diameter (Tenover *et al.*, 1994; Blanc *et al.*, 1994) or by the automated measurement of the minimal inhibitory concentration (Sanders *et al.*, 2001). For long-term epidemiological issues, antibiotic susceptibility typing has a relatively limited value because of selection factors for resistance and variability of the resistance traits. Various mechanisms may render strain antibiotic resistance (McGowan, 1991). These include single nucleotide mutations (Aubry-Damon *et al.*, 1998; Tanaka *et al.*, 2000), and the acquisition of resistance genes or resistance islands via mobile genetic elements (Kuroda *et al.*, 2001). In the latter case, resistance to multiple antimicrobial agents can be acquired simultaneously. These mobile DNA elements may be lost by the strain in the absence of specific selective antimicrobial pressure (Meyers *et al.*, 1976; Mayer, 1988). New resistance markers are often related to clonal expansion of an existing epidemic MRSA strain after environmental adaptation (Leski *et al.*, 1999; Hiramatsu *et al.*, 2001).

Multilocus Enzyme Electrophoresis

Multilocus enzyme electrophoresis (MLEE) analyses the differences in electrophoretic mobility of a number of bacterial house-keeping proteins, which is based on mutations present in the appropriate genes (Selander *et al.*, 1986). MLEE has proved to be only moderately discriminatory for the characterization of *S. aureus* isolates (Tenover *et al.*, 1994). In addition, the study of Tenover *et al.* (1994) also showed inconsistent MLEE results among outbreak strains. MLEE is amenable to the analysis of large strain collections and is used most effectively to study the population structure of (methicillin-resistant) *S. aureus* (Musser and Selander, 1990; Musser and Kapur, 1992; Kapur *et al.*, 1995). A study on the population structure of MRSA revealed completely different electrophoretic types and this finding supports the hypothesis that multiple phases of horizontal transmission and recombination of *mec* DNA has occurred (Musser and Kapur, 1992). Analysis of the genetic structure of the *S. aureus* population resulted in the existence of predominant types, corresponding to

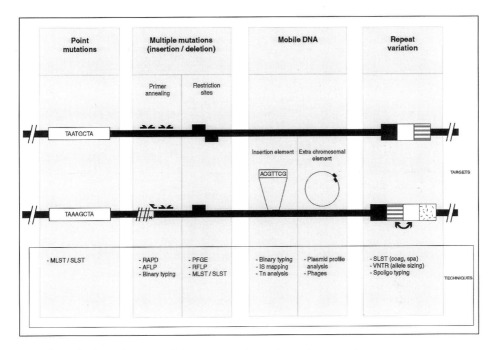

Figure 1. The molecular basis for the comparison of bacterial genomic DNA molecules: targets and techniques. The currently developed genotyping approaches for the discrimination of bacterial strains measure variability in single nucleotides, insertion or deletion of DNA fragments, presence or absence of (extra) chromosomal mobile DNA elements (plasmids, transposons, insertion sequences, phages, pathogenicity- and resistance islands), and polymorphism in the frequency of DNA repeat sequences. The different strategies to detect the different targets are: MLST, multi-locus sequence typing; SLST, single-locus sequence typing; RAPD, randomly amplified polymorphic DNA; AFLP, amplified fragment length polymorphism; PFGE, pulsed-field gel electrophoresis; RFLP, restriction fragment length polymorphism; IS, insertion sequence; Tn, transposon; VNTR, variable number of tandem repeats.

the criteria posed for an epidemic population structure (Maynard Smith *et al.*, 1993). Another study revealed the host-specificity of *S. aureus* strains (Kapur *et al.*, 1995).

Genotyping of *S. aureus* Strains

The advances of molecular biology have resulted in the development of multiple DNA-based strain typing strategies. The molecular basis of the different techniques for discriminating individual DNA molecules and the respective targets are summarized in Figure 1.

Over the last two decades DNA-based technologies have been introduced and are increasingly being used in clinical laboratories, which is reflected by

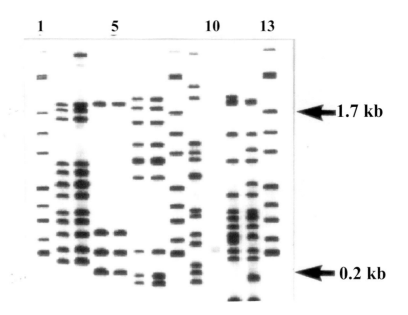

Figure 2. Southern hybridization analysis. Genomic DNA of 13 MRSA strains were digested with a fine-cutting restriction enzyme. Restriction fragments were separated by size through agarose gel electrophoresis and subsequently transferred onto a nylon membrane. The immobilized DNA restriction fragments are hybridized with a specific probe and hybrids were detected using a radio-active label. DNA fragment sizes are indicated on the right.

the number of papers reporting on *S. aureus* epidemiology, MRSA in particular (van Belkum, 2000). Over time, several stages of molecular typing methods have found their application in the analysis of *S. aureus* collections (Goering, 2000). These laboratory developments are reviewed chronologically here.

First-phase Molecular Typing: Plasmid Pofile Analysis

The first DNA-based techniques applied to epidemiological studies of *S. aureus* involved the analysis of plasmids; plasmid analysis was introduced in the mid-70s (Meyers *et al.*, 1976; McGowan *et al.*, 1979; Locksley *et al.*, 1982). Bacterial plasmids are autonomously replicating extra-chromosomal elements, distinct from the chromosome. The analysis of plasmids is a technically simple process. Although plasmids are present in more than 90 percent of MRSA strains, approximately 50 percent of methicillin-susceptible *S. aureus* isolates lack them and can, therefore, not be typed by this approach (Coia *et al.*, 1988; Hartstein *et al.*, 1989; Trilla *et al.*, 1993; Hartstein *et al.*, 1995). Also, the reproducibility of plasmid profiling is confounded by structure-variability of the plasmid itself (supercoiled, nicked, linear and oligomeric). This problem can be circumvented by the digestion of the plasmids into restriction fragments and analyzing their numbers and sizes. The fundamental drawbacks have limited the application of plasmid analysis and the method has only proven effective for evaluating isolates under restricted temporal and geographical conditions, such as during an acute outbreak episode in a single hospital.

Second-phase Molecular Typing: Southern Hybridization Analysis of Digested Chromosomal DNA.

The bacterial chromosome is the prime target molecule for the measurement of relationship between bacterial cells. In the mid-70s Southern hybridization (Southern, 1975) became available as a tool for epidemiological studies. Classical Southern blot analysis detects only specific restriction fragments carrying DNA sequences homologous to the probe used. The choice of the probe is a critical consideration with respect to typeability and discriminatory power and is directly related to the frequency with which the detected restriction fragments vary in number or size, or both (Figure 2). The best-known hybridization-mediated typing procedure is ribotyping. DNA probes corresponding to (parts of) ribosomal genes are used to highlight polymorphisms (Stull *et al.*, 1988; Wada *et al.*, 1993; Greisen *et al.*, 1994; Anthony *et al.*, 2000). All staphylococci carry five to seven ribosomal operons and are, therefore, typeable with this method (Blumberg *et al.*, 1992; Arbeit, 1997). The complete ribotyping procedure has recently been automated and in the case of MRSA the results have been coupled to a database management system (Diekema *et al.*, 2000). This library system should facilitate intercenter data exchange and ongoing multicenter studies explore these possibilities, e.g. GENE (Genetic Epidemiology Network for Europe, S. Brisse, Utrecht, the Netherlands; Qualicon Riboprinter as core method), an EU sponsored concerted action. Polymorphism in the ribosomal operons, as defined by for instance ribotyping, can also be identified through PCR-mediated procedures (Stull *et al.*, 1988; Prevost *et al.*, 1992).

A wide variety of DNA probes homologous to mobile genetic elements has been used for the epidemiological analysis of (methicillin-resistant) *S. aureus*. These probes target mobile genetic elements including transposons, Tn*554* (Kreiswirth *et al.*, 1995); or insertion sequences, IS*256* (Morvan *et al.*, 1997) and IS*257*/IS*431* (Yoshida *et al.*, 1997). In addition, different virulence genes such as those encoding coagulase (Goh *et al.*, 1992; Schwarzkopf and Karch, 1994; Hookey *et al.*, 1999a), and protein A (Frenay *et al.*, 1994, Shopsin *et al.*, 1999) have been used as targets for DNA probing. Genes encoding virulence factors are typically present only as a single copy on the genome; consequently, proper strain discrimination often requires probing for multiple different genetic loci (Kreiswirth *et al.*, 1993; Speaker *et al.*, 1994; Tenover *et al.*, 1994). The use of multiple probes, including virulence factors (Monzon-Moreno *et al.*, 1991; Smeltzer *et al.*, 1996; Hoefnagels-Schuermans, 1997;

Smeltzer *et al.*, 1997;) or antibiotic resistance gene sequences (Monzon-Moreno *et al.*, 1991; Kreiswirth *et al.*, 1994) in individual hybridization reactions generates enhanced resolution. Repetitive elements on the staphylococcal genome may expand the specificity and accuracy of strain typing (Wei *et al.*, 1992; Cramton *et al.*, 2000). Finally, the methicillin-resistance encoding *mec* region provides a number of possibilities for deducing DNA probes and accessing genetic polymorphism (Kreiswirth *et al.*, 1995). Application of PCR multiplexing or micro-array techniques using different probes may facilitate epidemiological studies in the future.

Third-phase Molecular Typing: PCR-based Techniques and Pulsed-field Gel Electrophoresis

PCR-based Typing Systems

The polymerase chain reaction (PCR) was developed in the mid-80s and can now be considered one of the major biological and technical innovations of the 20th century (Mullis *et al.*, 1986). The essential property of PCR is the ability to exponentially replicate (amplify) parts of a given template DNA genome, leading to the accumulation of a huge number of copies (amplicon) of the original nucleic acid fragment. Several different approaches have been proposed.

Restriction Digestion of PCR Products

The PCR products (amplicons) can be digested with specific DNA-splitting enzymes, called restriction endonuclease(s). The amplicon generally contains restriction sites for several of these enzymes. The DNA fragment length between the restriction sites is variable. Restriction fragment length polymorphism (RFLP) is analyzed by gel electrophoresis. DNA polymorphisms within the coagulase gene, the rRNA operon or the intergenic spacer region between 16S and 23S rRNA genes of *S. aureus* genomes are targets for RFLP analysis (Schwarzkopf and Karch, 1994; 1999; Kumari *et al.*, 1997; Hookey *et al.*, 1999a). Like most of the staphylococcal surface proteins, sharing common structural features, the extracellular part of the *S. aureus* coagulase displays a high level of variability, based on the presence of a variable number of tandem repeats. The repeat sequences of the coagulase gene are similar and differ in the presence or absence of an *Alu*I restriction site. The discrimination of strains with this technique is moderate (Goh *et al.*, 1992; Tenover *et al.*, 1994). The resolution can be improved by increasing the number of loci analyzed, or by increasing the number of restriction enzymes per locus analyzed (Calderwood *et al.*, 1996). This approach is analogous to multilocus enzyme electrophoresis (MLEE) and the results are suitable for population genetic studies.

PCR Based on Repetitive Chromosomal Sequences

Short extragenic repetitive sequences, originally identified in *Enterobacteriaceae* can be used as templates for PCR (Versalovic *et al.*, 1991; Woods *et al.*, 1992). Repetitive interspersed sequences can be found in most (if not all) bacteria and are scattered around the bacterial genome. These elements can serve as primer sites for genomic DNA amplification. Several families of repetitive sequences have been studied in detail, including the repetitive palindromic (REP) sequence (van der Zee *et al.*, 1999; Deplano *et al.*, 2000) the enterobacterial repetitive intergenic consensus (ERIC) sequence (van Belkum *et al.*, 1994; Witte *et al.*, 1997) or the BOX element (Versalovic *et al.*, 1991). These can give rise to a PCR product called an inter-repeat fragment. Several studies using primers that target such repetitive sequences have demonstrated only a moderate resolution of this typing strategy among MRSA strains (Struelens *et al.*, 1993; van Belkum *et al.*, 1993; Del Vecchio *et al.*, 1995). Another sort of repetitive sequence analysis by PCR is that of highly polymorphic short-sequence direct DNA repeats in prokaryotic genomes (van Belkum *et al.*, 1998). In *S. aureus*, genes that encode some of the surface proteins that recognize adhesive matrix molecules such as the protein A or the coagulase genes, contain a variable number of contiguous repetitive DNA elements. The bordering sequences of these direct-repeat sequences can form a template for PCR primers. The size variation of the amplicon reflects the number of direct-repeats units and can be established by agarose gel electrophoresis (Goh *et al.*, 1992; Frenay *et al.*, 1994).

Arbitrarily Primed PCR

Arbitrarily primed PCR (AP-PCR) was first described in the early nineties (Williams *et al.*, 1990; Welsh and McClelland, 1991) and appeared to be discriminatory for MRSA strains (van Belkum *et al.*, 1993). The discrimination level obtained with AP-PCR, also known as randomly amplified polymorphic DNA analysis (RAPD) is based on short primers (10 bp). These oligo's are used under low stringency of amplification conditions. The genetic organization of the *S. aureus* genome among different lineages is reflected by the variable size and numbers of amplified fragments (Figure 3). The inter-laboratory reproducibility is moderate (van Belkum *et al.*, 1995). Standardization with help of sophisticated equipment will optimize the reproducibility of this technique (Olmos *et al.*, 1998). PCR fingerprinting provides a generally applicable typing procedure for *ad hoc* epidemiological diagnostics and complies with most of the convenience criteria.

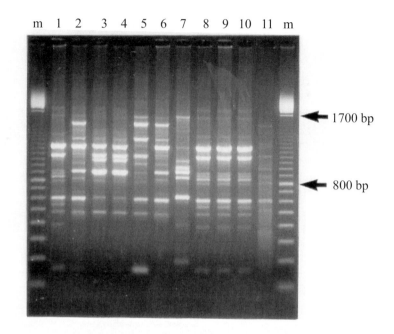

Figure 3. Characterization of 11 MRSA strains with arbitrarily primed PCR (AP-PCR). The figure represents an agarose gel showing the amplification products using PCR with a random primer. An amplified fragment that can be identified on an agarose gel, is generated when the primer is binding on opposite DNA strands in proximity to one another. Size of the amplicons is indicated on the right.

Amplified Fragment Length Polymorphism Analysis

In the mid-nineties amplified fragment length polymorphism analysis (AFLP) was designed as a typing tool for microorganisms (Zabeau and Vos, 1993; Vos *et al.*, 1995). AFLP belongs to the category of selective restriction fragment amplification techniques, based on ligation of synthetic adapters, i.e. linkers and indexers, to genomic restriction fragments followed by a PCR-based amplification with adapter-specific primers. The amplified products are visualized by DNA electrophoresis (Figure 4). It can be calculated that AFLP covers more than 40% of the total bacterial genome (few megabases), when a panel of approximately 96 individual PCR reactions is used. In a recent study, fluorescent AFLP (fAFLP) is demonstrated to be of high discriminatory value among epidemic MRSA strains (Grady *et al.*, 1999). To date, the AFLP technique is developed into a highly standardized, robust and automated technique (Savelkoul *et al.*, 1999) and has the potential for surveillance of MRSA epidemiology on national and international levels. However, interpretation criteria (Hookey *et al.*, 1999b) and intercenter validation of the reproducibility of AFLP for MRSA are urgently needed (Fry *et al.*, 1999, 2000).

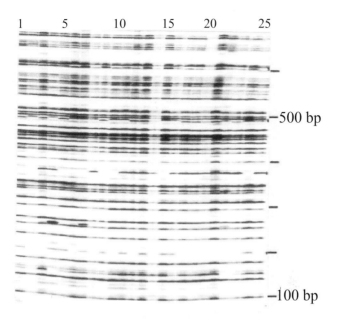

Figure 4. Example of AFLP patterns for 25 MRSA strains. The patterns are the result of amplification of templates generated after restriction with *Mbo*I and *Csp6*I and ligation with sequence specific oligonucleotide adapters. DNA fragment sizes are indicated on the right.

Pulsed-field Gel Electrophoresis

Restriction endonucleases that recognize only a few specific sites in bacterial genomes have been used since the late 70s. Consequently, exposure of DNA to those enzymes yielded large fragments, called macrorestriction fragments. Initially, these fragments were too large to be separated by conventional agarose gel electrophoresis. However, in 1984 this problem was solved with the introduction of pulsed-field gel electrophoresis (PFGE) by Schwartz and Cantor (1984). During the PFGE procedure, the orientation of the electric field across the gel is changed periodically. The separation of the DNA fragments by PFGE is primarily based on the time needed by the DNA molecules to reorient themselves in this gel, rather than the speed by which they can migrate in it (Figure 5). PFGE has often been applied for the comparison of bacterial genomes, sometimes combined with the use of probes (Allardet-Servent *et al.*, 1989; Goering and Duensing, 1990; Prevost *et al.*, 1991). PFGE is now generally accepted as the current "gold standard" for typing MRSA as well as many other bacterial species (Prevost *et al.*, 1992; Schlichting *et al.*, 1993; Leonard *et al.*, 1995; Liu *et al.*, 1996). However, PFGE generates complex banding patterns and internationally accepted guidelines for data interpretation

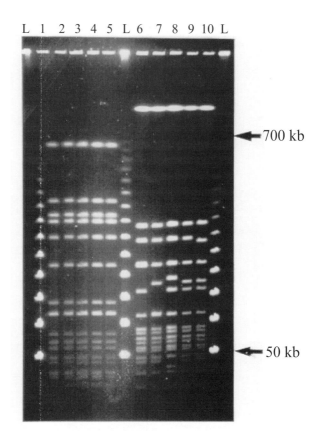

Figure 5. Representative example of PFGE results obtained after macrorestriction analysis of chromosomal DNA obtained from 5 outbreak MRSA strains (1 to 5) and 5 epidemiologically related MRSA strains (6 to 10). The bacterial DNA is cleaved with the rare-cutter restriction enzyme *Sma*I followed by an electrophoresis protocol, in which the very large DNA fragments are separated on orientation time by current switching. Size markers, indicated as L, were used and the sizes of fragments are depicted on the right.

were drawn up (Tenover *et al.*, 1995; Cookson *et al.*, 1996). Nevertheless, care has to be taken since the intercenter reproducibility of PFGE remains moderate (van Belkum *et al.*, 1998). Recently, diverse multinational European groups cooperated to establish a normalized procedure on the optimization of PFGE and a good level of reproducibility was reached, enabling multicenter comparison of PFGE data (Chung *et al.*, 2000; Murchan *et al.*, 2000, 2003).

Fourth-phase Molecular Typing: Sequence Typing and Probe-Mediated Typing.

Sequence Typing

Comparison of nucleic acid sequences is the most stringent method by which potential relatedness among strains can be defined. However, sequencing of whole genomes is not yet feasible when studying large collections of strains within a species. The challenge for sequence-based typing, therefore, is to identify region(s) within the genome, which exhibit variable and conserved sequences that can be sequenced efficiently. An elegant strategy has been the classification of bacterial isolates on the basis of sequences of internal fragments of six or seven so-called housekeeping genes (Maiden *et al.*, 1998). Housekeeping genes are those genes that encode the many proteins that are essential for cell viability. For each gene fragment, the different sequences are assigned to distinct allele identification numbers and the combination of the numbers defined for all gene fragments generates the sequence type (ST). Isolates with the same allelic profile can be considered clonally related. Such typing is called multilocus sequence typing (MLST) (Maiden *et al.*, 1998; Enright and Spratt, 1999; Spratt, 1999). MLST data can be conveniently stored in a computer and comparison of results between different laboratories is possible via the Internet (Spratt, 1999). MLST has already been developed for the identification of virulent clones of *Neisseria meningitidis* (Maiden *et al.*, 1998), *Streptococcus pneumoniae* (Enright and Spratt, 1998; Enright *et al.*, 1999; Spratt, 1999) and (methicillin-resistant) *Staphylococcus aureus* (Enright *et al.*, 2000). A new development was the establishment of a MRSA database. Despite of the limited age of MRSA, significant allelic polymorphism was defined and these STs correspond with PFGE profiles (Enright *et al.*, 2000). This suggests that MLST is a highly discriminatory typing method for the identification of MRSA lineages. MLST is thought to be technically very demanding and the technique is suited more to investigate the bacterial phylogeny and evolution of population lineages than for typing many strains in hospital outbreaks and epidemics (Enright *et al.*, 2002).

Probe-mediated Typing

In the last decade, nucleic acid probe technology has been introduced in clinical diagnostics for staphylococcal species identification (Goh *et al.*, 1996; Martineau *et al.*, 1998; Krimmer *et al.*, 1999). The application of such probes in typing systems for epidemiological studies should be considered seriously, since these approaches circumvent the obstacles surrounding banding pattern standardization and interpretation. Although several have been developed (Kamerbeek *et al.*, 1997; Molhuizen *et al.*, 1998; Allerberger and Fritschel, 1999), reliable probe-based microbial typing systems are not yet commonplace in microbiological practice. Gene probes for the study of *S. aureus* epidemiology have been described (Kreiswirth *et al.*, 1994), but these probe-mediated typing systems seem to be rather cumbersome and again give rise to banding patterns. It is known that DNA probes can be generated with the aid of RAPD analysis (Fonstein and Haselkorn, 1995). When strains are typed through RAPD analysis, DNA fingerprints are being generated. These banding patterns comprise of DNA fragments that can be characteristic for the genus, the species or the strain. These distinct but frequently not identified amplicons have already been used for the identification of fungal (Lanfranco *et al.*, 1993; Mehling *et al.*, 1995), archaeal (Martinez-Murcia and Rodriguez-Valera, 1994) and bacterial genera (Giesendorf *et al.*, 1993; Manulis *et al.*, 1994), or for the discrimination on the species level (Giesendorf *et al.*, 1996) and, most importantly, for the discrimination among different strains within fungal (Dobrowolski and O'Brien, 1993; Vazquez *et al.*, 1994; Bidochka *et al.*, 1995) or bacterial species (Vandenesch *et al.*, 1995; Erlandson *et al.*, 1997; Matheson *et al.*, 1997; Hayford *et al.* 1999).

For *S. aureus* strains it was shown that a distinct RAPD-generated DNA fragment could be used as a virulence-associated DNA probe (van Belkum *et al.*, 1994). These observations resulted in the development of the so-called binary typing system for the identification of (methicillin-resistant) *S. aureus* strains. A set of *S. aureus* specific DNA probes was generated from randomly amplified PCR products. Each DNA probe recognizes only a specific fraction of the *S. aureus* genome. Hybridization assays with the individual DNA probes reveal a simple yes or no result: the DNA probe hybridizes or not to the DNA of a certain *S. aureus* strain. A total number of twelve differentially reacting DNA elements have been isolated. This implies that direct hybridization of the individual probes with genomic DNA of a distinct *S. aureus* strain, will provide this strain with twelve yes/no characters, the 12-digit binary code (van Leeuwen *et al.*, 1996, 1999). The codes are stable within a strain (technically reproducible), constant within an outbreak strain (biologically reproducible), and unique for epidemiologically unrelated strains (resolution) (van Belkum *et al.*, 1997; van Leeuwen *et al.*, 1998; van Leeuwen *et al.*, 1999; Zadoks *et al.*, 2000). The codes can be stored in a computer database and the binary codes can be exchanged among diverse laboratories (van Leeuwen *et al.*, 2002). Currently, binary typing (BT) is being translated into a simple reversed hybridization protocol (van Leeuwen *et al.* 2001). Its reproducibility has already been demonstrated in a multicenter study (van Leeuwen *et al.*, 2002). In a recent study, the strain-specific DNA probes comprising the BT assay were extended with DNA probes, obtained from PCR products, encoding putative virulence factors. They demonstrated that other genetic markers than exfoliative toxine A were involved in the development of bullous impetigo in children (Koning *et al.*, 2002).

Criteria for the Interpretation of Typing Results

Theoretically, strain typing simply identifies an outbreak strain and differentiates among non-related strains. In practice, the interpretation of the experimental data leading to correct identification is complex. This is based on technical factors relating to the typing method used or by the fact that an epidemic strain can evolve during an ongoing outbreak and may demonstrate limited genetic variability. A recent study showed that MRSA strains produce PFGE patterns that were relatively stable over periods of weeks to months (Blanc *et al.*, 2001). Interpretation of strain typing results has to distinguish the diverse distances between the strains from the level of micro-evolution (takes place over days/months during the infectious cycle of a pathogen in a host) within outbreak strains to major differences among strains as a consequence of macro-evolution (spans millions of years over global and ecological range of the organism) (Struelens *et al.*, 1998). Interpretation criteria should provide clear guidelines for the unambiguous determination of the genetic variation level, whether a strain is unique or a component of an outbreak.

The majority of typing methods reviewed here, analyses a relatively small part of the overall bacterial genome. Therefore, identical genotypes have to be classified as "indistinguishable" and not "identical" (Tenover *et al.*, 1997). Tenover *et al.* (1997) translated the number of genetic events into strain (un)relatedness from results, obtained with so-called image-based typing techniques (PFGE, RAPD analysis, RFLP). The same interpretation criteria were applied for MLST typing (Enright *et al.*, 2000). Essentially, most of the image-based techniques generate complex banding patterns and the interpretation remains speculative. For a more precise definition of strain relatedness, the results obtained with image-based typing systems, can be compared with computer-based software. Analogue peak patterns will be translated into numerical patterns by mathematical calculation. Currently, approximately 200 phylogeny software programs are commercially available, including for instance GelCompar (Garaizar *et al.*, 2000), PHYLIP (van Belkum *et al.*, 1997), AMBIS (Smith, 1985), BioImage (Gerner-Smidt *et al.*, 1998), Dendron (Eribe and Olsen, 2000), Taxotron, (Meugnier *et al.*, 1996), Molecular Analyst (Cloak and Fratamico, 2002) and Bionumerics, a biological data analysis software package with a wide variety of applications (Klein *et al.*, 2000; Gillman *et al.*, 2001). A disadvantage of these bioanalysis software products is the fact that election of bands from fingerprints and normalization between gels had to be done manually, leading to subjective bias of the user. Currently, there are no methods for solving these problems.

Comparative Analysis of Typing Methods

Unfortunately, for the characterization of *S. aureus* strains no universally accepted epidemiological typing tool is available at the moment (van Belkum, 2000). Over the years, many studies in which multiple laboratory techniques were compared and evaluated in a single laboratory by screening of large collections of *S. aureus* strains have been published (Schlichting *et al.*, 1993; Schwarzkopf and Karch, 1994; Tenover *et al.*, 1994; Jorgensen *et al.*, 1996; Kumari *et al.*, 1997; Yoshida *et al.*, 1997; Schmitz *et al.*, 1998; Walker *et al.*, 1999; van der Zee *et al.*, 1999; Galdbart *et al.*, 2000). Various multicenter studies have been published describing problems concerning the performance of such technique (van Belkum *et al.*, 1995, 1998) or interpretation of experimental results (Tenover *et al.*, 1995; Cookson *et al.*, 1996). There is no consensus as to the optimal procedure for typing staphylococci. Most of the image-based techniques lack interlaboratory reproducibility. The analyses of the results are subjective and for reasons mentioned it is not possible to construct databases that can be freely exchanged internationally. This indicates that, as yet, there is no typing system that is suitable for the establishment of networks and it is foreseen that all procedures that generate fingerprints will in the end be displaced by procedures that produce a binary output (van Belkum, 2000). The advantage of these latter approaches is simplification of database management, comparison of results and intercenter in-silico data exchange via computer network systems.

Functional Typing

Typing methods can be used for comparison on the strain level as a tool for the detection of genetic determinants. Moreover, functional factors can be determined with these methods for the definition of biologically interesting features, such as pathogen-host interaction, virulence- (pathotypes) or resistance (resistotypes) factors. Currently, the typing differences on the gene or expression level can be identified from data-output obtained from genome sequencing (Figure 6).

Multiplex PCR

Recently, the development of multiplex PCR techniques were applied for the identification of multiple *S. aureus* virulence factors, to determine the relationship between the genetic background of *S. aureus*, the distribution of virulence factors among *S. aureus* strains, virulence regulator mechanisms and human diseases. In a recent study it was suggested that the core genome or the presence of accessory genetic elements within a strain may determine the gain and loss of DNA elements encoding virulence factors (Moore and

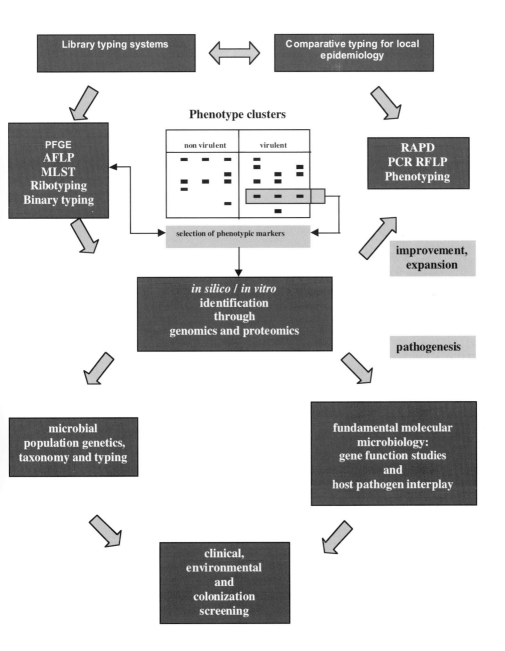

Figure 6. Accompanying downstream applications of *S. aureus* strain typing. Relevant markers can be selected from a fingerprint for the definition of biological features, such as host-pathogen interaction, virulence or resistance. Identification of the primary structure of the markers on the nucleotide or on the amino acid level allow the development of new assays that may reveal population genetic features (see figure 6a).

Lindsay, 2001). Gravet *et al.* concluded that the presence of epidermolysins and leukotoxins in *S. aureus* were correlated with the clinical syndrome; impetigo (Gravet *et al.*, 2001). This observation was confirmed by other studies (Kanzaki *et al.*, 1997; Koning *et al.*, 2002).

Microarrays

The introduction of high throughput systems such as DNA-microarrays may speed up the study of functional typing. Fitzgerald *et al.* (2001) developed a microarray that represents >90% of the *S. aureus* genome and, based on comparative analysis, they showed that >20% of the genome comprises dispensable DNA elements most of it identified as genes encoding putative virulence factors or antibiotic resistance. These findings were supported by the release of two *S. aureus* genome sequences, which corroborated that horizontal transmission of genes plays a fundamental role in the evolution of *S. aureus*. In this study it was also demonstrated that an epidemic rise of a virulent strain was caused by environmental (host) adaptation. Another study indicated that factors, promoting ecological fitness will increase virulence and these virulent clones are capable to disseminate easily (Day *et al.*, 2001, 2002).

Whole Genome Sequencing

Whole genomes of bacteria are currently being sequenced at high rates, and information can be derived from analysis and comparison of these chromosomes. Essential paralogous regions as well as narrowly distributed gene families can be identified. The latter group may be genus, species, or even strain specific. For instance, the genome of *Mycoplasma genitalium* commits about 5% of its content to a single species-specific domain, encoding an adhesin gene (Fraser *et al.*, 1995). Another type of DNA variability was observed after the completion of the *Haemophilus influenzae* DNA sequence (Fleischmann *et al.*, 1995). Repeats in the genes encoding enzymes involved in lipopolysaccharide biosynthesis and iron acquisition and a gene encoding an adhesin display clear heterogeneity (van Belkum, 1999; Hood *et al.*, 1996). The *E. coli* genome highlights novel insertion sequence elements, phage remnants, and many DNA fragments of unusual composition, indicating genome plasticity and horizontal gene transfer (Blattner *et al.*, 1997). Many bacterial virulence genes are found as discrete DNA fragments, present in pathogenic organisms but absent from nonpathogenic members of the same genus or species, e.g., the "pathogenicity islands" of uropathogenic *E. coli* or enteropathogenic *E. coli* (Brikun *et al.*, 1994, Blattner *et al.*, 1997). The *B. subtilis* genome contains phage-type elements as well (Kunst *et al.*, 1997). In April 2002, whole genome sequences of two genetically related MRSA strains (N315 and Mu50) were released in the public domain (Kuroda *et al.*, 2001). Unraveling the genome sequence of both *S. aureus* strains, the enormous flexibility was demonstrated by the presence of mobile elements encoding a number of resistance determinants and virulence factors (resistance- and pathogenicity islands), transposons, plasmids, phages and putative alien genes disseminated all over the chromosome (reviewed in Table 4, Kuroda *et al.* (2001)). These elements form the basis of multiple new epidemiological targets for typing methods.

Staphylococcal Chromosome Cassette *mec* (SSC*mec*)

Historically, the origins of MRSA strains, still poorly understood, were explained by two divergent hypotheses. The mobility of the *mecA* gene in combination with the genetic diversity of MRSA strains, led to the statement from Musser *et al.* that several MRSA strains evolved independently by horizontal transmission of the *mecA* gene into distinct methicillin-susceptible *S. aureus* clones (Musser and Kapur, 1992). This finding was supported by the work of Hiramatsu *et al.* (2001) and Enright *et al.* (2002). They could infer the original MRSA clone from which a MRSA strain is derived, i.e. the ancestral *S. aureus* chromosome type that served as the *SSCmec* recipient at the moment of "birth" of a MRSA clone. They described the presence of at least five different MRSA clonotypes or clonal clusters, signifying that a MRSA clone was generated by intergration of a *SSCmec* type into a distinct genotype chromosome. In contrast, Kreiswirth interpreted data obtained from RFLP analysis of MRSA genomes with *mecA* end Tn*554* probes, that MRSA evolved from a single ancestral methicillin-susceptible clone that required the *mecA* gene and subsequently evolved into diverse clones (Kreiswirth *et al.*, 1993). Oliveira *et al.* (2000) concluded that the downstream region of the *mecA* gene, reflecting *mecA* polymorphism, do not show significant variation. Crisostomo *et al.* (2001) demonstrated the relatedness of early methicillin-susceptible, methicillin-resistant *S. aureus* strains and contemporary MRSA strains. Both studies support the theory of Kreiswirth.

Concluding Remarks

In the future, microbiological typing and identification procedures that are based on the generation of DNA banding patterns, the image-based methods, will be replaced by techniques that produce a binary output. These prospective approaches will depend on probe-mediated identification or primary DNA sequence elucidation. In the coming years research should identify the optimal system for *S. aureus*. Currently, comparative typing methods are used for ad-hoc outbreak studies of limited numbers of strains. Long-term studies, such as continuous surveillance of pathogenic bacteria in specific human populations, require standardized high-throughput methods, the so-called library typing systems which use an uniform nomenclature (Struelens *et al.*, 1998). In high-density micro-array systems, amplification of polymorphic genome regions is followed by solid-phase hybridization with a reference panel of thousands of sequence-variant specific probes, immobilized through chemical or physical procedures (Marshall, and Hodgson, 1998). The resulting binary typing data allow determination of reproducible and extensive numeric profiles. However, interpretation of typing results for large-scale surveillance purposes requires a better understanding of the population structure and micro-evolution of most bacterial pathogens.

The importance of high-density DNA probe array technology has been demonstrated in case of the species identification and determination of rifampicin resistance of *Mycobacteria* spp. (Troesch *et al.*, 1999). Diverse elements that have been identified in the staphylococcal genome and applied in the current more superficial systems, can be addressed as potential targets for the development of probes. House-keeping genes (Enright and Spratt, 1998; Enright and Spratt, 1999; Enright *et al.*, 2000), repeat elements and bordering sequences (Versalovic *et al.*, 1991; Goh *et al.*, 1992; Woods *et al.*, 1992; Frenay *et al.*, 1994; van der Zee, 1999; Cramton *et al.*, 2000), binary probes (van Leeuwen *et al.*, 1996, 1998, 1999) and mobile genetic elements such as plasmid-specific genes, insertion sequences, transposons (Kreiswirth *et al.*, 1994) can be translated into immobilized probes in DNA chip-mediated testing for genetic variability among the staphylococcal genomes. The release of the whole genome sequences from two genetically related MRSA strains generated a large amount of additional nucleic acid targets within one year: much of this most probably, constitutes additional candidates for the epidemiological characterization of MRSA (Kuroda *et al.*, 2001).

Shortly after the introduction of methicillin in 1959, a drug for the treatment of penicillin-resistant *S. aureus* infections, the first cases of isolates that acquired methicillin resistance were reported (Jevons, 1961). To date, the prevalence of MRSA is still increasing. In the intervening period an array of typing techniques have been developed and used, but not a single molecular technique appeared uniformly applicable for the characterization of *S. aureus*. Some image-based typing methods, such as PFGE, have been used for large multi-center studies. Globally, several networks were developed for the validation and characterization of these technologies to obtain inter-center data exchange. The main outcome of these studies was that the optimal *S. aureus* procedure has yet to be developed, be it that MLST is a very promising candidate-technology (Enright *et al.*, 2000; Dingle *et al.*, 2001; Enright *et al.*, 2002). Research in the near future will have to demonstrate whether this technology, which is currently not well available to the routine diagnostic medical microbiology laboratory. It remains to be determined whether MLST is also suited for ad-hoc nosocomial epidemiological studies on MSSA and MRSA. Until then, personal preferences of the researchers involved will remain the prime determinant for the choice of a *S. aureus* typing system.

Acknowledgements

I am very grateful to Alex van Belkum for critical review of the manuscript.

References

Allardet-Servent, A., Bouziges, N., Carles-Nurit, M.J., Bourg, G., Gouby, A. and Ramuz, M. 1989. Use of low-frequency-cleavage restriction endonucleases for DNA analysis in epidemiological investigations of nosocomial bacterial infections. J. Clin. Microbiol. 27: 2057-2061.

Allerberger, F. and Fritschel, S.J. 1999. Use of automated ribotyping of Austrian Listeria monocytogenes isolates to support epidemiological typing. J. Microbiol. Methods 35: 237-244.

Andersen, B.M., Lindemann, R., Bergh, K., Nesheim, B.I., Syversen, G., Solheim, N. and Laugerud, F. 2002. Spread of methicillin-resistant *Staphylococcus aureus* in a neonatal intensive unit associated with understaffing, overcrowding and mixing of patients. J. Hosp. Infect. 50: 18-24.

Anthony, R.M., Brown, T.J. and French, G.L. 2000. Rapid diagnosis of bacteremia by universal amplification of 23S ribosomal DNA followed by hybridization to an oligonucleotide array. J. Clin. Microbiol. 38: 781-788.

Arbeit, R.B. 1997. Laboratory procedures for epidemiologic analysis. In: The staphylococci in human disease Crossley, K.B., and G. L. Gordon, ed., Churchill Livingstone, New York.

Arber, W. 1999. Involvement of gene products in bacterial evolution. Ann. N. Y. Acad. Sci. 870: 36-44.

Aubry-Damon, H., Soussy, C.J. and Courvalin, P. 1998. Characterization of mutations in the *rpoB* gene that confer rifampicin resistance in *Staphylococcus aureus*. Antimicrob. Agents Chemother. 42: 2590-2594.

Bidochka, M.J., Walsh, S.R., Ramos, M.E., Leger, R.J., Silver, J.C. and Roberts, D.W. 1995. Pathotypes in the *Entomophaga grylli* species complex of grasshopper pathogens differentiated with random amplification of polymorphic DNA and cloned-DNA probes. Appl. Environ. Microbiol. 61: 556-560.

Blair, J.E. and Williams, R.E.O. 1961. Phage typing of staphylococci. Bull. W.H.O. 23: 771-784.

Blanc, D.S., Petignat, C., Moreillon, P., Wenger, A., Bille, J. and Francioli, P. 1994. Quantitative antibiogram typing as a typing method for the prospective epidemiological surveillance and control of MRSA: comparison with molecular typing. Infect. Control Hosp. Epidemiol. 17: 654-659.

Blanc, D.S., Struelens, M., Deplano, A., De Ryck, R., Hauser, P.M., Petignat, C. and Francioli, P. 2001. Epidemiological validation of pulsed-field gel electrophoresis patterns for methicillin-resistant *Staphylococcus aureus*. J. Clin. Microbiol. 39: 3442-3445.

Blattner, F.R., Plunkett, G., 3rd, Bloch, C.A., Perna, N.T., Burland, V., Riley, M., Collado-Vides, J., Glasner, J.D., Rode, C.K., Mayhew, G.F., Gregor, J., Davis, N.W., Kirkpatrick, H.A., Goeden, M.A., Rose, D.J., Mau, B. and Shao, Y. 1997. The complete genome sequence of *Escherichia coli* K-12. Science 277: 1453-1474.

Blumberg, H.M., Rimland, D., Kiehlbauch, J.A., Terry, P.M. and Wachsmuth, I.K. 1992. Epidemiologic typing of *Staphylococcus aureus* by DNA restriction fragment length polymorphisms of rRNA genes: elucidation of the clonal nature of a group of bacteriophage-nontypeable, ciprofloxacin-resistant, methicillin-susceptible *S. aureus* isolates. J. Clin. Microbiol. 30: 362-369.

Brikun, I., Suziedelis, K. and Berg, D.E. 1994. DNA sequence divergence among derivatives of *Escherichia coli* K-12 detected by arbitrary primer PCR (random amplified polymorphic DNA) fingerprinting. J. Bacteriol. 176: 1673-1682.

Calderwood, S.B., Baker, M.A., Carroll, P.A., Michel, J.L., Arbeit, R.D. and Ausubel, F.M. 1996. Use of cleaved amplified polymorphic sequences to distinguish strains of *Staphylococcus epidermidis*. J. Clin. Microbiol. 34: 2860-2865.

Chung, M., de Lencastre, H., Matthews, P., Tomasz, A., Adamsson, I., Aries de Sousa, M., Camou, T., Cocuzza, C., Corso, A., Couto, I., Dominguez, A., Gniadkowski, M., Goering, R., Gomes, A., Kikuchi, K., Marchese, A., Mato, R., Melter, O., Oliveira, D., Palacio, R., Sa-Leao, R., Santos Sanches, I., Song, J.H., Tassios, P.T. and Villari, P. 2000. Molecular typing of methicillin-resistant *Staphylococcus aureus* by pulsed-field gel electrophoresis: comparison of results obtained in a multilaboratory effort using identical protocols and MRSA strains. Microb. Drug Resist. 6: 189-198.

Clink, J. and Pennington, T.H. 1987. Staphylococcal whole-cell polypeptide analysis: evaluation as a taxonomic and typing tool. J. Med. Microbiol. 23: 41-44.

Cloak, O.M. and Fratamico, P.M. 2002. A multiplex polymerase chain reaction for the differentiation of *Campylobacter jejuni* and *Campylobacter coli* from a swine processing facility and characterization of isolates by pulsed-field gel electrophoresis and antibiotic resistance profilest. J. Food Prot. 65: 266-273.

Coia, J.E., Noor-Hussain, I. and Platt, D.J. 1988. Plasmid profiles and restriction enzyme fragmentation patterns of plasmids of methicillin-sensitive and methicillin-resistant isolates of *Staphylococcus aureus* from hospital and the community. J. Med. Microbiol. 27: 271-276.

Cookson, B.D., Aparicio, P., Deplano, A., Struelens, M., Goering, R. and Marples, R. 1996. Inter-centre comparison of pulsed-field gel electrophoresis for the typing of methicillin-resistant *Staphylococcus aureus*. J. Med. Microbiol. 44: 179-184.

Costas, M., Cookson, B.D., Talsania, H.G. and Owen, R.J. 1989. Numerical analysis of electrophoretic protein patterns of methicillin- resistant strains of *Staphylococcus aureus*. J. Clin. Microbiol. 27: 2574-2581.

Cramton, S.E., Schnell, N.F., Gotz, F. and Bruckner, R. 2000. Identification of a new repetitive element in *Staphylococcus aureus*. Infect. Immun. 68: 2344-2348.

Crisostomo, M.I., Westh, H., Tomasz, A., Chung, M., Oliveira, D.C. and de Lencastre, H. 2001. The evolution of methicillin resistance in *Staphylococcus aureus*: similarity of genetic backgrounds in historically early methicillin-susceptible and -resistant isolates and contemporary epidemic clones. Proc. Natl. Acad. Sci. USA 98: 9865-9870.

Day, N.P.J., Moore, C.E., Enright, M.C., Berendt, A.R., Maynard Smith, J., Murphy, M.F., Peacock, S.J., Spratt, B.G. and Feil, E.J. 2001. A link between virulence and ecological abundance in natural populations of *Staphylococcus aureus*. Science 292: 114-116.

Day, N.P.J., Moore, C.E., Enright, M.C., Berendt, A.R., Maynard Smith, J., Murphy, M.F., Peacock, S.J., Spratt, B.G. and Feil, E.J. 2002. Retraction. Science 295: 971.

Del Vecchio, V.G., Petroziello, J.M., Gress, M.J., McCleskey, F.K., Melcher, G.P., Crouch, H.K. and Lupski, J.R. 1995. Molecular genotyping of methicillin-resistant *Staphylococcus aureus* via fluorophore-enhanced repetitive-sequence PCR. J. Clin. Microbiol. 33: 2141-2144.

Deplano, A., Schuermans, A., Van Eldere, J., Witte, W., Meugnier, H., Etienne, J., Grundmann, H., Jonas, D., Noordhoek, G.T., Dijkstra, J., van Belkum, A., van Leeuwen, W., Tassios, P.T., Legakis, N.J., van der Zee, A., Bergmans, A., Blanc, D.S., Tenover, F.C., Cookson, B.C., O'Neil, G. and Struelens, M.J. 2000. Multicenter evaluation of epidemiological typing of methicillin- resistant *Staphylococcus aureus* strains by repetitive-element PCR analysis. The European Study Group on Epidemiological Markers of the ESCMID. J. Clin. Microbiol. 38: 3527-3533.

Diekema, D.J., Pfaller, A., Turnidge, J., Verhoef, J., Bell, J., Fluit, A.C., Van Doern, G. and Jones, R.N., and the SENTRY Participants Group. 2000. Genetic relatedness of multidrug-resistant *Staphylococcus aureus* bloodstream isolates from SENTRY antimicrobial resistance surveillance centers worldwide, 1998. Microb. Drug Resist. 6: 213-221.

Dingle, K.E., Colles, F.M., Wareing, D.R., Ure, R., Fox, A.J., Bolton, F.E., Bootsma, H.J., Willems, R.J., Urwin, R. and Maiden, M.C. 2001. Multilocus sequence typing system for *Campylobacter jejuni*. J. Clin. Microbiol. 39: 14-23.

Dobrowolski, M.P. and O'Brien, P.A. 1993. Use of RAPD-PCR to isolate a species specific DNA probe for *Phytophthora cinnamomi*. FEMS Microbiol. Lett. 113: 43-47.

Enright, M.C., Day, N.P.J., Davies, C.E., Peacock, S.J. and Spratt, B.G. 2000. Multilocus sequence typing for characterization of methicillin-resistant and methicillin-susceptible clones of *Staphylococcus aureus*. J. Clin. Microbiol. 38: 1008-1015.

Enright, M.C., Fenoll, A., Griffiths, D. and Spratt, B.G. 1999. The three major Spanish clones of penicillin-resistant *Streptococcus pneumoniae* are the most common clones recovered in recent cases of meningitis in Spain. J. Clin. Microbiol. 37: 3210-3216.

Enright, M.C., Robinson, D.A., Randle, G., Feil, E.J., Grundmann, H. and Spratt, B.G. 2002. The evolutionary history of methicillin-resistant *Staphylococcus aureus* (MRSA). Proc. Natl. Acad. Sci. USA 99: 7687-7692.

Enright, M.C. and Spratt, B.G. 1998. A multilocus sequence typing scheme for *Streptococcus pneumoniae*: identification of clones associated with serious invasive disease. Microbiology 144: 3049-3060.

Enright, M.C. and Spratt, B.G. 1999. Multilocus sequence typing. Trends Microbiol. 7: 482-487.

Eribe, E.R. and Olsen, I. 2000. Strain differentiation in *Bacteroides fragilis* by RAPD and Dendron computer-assisted gel analysis. Apmis 108: 676-684.

Erlandson, K. and Batt, C.A. 1997. Strain-specific differentiation of lactococci in mixed starter culture populations using randomly amplified polymorphic DNA-derived probes. Appl. Environ. Microbiol. 63: 2702-2707.

Fitzgerald, J.R., Sturdevant, D., Mackie, S.M., Gill, S.R. and Musser, J.M. 2001. Evolutionary genomics of *Staphylococcus aureus*: Insights into the origin of methicillin-resistant strains and the toxic shock syndrome epidemic. Proc. Natl. Acad. Sci. USA 98: 8821-8826.

Fleischmann, R.D., Adams, M.D., White, O., Clayton, R.A., Kirkness, E.F., Kerlavage, A.R., Bult, C.J., Tomb, J.F., Dougherty, B.A., Merrick, J.M., McKenney, K., Sutton, G., FitzHugh, W., Fields, C., Gocayne, J.D., Scott, J., Shirley, R., Liu, L., Glodek, A., Kelley, J.M., Weidman, J.F., Phillips, C.A., Spriggs, T., Hedblom,. E., Cotton, M.d., Utterback, T.R., Hanna, M.C., Nguyen, D.T., Saudek, D.M., Brandon, R.C., Fine, L.D., Fritchman, J.L., Fuhrman, J.L., Geohagen, N.S.M., Gnehm, C.L., McDonald L.A., Small, K.V., Fraser, C.M., Smith, H.O., and Venter, J.C. 1995. Whole-genome random sequencing and assembly of *Haemophilus influenzae* Rd. Science 269: 496-512.

Fonstein, M. and Haselkorn, R. 1995. Physical mapping of bacterial genomes. J. Bacteriol. 177: 3361-3369.

Fraser, C.M., Gocayne, J.D., White, O., Adams, M.D., Clayton, R.A., Fleischmann, R.D., Bult, C.J., Kerlavage, A.R., Sutton, G., Kelley, J.M., Fritchman, J.L., Weidman, J.F., Small, K.V., Sandusky, M., Fuhrman, J., Nguyen, D., Utterback, T.R., Saudeck, D.M., Phillips, C.A., Merrick, J.M., Tomb, J., Dougherty, B.A., Bott, K.F., Hu., P., Lucier, T.S., Peterson, S.N., Smith, H.O., Hutchinson III, C.A. and Venter, J.C. 1995. The minimal gene complement of *Mycoplasma genitalium*. Science 270: 397-403.

Frenay, H.M., Theelen, J.P., Schouls, L.M., Vandenbroucke-Grauls, C.M., Verhoef, J., van Leeuwen, W.J. and Mooi, F.R. 1994. Discrimination of epidemic and nonepidemic methicillin-resistant *Staphylococcus aureus* strains on the basis of protein A gene polymorphism. J. Clin. Microbiol. 32: 846-847.

Fry, N.K., Alexiou-Daniel, S., Bangsborg, J.M., Bernander, S., Castellani Pastoris, M., Etienne, J., Forsblom, B., Gaia, V., Helbig, J.H., Lindsay, D., Christian Luck, P., Pelaz, C., Uldum, S.A. and Harrison, T.G. 1999. A multicenter evaluation of genotypic methods for the epidemiologic typing of *Legionella pneumophila* serogroup 1: results from a pan-European study. Clin. Microbiol. Infect. 5: 462-477.

Fry, N.K., Bangsborg, J.M., Bernander, S., Etienne, J., Forsblom, B., Gaia, V., Hasenberger, P., Lindsay, D., Papoutsi, A., Pelaz, C., Struelens, M., Uldum, S.A., Visca, P. and Harrison, T.G. 2000. Assessment of intercentre reproducibility and epidemiological concordance of *Legionella pneumophila* serogroup 1 genotyping by amplified fragment length polymorphism analysis. Eur. J. Clin. Microbiol. Infect. Dis. 19: 773-780.

Galdbart, J.O., Morvan, A. and El Solh, N. 2000. Phenotypic and molecular typing of nosocomial methicillin-resistant *Staphylococcus aureus* strains susceptible to gentamicin isolated in France from 1995 to 1997. J. Clin. Microbiol. 38: 185-190.

Garaizar, J., Lopez-Molina, N., Laconcha, I., Lau Baggesen, D., Rementeria, A., Vivanco, A., Audicana, A. and Perales, I. 2000. Suitability of PCR fingerprinting, infrequent-restriction-site PCR, and pulsed-field gel electrophoresis, combined with computerized gel analysis, in library typing of *Salmonella enterica* serovar enteritidis. Appl. Environ. Microbiol. 66: 5273-5281.

Gaston, M.A., Duff, P.S., Naidoo, J., Ellis, K., Roberts, J.I., Richardson, J.F., Marples, R.R. and Cooke, E.M. 1988. Evaluation of electrophoretic methods for typing methicillin-resistant *Staphylococcus aureus*. J. Med. Microbiol. 26: 189-197.

Gerner-Smidt, P., Graves, L.M., Hunter, S. and Swaminathan, B. 1998. Computerized analysis of restriction fragment length polymorphism patterns: comparative evaluation of two commercial software packages. J. Clin. Microbiol. 36: 1318-1323.

Giesendorf, B.A., Quint, W.G., Vandamme, P. and van Belkum, A. 1996. Generation of DNA probes for detection of microorganisms by polymerase chain reaction fingerprinting. Zentralbl. Bakteriol. 283: 417-430.

Giesendorf, B.A., van Belkum, A., Koeken, A., Stegeman, H., Henkens, M.H., van der Plas, J., Goossens, H., Niesters, H.G. and Quint, W.G. 1993. Development of species-specific DNA probes for *Campylobacter jejuni*, *Campylobacter coli*, and *Campylobacter lari* by polymerase chain reaction fingerprinting. J. Clin. Microbiol. 31: 1541-1546.

Gillet, Y., Issartel, B., Vanhems, P., Fournet, J.C., Lina, G., Bes, M., Vandenesch, F., Piemont, Y., Brousse, N., Floret, D. and Etienne, J. 2002. Association between *Staphylococcus aureus* strains carrying gene for Panton-Valentine leukocidin and highly lethal necrotising pneumonia in young immunocompetent patients. Lancet 359: 753-759.

Gillman, L.M., Gunton, J., Turenne, C.Y., Wolfe, J. and Kabani, A.M. 2001. Identification of *Mycobacterium* species by multiple-fluorescence PCR-single-strand conformation polymorphism analysis of the 16S rRNA gene. J. Clin. Microbiol. 39: 3085-3091.

Goering, R.V. 2000. The molecular epidemiology of nosocomial infection: past, present and future. Rev. Med. Microbiol. 11: 145-152.

Goering, R.V. and Duensing, T.D. 1990. Rapid field inversion gel electrophoresis in combination with an rRNA gene probe in the epidemiological evaluation of staphylococci. J. Clin. Microbiol. 28: 426-429.

Goh, S.H., Byrne, S.K., Zhang, J.L. and Chow, A.W. 1992. Molecular typing of *Staphylococcus aureus* on the basis of coagulase gene polymorphisms. J. Clin. Microbiol. 30: 1642-1645.

Goh, S.H., Potter, S., Wood, J.O., Hemmingsen, S.M., Reynolds, R.P. and Chow, A.W. 1996. HSP60 gene sequences as universal targets for microbial species identification: studies with coagulase-negative staphylococci. J. Clin. Microbiol. 34: 818-823.

Grady, R., Desai, M., Neill., O., Cookson, B. and Stanley, J. 1999. Genotyping of epidemic methicillin-resistant *Staphylococcus aureus* phage type 15 isolates by fluorescent amplified length polymorphism analysis. J. Clin. Microbiol. 37: 3198-3203.

Gravet, A., Couppie, P., Meunier, O., Clyti, E., Moreau, B., Pradinaud, R., Monteil, H. and Prevost, G. 2001. *Staphylococcus aureus* isolated in cases of impetigo produces both epidermolysin A or B and LukE-LukD in 78% of 131 retrospective and prospective cases. J. Clin. Microbiol. 39: 4349-4356.

Greisen, K., Loeffelholz, M., Purohit, A. and Leong, D. 1994. PCR primers and probes for the 16S rRNA gene of most species of pathogenic bacteria, including bacteria found in cerebrospinal fluid. J. Clin. Microbiol. 32: 335-351.

Hartstein, A.I., Morthland, V.H., Eng, S., Archer, G.L., Schoenknecht, F.D. and Rashad, A.L. 1989. Restriction enzyme analysis of plasmid DNA and bacteriophage typing of paired *Staphylococcus aureus* blood culture isolates. J. Clin. Microbiol. 27: 1874-1879.

Hartstein, A.I., Phelps, C.L., Kwok, R.Y. and Mulligan, M.E. 1995. *In vivo* stability and discriminatory power of methicillin-resistant *Staphylococcus aureus* typing by restriction endonuclease analysis of plasmid DNA compared with those of other molecular methods. J. Clin. Microbiol. 33: 2022-2026.

Hayford, A.E., Petersen, A., Vogensen, F.K. and Jakobsen, M. 1999. Use of conserved randomly amplified polymorphic DNA (RAPD) fragments and RAPD pattern for characterization of *Lactobacillus fermentum* in Ghanaian fermented maize dough. Appl. Environ. Microbiol. 65: 3213-3221.

Hiramatsu, K., Cui, L., Kuroda, M. and Ito, T. 2001. The emergence and evolution of methicillin-resistant *Staphylococcus aureus*. Trends Microbiol. 9: 486-493.

Hoefnagels-Schuermans, A., Peetermans, W.E., Struelens, M.J., Van Lierde, S. and Van Eldere, J. 1997. Clonal analysis and identification of epidemic strains of methicillin-resistant *Staphylococcus aureus* by antibiotyping and determination of protein A gene and coagulase gene polymorphisms. J. Clin. Microbiol. 35: 2514-2520.

Hood, D.W., Deadman, M.E., Allen, T., Masoud, H., Martin, A., Brisson, J.R., Fleischmann, R., Venter, J.C., Richards, J.C. and Moxon, E.R. 1996. Use of the complete genome sequence information of *Haemophilus influenzae* strain Rd to investigate lipopolysaccharide biosynthesis. Mol. Microbiol. 22: 951-965.

Hookey, J.V., Edwards, V., Cookson, B.D. and Richardson, J.F. 1999a. PCR-RFLP analysis of the coagulase gene of *Staphylococcus aureus*: application to the differentiation of epidemic and sporadic methicillin-resistant strains. J. Hosp. Infect. 42: 205-212.

Hookey, J.V., Edwards, V., Patel, S., Richardson, J.F. and Cookson, B.D. 1999b. Use of fluorescent amplified fragment length polymorphism Methods (fAFLP) to characterise methicillin-resistant *Staphylococcus aureus*. J. Microbiol. 37: 7-15.

Hunter, P.R. 1990. Reproducibility and indices of discriminatory power of microbial typing methods. J. Clin. Microbiol. 28: 1903-1905.

Hunter, P.R. and Gaston, M.A. 1988. Numerical index of the discriminatory ability of typing systems: an application of Simpson's index of diversity. J. Clin. Microbiol. 26: 2465-2466.

Jevons, M.P. 1961. "Celbenin"-resistant staphylococci. British Medical Journal 1: 124-125.

Jorgensen, M., Givney, R., Pegler, M., Vickery, A. and Funnell, G. 1996. Typing multidrug-resistant *Staphylococcus aureus*: conflicting epidemiological data produced by genotypic and phenotypic methods clarified by phylogenetic analysis. J. Clin. Microbiol. 34: 398-403.

Kamerbeek, J., Schouls, L., Kolk, A., van Agterveld, M., van Soolingen, D., Kuijper, S., Bunschoten, A., Molhuizen, H., Shaw, R., Goyal, M. and van Embden, J. 1997. Simultaneous detection and strain differentiation of *Mycobacterium tuberculosis* for diagnosis and epidemiology. J. Clin. Microbiol. 35: 907-914.

Kanzaki, H., Ueda, M., Morishita, Y., Akiyama, H., Arata, J. and Kanzaki, S. 1997. Producibility of exfoliative toxin and staphylococcal coagulase types of *Staphylococcus aureus* strains isolated from skin infections and atopic dermatitis. Dermatology 195: 6-9.

Kapur, V., Sischo, W.M., Greer, R.S., Whittam, T.S. and Musser, J.M. 1995. Molecular population genetic analysis of *Staphylococcus aureus* recovered from cows. J. Clin. Microbiol. 33: 376-380.

Klein, P.E., Klein, R.R., Cartinhour, S.W., Ulanch, P.E., Dong, J., Obert, J.A., Morishige, D.T., Schlueter, S.D., Childs, K.L., Ale, M. and Mullet, J.E. 2000. A high-throughput AFLP-based method for constructing integrated genetic and physical maps: progress toward a sorghum genome map. Genome Res. 10: 789-807.

Kluytmans, J., van Leeuwen, W., Goessens, W., Hollis, R., Messer, S., Herwaldt, L., Bruining, H., Heck, M., Rost, J., van Leeuwen, N. 1995. Food-initiated outbreak of methicillin-resistant *Staphylococcus aureus* analyzed by pheno- and genotyping. J. Clin. Microbiol. 33: 1121-1128.

Koning, S., van Belkum, A., Snijders, S., van Leeuwen, W., Verbrugh, H.A., Nouwen, J., Op 't Veld, M., van Suijlekom-Smit, L.W.A., van der Wouden, J.C. and Verduin, C. 2003. Virulence of *Staphylococcus aureus* strains and severity of impetigo in children. J. Clin. Microbiol. In Press.

Kreiswirth, B., Kornblum, J., Arbeit, R.D., Eisner, W., Maslow, J.N., McGeer, A., Low, D.E. and Novick, R.P. 1993. Evidence for a clonal origin of methicillin resistance in *Staphylococcus aureus*. Science 259: 227-230.

Kreiswirth, B.N., Lutwick, S.M., Chapnick, E.K., Gradon, J.D., Lutwick, L.I., Sepkowitz, D.V., Eisner, W. and Levi, M.H. 1995. Tracing the spread of methicillin-resistant *Staphylococcus aureus* by Southern blot hybridization using gene-specific probes of mec and Tn*554*. Microb. Drug Resist. 1: 307-313.

Kreiswirth, B.N., McGeer, A., Kornblum, J., Simon, A.E., Eisner, W., Poon, R., Righter, J., Campbell, I. and Low, D.E. 1994. The use of variable gene probes to investigate a multihospital outbreak of MRSA. In: Molecular biology of the Staphylococci. Novick, R.P., ed., pp. 521-530, VCH Publishers, New York.

Krimmer, V., Merkert, H., von Eiff, C., Frosch, M., Eulert, J., Lohr, J.F., Hacker, J. and Ziebuhr, W. 1999. Detection of *Staphylococcus aureus* and *Staphylococcus epidermidis* in clinical samples by 16S rRNA-directed in situ hybridization. J. Clin. Microbiol. 37: 2667-2673.

Kumari, D.N., Keer, V., Hawkey, P.M., Parnell, P., Joseph, N., Richardson, J.F. and Cookson, B. 1997. Comparison and application of ribosome spacer DNA amplicon polymorphisms and pulsed-field gel electrophoresis for differentiation of methicillin-resistant *Staphylococcus aureus* strains. J. Clin. Microbiol. 35: 881-885.

Kunst, F., Ogasawara, N., Moszer, I., Albertini, A.M., Alloni, G., Azevedo, V., Bertero, M.G., Bessieres, P., Bolotin, A., Borchert, S., Borriss, R., Boursier, L., Brans, A., Braun, M., Brignell, S.C., Bron, S., Brouillet, S., Bruschi, C.V., Caldwell, B., Capuano, V., Carter, N.M., Choi, S.K., Codani, J.J., Connerton, I.F., Cummings, N.J., Daniel, R.A., Denizoti, F., Devine, K.M., Düsterhöft, A., Ehrlich, S.D., Emmerson, P.T., Entian, K.D., Errington, J.,Fabret, C., Ferrari, E., Foulger, D., Fritz, C., Fujita, M., Fujita, Y., Fuma, S., Galizzi, A., Galleron, N., Ghim, S.-Y., Glaser, P.,Goffeau, A., Golightly, E.J., Grandi, G., Guiseppi, G., Guy, B.J., Haga, K., Haiech, J., Harwood, C.R., Hénaut, A., Hilbert, H., Holsappel, S., Hosono, S., Hullo, M.-F., Itaya, M., Jones, L., Joris, B., Karamata, D., Kasahara, Y., Klaerr-

Blanchard, M., Klein, C., Kobayashi, Y., Koetter, P., Koningstein, G., Krogh, S., Kumano, M., Kurita, K., Lapidus, A., Lardinois, S., Lauber, J. Lazarevic, V., Lee, S.-M., Levine, A., Liu, H., Masuda, S., Mauël, C., Médigue, C., Medina, N., Mellado, R.P., Mizuno, M., Moestl, D., Nakai, S., Noback, M., Noone, D., O'Reilly, M., Ogawa, Ogiwara, A., Oudega, B., Park, S.-H., Parro, V., Pohl, T.M., Portetelle, D., Porwollik, S., Prescott, A.M., Presecan, E., Pujic, P., Purnelle, B., Rapoport, G., Rey, M., Reynolds, S., Rieger, M., Rivolta, C., Rocha, E., Roche, B., Rose, M., Sadaie, Y. Sato, T., Scanlan, E., Schleich, S., Schroeter, R., Scoffone, F., Sekiguchi, J., Sekowska, A., Seror, S.J., Serror, P., Shin, B.-S., Soldo, B., Sorokin, A., Tacconi, E., Takagi, T., Takahashi, H., Takemaru, K., Takeuchi, M., Tamakoshi, A., Tanaka, T., Terpstra, P., Tognoni, A., Tosato, V., Uchiyama, S., Vandenbol, M., Vannier, F., Vassarotti, A., Viari, A., Wambutt, R., Wedler, E.,Wedler, H., Weitzenegger, T., Winters, P., Wipat, A., Yamamoto, H., Yamane, K., Yasumoto, K., Yata, K., Yoshida, K., Yoshikawa, H.-F., Zumstein, E., Yoshikawa, H. and Danchin, A. 1997. The complete genome sequence of the gram-positive bacterium *Bacillus subtilis*. Nature 390: 249-256.

Kuroda, M., Ohta, T., Uchiyama, I., Baba, T., Yuzawa, H., Kobayashi, I., Cui, L., Oguchi, A., Aoki, K., Nagai, Y., Lian, J., Ito, T., Kanamori, M., Matsumaru, H., Maruyama, A., Murakami, H., Hosoyama, A., Mizutani-Ui, Y., Takahashi, N.K., Sawano, T., Inoue, R., Kaito, C., Sekimizu, K., Hirakawa, H., Kuhara, S., Goto, S., Yabuzaki, J., Kanehisa, M., Yamashita, A., Oshima, K., Furuya, K., Yoshino, C., Shiba, T., Hattori, M., Ogasawara, N., Hayashi, H. and Hiramatsu, K. 2001. Whole genome sequencing of meticillin-resistant *Staphylococcus aureus*. Lancet 357: 1225-1240.

Lanfranco, L., Wyss, P., Marzachi, C. and Bonfante, P. 1993. DNA probes for identification of the ectomycorrhizal fungus *Tuber magnatum* Pico. FEMS Microbiol. Lett. 114: 245-251.

Leonard, R.B., Mayer, J., Sasser, M., Woods, M.L., Mooney, B.R., Brinton, B.G., Newcomb-Gayman, P.L. and Carroll, K.C. 1995. Comparison of MIDI Sherlock system and pulsed-field gel electrophoresis in characterizing strains of methicillin-resistant *Staphylococcus aureus* from a recent hospital outbreak. J. Clin. Microbiol. 33: 2723-2727.

Leski, T.A., Gniadkowski, M., Skoczynska, A., Stefaniuk, E., Trzcinski, K. and Hryniewicz, W. 1999. Outbreak of mupirocin-resistant staphylococci in a hospital in Warsaw, Poland, due to plasmid transmission and clonal spread of several strains. J. Clin. Microbiol. 37: 2781-2788.

Liu, P.Y., Shi, Z.Y., Lau, Y.J., Hu, B.S., Shyr, J.M., Tsai, W.S., Lin, Y.H. and Tseng, C.Y. 1996. Use of restriction endonuclease analysis of plasmids and pulsed-field gel electrophoresis to investigate outbreaks of methicillin-resistant *Staphylococcus aureus* infection. Clin. Infect. Dis. 22: 86-90.

Locksley, R.M., Cohen, M.L., Quinn, T.C., Tompkins, L.S., Coyle, M.B., Kirihara, J.M. and Counts, G.W. 1982. Multiple antibiotic-resistant *Staphylococcus aureus*: introduction, transmission, and evolution of nosocomial infection. Ann. Intern. Med. 97: 317-324.

Maiden, M.C., Bygraves, J.A., Feil, E., Morelli, G., Russell, J.E., Urwin, R., Zhang, Q., Zhou, J., Zurth, K., Caugant, D.A., Feavers, I.M., Achtman, M. and Spratt, B.G. 1998. Multilocus sequence typing: a portable approach to the identification of clones within populations of pathogenic microorganisms. Proc. Natl. Acad. Sci. USA 95: 3140-3145.

Manulis, S., Valinsky, L., Lichter, A. and Gabriel, D.W. 1994. Sensitive and specific detection of *Xanthomonas campestris* pv. pelargonii with DNA primers and probes identified by random amplified polymorphic DNA analysis. Appl. Environ. Microbiol. 60: 4094-4099.

Marshall, A. and Hodgson, J. 1998. DNA chips: an array of possibilities. Nat. Biotechnol. 16: 27-31.

Martineau, F., Picard, F.J., Roy, P.H., Ouellette, M. and Bergeron, M.G. 1998. Species-specific and ubiquitous-DNA-based assays for rapid identification of *Staphylococcus aureus*. J. Clin. Microbiol. 36: 618-623.

Martinez-Murcia, A.J. and Rodriguez-Valera, F. 1994. The use of arbitrarily primed PCR (AP-PCR) to develop taxa specific DNA probes of known sequence. FEMS Microbiol. Lett. 124: 265-269.

Maslow, J.N., Mulligan, M.E. and Arbeit, R.D. 1993. Molecular epidemiology: application of contemporary techniques to the typing of microorganisms. Clin. Infect. Dis. 17: 153-162; quiz 163-164.

Matheson, V.G., Munakata-Marr, J., Hopkins, G.D., McCarty, P.L., Tiedje, J.M. and Forney, L.J. 1997. A novel means to develop strain-specific DNA probes for detecting bacteria in the environment. Appl. Environ. Microbiol. 63: 2863-2869.

Mayer, L.W. 1988. Use of plasmid profiles in epidemiologic surveillance of disease outbreaks and in tracing the transmission of antibiotic resistance. Clin. Microbiol. Rev. 1: 228-243.

Maynard Smith, J., Feil, E.J. and Smith, N.H. 2000. Population structure and evolutionary dynamics of pathogenic bacteria. BioEssays 20: 1115-1122.

Maynard Smith, J., Smith, N.H., O' Rourke, M. and Spratt, B.G. 1993. How clonal are bacteria? Proc. Natl. Acad. Sci. USA 90: 4384-4388.

McGowan, J.E., Jr. 1991. Abrupt changes in antibiotic resistance. J. Hosp. Infect. 18 Suppl A: 202-210.

McGowan, J.E., Jr., Terry, P.M., Huang, T.S., Houk, C.L. and Davies, J. 1979. Nosocomial infections with gentamicin-resistant *Staphylococcus aureus*: plamid analysis as an epidemiologic tool. J. Infect. Dis. 140: 864-872.

Mehling, A., Wehmeier, U.F. and Piepersberg, W. 1995. Application of random amplified polymorphic DNA (RAPD) assays in identifying conserved regions of actinomycete genomes. FEMS Microbiol. Lett. 128: 119-125.

Meugnier, H., Bes, M., Vernozy-Rozand, C., Mazuy, C., Brun, Y., Freney, J. and Fleurette, J. 1996. Identification and ribotyping of *Staphylococcus xylosus* and *Staphylococcus equorum* strains isolated from goat milk and cheese. Int. J. Food Microbiol. 31: 325-331.

Meyers, J.A., Sanchez, D., Elwell, L.P. and Falkow, S. 1976. Simple agarose gel electrophoretic method for the identification and characterization of plasmid deoxyribonucleic acid. J. Bacteriol. 127: 1529-1537.

Molhuizen, H.O., Bunschoten, A.E., Schouls, L.M. and van Embden, J.D. 1998. Rapid detection and simultaneous strain differentiation of *Mycobacterium tuberculosis* complex bacteria by spoligotyping. Meth. Mol. Biol. 101: 381-394.

Monzon-Moreno, C., Aubert, S., Morvan, A. and El Solh, N. 1991. Usefulness of three probes in typing isolates of methicillin-resistant *Staphylococcus aureus* (MRSA). J. Med. Microbiol. 35: 80-88.

Moore, P.C. and Lindsay, J.A. 2001. Genetic variation among hospital isolates of methicillin-sensitive *Staphylococcus aureus*: evidence for horizontal transfer of virulence genes. J. Clin. Microbiol. 39: 2760-2767.

Morvan, A., Aubert, S., Godard, C. and El Solh, N. 1997. Contribution of a typing method based on IS*256* probing of *Sma*I-digested cellular DNA to discrimination of European phage type 77 methicillin-resistant *Staphylococcus aureus* strains. J. Clin. Microbiol. 35: 1415-1423.

Mulligan, M.E. and Arbeit, R.D. 1991. Epidemiologic and clinical utility of typing systems for differentiating among strains of methicillin-resistant *Staphylococcus aureus*. Infect. Control Hosp. Epidemiol. 12: 20-28.

Mullis, K., Faloona, F., Scharf, S., Saiki, R., Horn, G. and Erlich, H. 1986. Specific enzymatic amplification of DNA *in vitro*: the polymerase chain reaction. Cold Spring Harb. Symp. Quant. Biol. 51: 263-273.

Murchan, S., Carter, M. and Aucken, H.M. 2000. Strain identities of phage non-typable MRSA in the UK. J. Hosp. Infect. 46: 157-158.

Murchan, S., Kaufmann, M.E., Deplano, A., De Ryck, R., Elsberg-Vinn, C., Fussing, V., Rosdahl, V.T., Salmenlinna, S., Vuopio-Varkila, J., El Sohl, N., Cuny, C., Witte, W., Tassios, P., Legakis, N., van Leeuwen, W., van Belkum, A., Vindel, A., Laconcha, I., Garaizar, J., Haeggman, S., Hoffman, B.M., Ransjo, U., Olsson-Liljequist, B., Coombes, G. and Cookson, B. 2003. Harmonisation of pulsed-field gel electrophoresis for epidemiological typing of methicillin-resistant *Staphylococcus aureus* by consensus in 10 European centers and its use to plot the spread of related strains. J. Clin. Microbiol. In Press.

Murchan, S., Kaufman, M.E., Deplano, A., Struelens, M., Elsberg, C.S., Rosdahl, V.T., Salmenlinna, S., Vuopio-Varkila, J., El Solh, N., Cuny, C., Witte, W., Tassios, P.T., Legakis, N., Van Leeuwen, W.B., Van Belkum, A., Melo-Cristino, J., Vindel, A., Garaizar, J., Hoffman, B.M., Olsson-Liljequist, B., Ransjo, U. and Cookson, B.D. 2000. Harmony: Establishment of a collection and database of European epidemic MRSA (EMRSA) strains and harmonisation of pulsed-field gel electrophoresis (PFGE). In: 5th International Meeting on Bacterial Epidemiological Markers. p. P084: p.156.

Musser, J.M. and Kapur, V. 1992. Clonal analysis of methicillin-resistant *Staphylococcus aureus* strains from intercontinental sources: association of the *mec* gene with divergent phylogenetic lineages implies dissemination by horizontal transfer and recombination. J. Clin. Microbiol. 30: 2058-2063.

Musser, J.M. and Selander, R.K. 1990. Genetic analysis of natural populations of *Staphylococcus aureus*. In: Molecular Biology of the Staphylococci. Novick, R. and Skurray, R.A., eds. VCH, New York. p. 59.

Nakano, M., Miyazawa, H., Kawano, Y., Kawagishi, M., Torii, K., Hasegawa, T., Iinuma, Y. and Ohta, M. 2002. An outbreak of neonatal toxic shock syndrome-like exanthematous disease (NTED) caused by methicillin-resistant *Staphylococcus aureus* (MRSA) in a neonatal intensive care unit. Microbiol. Immunol. 46: 277-284.

Oliveira, D.C., Wu, S.W. and de Lencastre, H. 2000. Genetic organization of the downstream region of the *mecA* element in methicillin-resistant *Staphylococcus aureus* isolates carrying different polymorphisms of this region. Antimicrob. Agents Chemother. 44: 1906-1910.

Olmos, A., Camarena, J.J., Nogueira, J.M., Navarro, J.C., Risen, J. and Sanchez, R. 1998. Application of an optimized and highly discriminatory method based on arbitrarily primed PCR for epidemiologic analysis of methicillin-resistant *Staphylococcus aureus* nosocomial infections. J. Clin. Microbiol. 36: 1128-1134.

O'Neill, G.L., Murchan, S., Gil-Setas, A. and Aucken, H.M. 2001. Identification and characterization of phage variants of a strain of epidemic methicillin-resistant *Staphylococcus aureus* (EMRSA-15). J. Clin. Microbiol. 39: 1540-1548.

Prevost, G., Jaulhac, B. and Piemont, Y. 1992. DNA fingerprinting by pulsed-field gel electrophoresis is more effective than ribotyping in distinguishing among methicillin-resistant *Staphylococcus aureus* isolates. J. Clin. Microbiol. 30: 967-973.

Prevost, G., Pottecher, B., Dahlet, M., Bientz, M., Mantz, J.M. and Piemont, Y. 1991. Pulsed-field gel electrophoresis as a new epidemiological tool for monitoring methicillin-resistant *Staphylococcus aureus* in an intensive care unit. J. Hosp. Infect. 17: 252-269.

Rosdahl, V.T., Witte, W., Musser, M. and Jarlov, J.O. 1994. *Staphylococcus aureus* strains of type 95. Spread of a single clone. Epidemiol. Infect. 113: 463-470.

Sanders, C.C., Peyret, M., Moland, E.S., Cavalieri, S.J., Shubert, C., Thomson, K.S., Boeufgras, J.M. and Sanders, W.E., Jr. 2001. Potential impact of the VITEK 2 system and the Advanced Expert System on the clinical laboratory of a university-based hospital. J. Clin. Microbiol. 39: 2379-2385.

Savelkoul, P.H.M., Aarts, H.J.M., de Haas, J., Dijkshoorn, L., Duim, B., Otsen, M., Rademaker, J.L.W., Schouls, L. and Lenstra, J.A. 1999. Amplified-fragment length polymorphism analysis: the state of an art. J. Clin. Microbiol. 37: 3083-3091.

Schlichting, C., Branger, C., Fournier, J.M., Witte, W., Boutonnier, A., Wolz, C., Goullet, P. and Doring, G. 1993. Typing of *Staphylococcus aureus* by pulsed-field gel electrophoresis, zymotyping, capsular typing, and phage typing: resolution of clonal relationships. J. Clin. Microbiol. 31: 227-232.

Schmitz, F.J., Steiert, M., Tichy, H.V., Hofmann, B., Verhoef, J., Heinz, H.P., Kohrer, K. and Jones, M.E. 1998. Typing of methicillin-resistant *Staphylococcus aureus* isolates from Dusseldorf by six genotypic methods. J. Med. Microbiol. 47: 341-351.

Schwartz, D.C. and Cantor, C.R. 1984. Separation of yeast chromosome-sized DNAs by pulsed field gradient gel electrophoresis. Cell 37: 67-75.

Schwarzkopf, A. and Karch, H. 1994. Genetic variation in *Staphylococcus aureus* coagulase genes: potential and limits for use as epidemiological marker. J. Clin. Microbiol. 32: 2407-2412.

Selander, R.K., Caugant, D.A., Ochman, H., Musser, J.M., Gilmour, M.N. and Whittam, T.S. 1986. Methods of multilocus enzyme electrophoresis for bacterial population genetics and systematics. Appl. Environ. Microbiol. 51: 873-884.

Selander, R.K., Musser, J.M., Caugant, D.A., Gilmour, M.N. and Whittam, T.S. 1987. Population genetics of pathogenic bacteria. Microbiol. Pathog. 3: 1-7.

Shopsin, B., Gomez, M., Montgomery, S.O., Smith, D.H., Waddington, M., Dodge, D.E., Bost, D.A., Riehman, M., Naidich, S. and Kreiswirth, B.N. 1999. Evaluation of protein A gene polymorphic region DNA sequencing for typing of *Staphylococcus aureus* strains. J. Clin. Microbiol. 37: 3556-3563.

Smeltzer, M.S., Gillaspy, A.F., Pratt, F.L. and Thames, M.D. 1997. Comparative evaluation of use of *cna*, *fnbA*, *fnbB*, and *hlb* for genomic fingerprinting in the epidemiological typing of *Staphylococcus aureus*. J. Clin. Microbiol. 35: 2444-2449.

Smeltzer, M.S., Pratt, F.L., Gillaspy, A.F. and Young, L.A. 1996. Genomic fingerprinting for epidemiological differentiation of *Staphylococcus aureus* clinical isolates. J. Clin. Microbiol. 34: 1364-1372.

Smith, I. 1985. The AMBIS beta scanning system. Bioessays 3: 225-229.

Southern, E.M. 1975. Detection of specific sequences among DNA fragments separated by gel electrophoresis. J. Mol. Biol. 98: 503-517.

Speaker, M.G., Milch, F.A. and Shah, M.K., *et al.* 1994. Role of external bacterial flora in the pathogenesis of acute postoperative endophthalmitis. Ophthalmol. 98: 639.

Spratt, B.G. 1999. Multilocus sequence typing: molecular typing of bacterial pathogens in an era of rapid DNA sequencing and the Internet. Curr. Opin. Microbiol. 2: 312-316.

Struelens, M.J., and the members of the European Study Group on Epidemiological Markers (ESGEM)of the European Society for Clinical Microbiology and Infectious Diseases (ESCMID). 1996. Consensus guidelines for appropriate use and evaluation of microbial epidemiologic typing systems. Clin. Microbiol. Infect. 2: 2-11.

Struelens, M.J., Bax, R., Deplano, A., Quint, W.G. and Van Belkum, A. 1993. Concordant clonal delineation of methicillin-resistant *Staphylococcus aureus* by macrorestriction analysis and polymerase chain reaction genome fingerprinting. J. Clin. Microbiol. 31: 1964-1970 (erratum 1932:1134).

Struelens, M.J., de Gheldre, Y. and Deplano, A. 1998. Comparative and library epidemiological typing systems: outbreak investigations versus surveillance systems. Infect. Control Hosp. Epidemiol. 19: 565-569.

Stull, T.L., LiPuma, J.J. and Edlind, T.D. 1988. A broad-spectrum probe for molecular epidemiology of bacteria: ribosomal RNA. J. Infect. Dis. 157: 280-286.

Tambic, A., Power, E.G., Talsania, H., Anthony, R.M. and French, G.L. 1997. Aanlysis of an outbreak of non-phage-typeable methicillin-resistant *Staphylococcus aureus* by using a randomly amplified polymorphic DNA assay. J. Clin. Microbiol. 35: 3092 -3097.

Tambic-Andrasevic, A., Power, E.G.M., Anthony, R.M., Kalenic, S. and French, G.L. 1999. Failure of bacteriophage typing to detect an inter-hospital outbreak of methicillin-resistant *Staphylococcus aureus* (MRSA) in Zagreb subsequently identified by random amplification of polymorphic DNA (RAPD) and pulsed-field gel electrophoresis (PFGE). Clin. Microbiol. Infect. 5: 634-642.

Tanaka, M., Wang, T., Onodera, Y., Uchida, Y. and Sato, K. 2000. Mechanism of quinolone resistance in *Staphylococcus aureus*. J. Infect. Chemother. 6: 131-139.

Tenover, F.C., Arbeit, R., Archer, G., Biddle, J., Byrne, S., Goering, R., Hancock, G., Hebert, G.A., Hill, B., Hollis, R. and *et al.* 1994. Comparison of traditional and molecular methods of typing isolates of *Staphylococcus aureus*. J. Clin. Microbiol. 32: 407-415.

Tenover, F.C., Arbeit, R.B. and Goering, R.V., and the Molecular Typing Working Group of the Society for Healthcare Epidemiology of America. 1997. How to select and interpret molecular strain typing methods for epidemiological studies of bacterial infections: a review for healthcare epidemiologists. Infect. Contr. Hosp. Epidemiol. 18: 426-439.

Tenover, F.C., Arbeit, R.D., Goering, R.V., Mickelsen, P.A., Murray, B.E., Persing, D.H. and Swaminathan, B. 1995. Interpreting chromosomal DNA restriction patterns produced by pulsed-field gel electrophoresis: criteria for bacterial strain typing. J. Clin. Microbiol. 33: 2233-2239.

Trilla, A., Nettleman, M.D., Hollis, R.J., Fredrickson, M., Wenzel, R.P. and Pfaller, M.A. 1993. Restriction endonuclease analysis of plasmid DNA from methicillin-resistant *Staphylococcus aureus:* clinical application over a three-year period. Infect. Contr. Hosp. Epidemiol. 14: 29-35.

Troesch, A., Nguyen, H., Miyada, C.G., Desvarenne, S., Gingeras, T.R., Kaplan, P.M., Cros, P. and Mabilat, C. 1999. *Mycobacterium* species identification and rifampin resistance testing with high-density DNA probe arrays. J. Clin. Microbiol. 37: 49-55.

van Belkum, A. 1999. Short sequence repeats in microbial pathogenesis and evolution. Cell. Mol. Life Sci. 56: 729-734.

van Belkum, A. 2000. Molecular epidemiology of methicillin-resistant *Staphylococcus aureus* strains: state of affairs and tomorrow's possibilities. Microb. Drug Resist. 6: 173 - 188.

van Belkum, A., Bax, R., Peerbooms, P., Goessens, W.H., van Leeuwen, N. and Quint, W.G. 1993. Comparison of phage typing and DNA fingerprinting by polymerase chain reaction for discrimination of methicillin-resistant *Staphylococcus aureus* strains. J. Clin. Microbiol. 31: 798-803.

van Belkum, A., Bax, R., Van Straaten, P.C.J., Quint, W.G.V. and Veringa, E. 1994. PCR fingerprinting for epidemiological studies of *Staphylococcus aureus*. J. Microbiol. Meth. 20: 235-247.

van Belkum, A., Kluytmans, J., van Leeuwen, W., Bax, R., Quint, W., Peters, E., Fluit, A., Vandenbroucke-Grauls, C., van den Brule, A., Koeleman, H. *et al*. 1995. Multicenter evaluation of arbitrarily primed PCR for typing of *Staphylococcus aureus* strains. J. Clin. Microbiol. 33: 1537-1547.

van Belkum, A., Scherer, S., van Alphen, L. and Verbrugh, H. 1998. Short-sequence DNA repeats in prokaryotic genomes. Microb. Mol. Biol. Rev. 62: 275-293.

van Belkum, A., van Leeuwen, W., Kaufmann, M.E., Cookson, B., Forey, F., Etienne, J., Goering, R., Tenover, F., Steward, C., O'Brien, F., Grubb, W., Tassios, P., Legakis, N., Morvan, A., El Solh, N., de Ryck, R., Struelens, M., Salmenlinna, S., Vuopio-Varkila, J., Kooistra, M., Talens, A., Witte, W. and Verbrugh, H. 1998. Assessment of resolution and intercenter reproducibility of results of genotyping *Staphylococcus aureus* by pulsed-field gel electrophoresis of *Sma*I macrorestriction fragments: a multicenter study. J. Clin. Microbiol. 36: 1653-1659.

van Belkum, A., van Leeuwen, W., Verkooyen, R., Sacilik, S.C., Cokmus, C. and Verbrugh, H. 1997. Dissemination of a single clone of methicillin-resistant *Staphylococcus aureus* among Turkish hospitals. J. Clin. Microbiol. 35: 978-981.

van der Zee, A., Verbakel, H., van Zon, J.C., Frenay, I., van Belkum, A., Peeters, M., Buiting, A. and Bergmans, A. 1999. Molecular genotyping of *Staphylococcus aureus* strains: comparison of repetitive element sequence-based PCR with various typing methods and isolation of a novel epidemicity marker. J. Clin. Microbiol. 37: 342-349.

van Leeuwen, W., Sijmons, M., Sluijs, J., Verbrugh, H. and van Belkum, A. 1996. On the nature and use of randomly amplified DNA from *Staphylococcus aureus*. J. Clin. Microbiol. 34: 2770-2777.

van Leeuwen, W., Snoeyers, S., van der Werken-Liebrechts, C., Tuip, A., van der Zee, A., Egberink, D., de Proost, M., Bik, E., Lunter, B., Kluytmans, J., Wannet, W., Noordhoek, G., Mulder, S., Renders, N., Boers, M., Zaat, B., van der Riet, D., Kooistra, M., Talens, A., Dijkshoorn, L., van der Reyden, T., Veenendaal, D., Bakker, N., Cookson, B., Lynch, A., Witte, W., Cuny, C., Blanc, D., Vernez, I., Hryniewicz, W., Fiet, J., Struelens, M., Deplano, A., Landegent, J., Verbrugh, H. and van Belkum, A. 2002. Assessment of inter-center reproducibility of the binary typing system for the characterization of *Staphylococcus aureus* strains. J. Microbiol. Meth. 51: 19-28.

van Leeuwen, W., van Belkum, A., Kreiswirth, B. and Verbrugh, H. 1998. Genetic diversification of methicillin-resistant *Staphylococcus aureus* as a function of prolonged geographic dissemination and as measured by binary typing and other genotyping methods. Res. Microbiol. 149: 497-507.

van Leeuwen, W., Verbrugh, H., van der Velden, J., van Leeuwen, N., Heck, M. and van Belkum, A. 1999. Validation of binary typing for *Staphylococcus aureus* strains. J. Clin. Microbiol. 37: 664-674.

van Leeuwen, W.B., Libregts, C., Schalk, M., Veuskens, J., Verbrugh, H. and Van Belkum, A. 2001. Binary typing of *Staphylococcus aureus* strains through reversed hybridization using digoxigenin-ULS^R labeled bacterial genomic DNA. J. Clin. Microbiol. 39: 328-331.

Vandenesch, F., Perrier-Gros-Claude, J.D., Bes, M., Fuhrmann, C., Delorme, V., Mouren, C. and Etienne, J. 1995. *Staphylococcus pasteuri*-specific oligonucleotide probes derived from a random amplified DNA fragment. FEMS Microbiol. Lett. 132: 147-152.

Vazquez, J.A., Arganoza, M.T., Vaishampayan, J., Vitzgerald, T., Zervos, M.J., Lipman, B. and Akins, R.A. 1994. Strain delineation of related and unrelated isolates of *Aspergillus flavus* with RAPD probes. Poster 226. In Infectious Diseases Society of America. p. 901.

Versalovic, J., Koeuth, T. and Lupski, J.R. 1991. Distribution of repetitive DNA sequences in eubacteria and application to fingerprinting of bacterial genomes. Nucleic Acids Res. 19: 6823-6831.

Vos, P., Hogers, R., Bleeker, M., Reijans, M., van de Lee, T., Hornes, M., Frijters, A., Pot, J., Peleman, J., Kuiper, M. *et al.* 1995. AFLP: a new technique for DNA fingerprinting. Nucleic Acids Res. 23: 4407-4414.

Voss, A., Milatovic, D., Wallrauch-Schwarz, C., Rosdahl, V.T. and Braveny, I. 1994. Methicillin-resistant *Staphylococcus aureus* in Europe. Eur. J. Clin. Microbiol. Infect. Dis. 13: 50-55.

Wada, A., Ohta, H., Kulthanan, K. and Hiramatsu, K. 1993. Molecular cloning and mapping of 16S-23S rRNA gene complexes of *Staphylococcus aureus*. J. Bacteriol. 175: 7483-7487.

Walker, J., Borrow, R., Goering, R.V., Egerton, S., Fox, A.J. and Oppenheim, B.A. 1999. Subtyping of methicillin-resistant *Staphylococcus aureus* isolates from the North-West of England: a comparison of standardised pulsed-field gel electrophoresis with bacteriophage typing including an inter-laboratory reproducibility study. J. Med. Microbiol. 48: 297-301.

Wei, M.Q., Groth, D., Mendis, A.H., Sampson, J., Wetherall, J.D. and Grubb, W.B. 1992. Typing of methicillin-resistant *Staphylococcus aureus* with an M13 repeat probe. J. Hosp. Infect. 20: 233-245.

Welsh, J. and McClelland, M. 1991. Genomic fingerprinting using arbitrarily primed PCR and a matrix of pairwise combinations of primers. Nucleic Acids Res. 19: 5275-5279.

Wielders, C.L.C., Vriens, M.R., Brisse, S., de Graaf-Miltenburg, L.A.M., Troelstra, A., Fleer, A., Schmitz, F.J., Verhoef, J. and Fluit, A.C. 2001. *In-vivo* transfer of *mecA* DNA to *Staphylocccus aureus*. Lancet 357: 1674-1675.

Williams, J.G., Kubelik, A.R., Livak, K.J., Rafalski, J.A. and Tingey, S.V. 1990. DNA polymorphisms amplified by arbitrary primers are useful as genetic markers. Nucleic Acids Res. 18: 6531-6535.

Witte, W., Kresken, M., Braulke, C. and Cuny, C. 1997. Increasing incidence and widespread dissemination of methicillin- resistant *Staphylococcus aureus* (MRSA) in hospitals in central Europe, with special reference to German hospitals. Clin. Microbiol. Infect. 3: 414-422.

Woods, C.R., Jr., Versalovic, J., Koeuth, T. and Lupski, J.R. 1992. Analysis of relationships among isolates of *Citrobacter diversus* by using DNA fingerprints generated by repetitive sequence-based primers in the polymerase chain reaction. J. Clin. Microbiol. 30: 2921-2929.

Yoshida, T., Kondo, N., Hanifah, Y.A. and Hiramatsu, K. 1997. Combined use of ribotyping, PFGE typing and IS*431* typing in the discrimination of nosocomial strains of methicillin-resistant *Staphylococcus aureus*. Microbiol. Immunol. 41: 687-695.

Zabeau, M. and Vos, P. 1993. Selective restriction fragment amplification: a general method for DNA fingerprinting. Publication 0 534 858 A1, bulletin 93/13, European Patent Office, Munich, Germany

Zadoks, R., van Leeuwen, W., Barkema, H., Sampimon, O., Verbrugh, H., Schukken, Y.H. and van Belkum, A. 2000. Application of pulsed-field gel electrophoresis and binary typing as tools in veterinary clinical microbiology and molecular epidemiologic analysis of bovine and human *Staphylococcus aureus* isolates. J. Clin. Microbiol. 38: 1931-1939.

From: *MRSA: Current Perspectives*
Edited by: A.C. Fluit and F.-J. Schmitz

Chapter 6

The Molecular Evolution of Methicillin-Resistant *Staphylococcus aureus*

J. Ross Fitzgerald and
James M. Musser

Abstract

Methicillin-resistant *Staphylococcus aureus* (MRSA) are a major problem in the treatment of nosocomial infections. Recent discoveries have contributed important insights into the molecular evolution and dissemination of MRSA. This chapter summarizes information regarding the nature of the chromosomal elements encoding resistance to methicillin, and the molecular evolution of the limited number of clones responsible for MRSA infections, worldwide. We also discuss data concerning the recent emergence of MRSA as a significant cause of community-acquired infections. Finally we discuss the implications of the availability of the complete genome sequence of MRSA strains, and suggest important directions for future research.

Introduction

In the early 1950s, plasmids encoding penicillinase spread rapidly among *Staphylococcus aureus* populations rendering them resistant to penicillin. A modified synthetic penicillin known as methicillin was introduced in 1959 as an alternative drug to treat infections caused by this pathogen (Figure 1). However, resistant strains emerged very rapidly, and the first methicillin-resistant *S. aureus* (MRSA) strain was identified in the United Kingdom in 1961 (Jevons, 1961). The first major outbreak in the USA was not reported until 1968 (Barrett *et al.*, 1968). Subsequently it was shown that MRSA strains had acquired a gene (*mecA*) encoding an altered penicillin-binding protein, which was responsible for methicillin resistance (Matsuhashi *et al.*, 1986). MRSA strains are now ubiquitous worldwide and are a significant healthcare problem (Carbon, 1999). A combination of lateral gene transfer of the *mecA* gene into distinct phylogenetic lines of *S. aureus*, and clonal dissemination, has been responsible for the emergence and spread of MRSA strains (Musser and Kapur, 1992; Ayliffe, 1997; Roman *et al.*, 1997; Witte *et al.*, 1997; Deplano *et al.*, 2000b; Fitzgerald *et al.*, 2001). In this chapter we summarize information bearing on the origin and nature of the chromosomal elements containing the *mecA* gene, and the limited number of genetic backgrounds that have acquired them.

The Genetic Elements Encoding Resistance to Methicillin

Staphylococcal Cassette Chromosome *mec* (SCC*mec*)

In *S. aureus*, resistance to methicillin and virtually all other β-lactam antibiotics is caused by the production of an altered penicillin-binding protein (PBP) designated PBP 2′. This novel PBP has significantly reduced binding affinity for β-lactam antibiotics. When PBP 2′ is produced, *S. aureus* can continue cell wall synthesis unabated, even when exposed to an otherwise inhibitory concentration of β-lactams. Recently, the crystal structure of PBP2′ has been solved and this has allowed dissection of the structural basis of methicillin resistance (Lim and Strynadka, 2002). PBP2′ is encoded by the *mecA* gene located in the chromosome of MRSA strains (Matsuhashi *et al.*, 1986). Several distinct genetic elements contain the *mecA* gene, and these have been well characterized recently by Hiramatsu and co-workers (Ito *et al.*, 1999; Hiramatsu *et al.*, 2001; Ito *et al.*, 2001). The staphylococcal cassette chromosome *mec* (SCC*mec*) is a 21-67 kb fragment of DNA that is integrated into the chromosome of MRSA near the origin of replication. SCC*mec* elements present in different strains vary in gene content (Fitzgerald *et al.*, 2001). Four distinct types of the SCC*mec* (I-IV) element have been described (Hiramatsu *et al.*, 2001; Ito *et al.*, 2001) (Figure 2). SCC*mec* I, II, and III are 39 kb, 52 kb, and

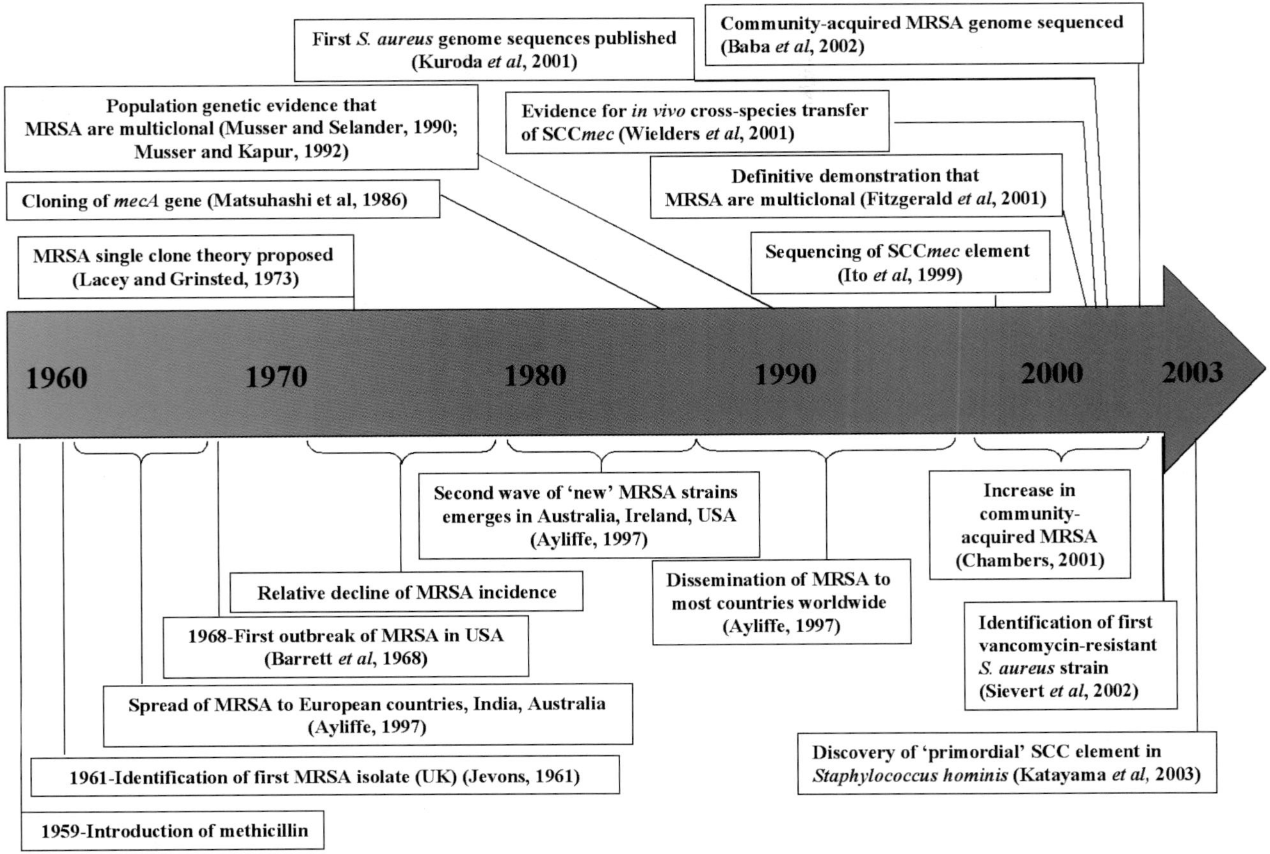

Figure 1. Chronology of MRSA emergence and discoveries

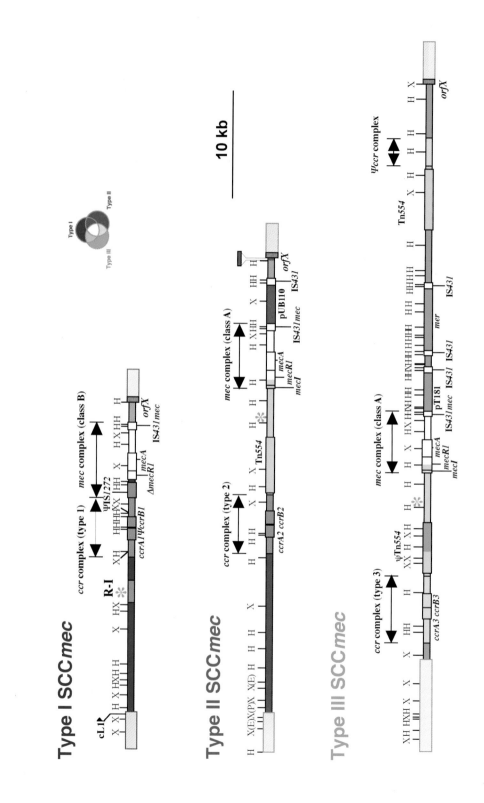

Type I SCCmec

cL.1
ccr complex (type 1)   mec complex (class B)
R-I
ccrA1 ΨccrB1   ΨIS1272   mecA   IS431mec
                          ΔmecR1   orfX

Type II SCCmec

ccr complex (type 2)   mec complex (class A)
ccrA2 ccrB2   Tn554   mecI   IS431mec   orfX
                        mecR1   pUB110
                        mecA

10 kb

Type III SCCmec

ccr complex (type 3)   mec complex (class A)   Ψccr complex
ΨTn554   ccrA3 ccrB3   mecI   IS431mec   IS431   Tn554   orfX
                        mecR1   pT181   mer
                        mecA   IS431

Type IV SCC*mec*

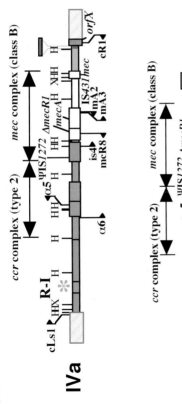
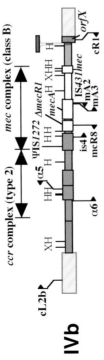

Figure 2. SCC*mec* types I-IV. SCC*mec* variants are 21-67 kb in size. All have conserved terminal inverted repeats and direct repeats located at the chromosomal integration sites, genes for chromosome recombinases (*ccrA* and *ccrB*) and conserved organization around the *mecA* gene. Colour coding represents regions of homology between different SCC*mec* types. Reproduced from Hiramatsu *et al.*, 2001.

67 kb in size, respectively. Analysis of each SCC*mec* type by Southern blot hybridization with probes specific for multiple genes in each SCC*mec* type revealed that 34 of 38 strains examined had either SCC*mec* I, II or III (Ito *et al.*, 2001). Two related subtypes (20.9 kb and 24.3 kb) of SCC*mec* type IV have been identified among recently emerged community-acquired MRSA strains (Hiramatsu *et al.*, 2001; Dunn *et al.*, 2002; Ma *et al.*, 2002).

Although the 4 SCC*mec* elements vary extensively in size and nucleotide sequence, they also have several features in common. All 4 SCC*mec* variants have conserved terminal inverted repeats and direct repeats located at the chromosomal integration sites, genes for cassette chromosome recombinases (*ccrA* and *ccrB*), which direct chromosomal integration and excision events of SCC*mec* (Katayama *et al.*, 2000), and conserved gene organization around the *mecA* gene. The overall G+C content of SCC*mec* II is 33% (Ito *et al.*, 1999), a value within the range of variation for the species *S. aureus* (32% to 36%). However, several regions in SCC*mec* II deviate significantly in GC content, suggesting relatively recent horizontal acquisition. For example, a group of ORFs homologous to the *Clostridium acetobutylicum* potassium ion transport complex operon (*kdp*), remnants of transposase genes, the *mec* regulator genes *mecI* and *mecR1*, and several genes encoding resistance to other antibiotics, all have markedly different GC contents compared to the *S. aureus* chromosome (Ito *et al.*, 1999).

The Chromosomal Integration Site of SCC*mec* and Evidence for Mobility

SCC*mec* is integrated in an open reading frame (*orfX*) encoding a hypothetical protein of unknown function that is well conserved among *S. aureus* clinical isolates (Ito *et al.*, 1999). The precise integration site within *orfX* is conserved among representative isolates containing SCC*mec* I, II, or III and has been identified to within 4 ambiguous bases (the left terminal nucleotide was estimated to be one of the 4 nucleotides TCTG, and the right terminal nucleotide was estimated to be one of the 4 nucleotides TTCT) (Ito *et al.*, 2001). Integration occurs 15 bp upstream from the *orfX* stop codon. Interestingly, these 15 bp are replaced by identical nucleotides present in the integrated SCC*mec* element, thus maintaining an intact *orfX* gene. Both ends of SCC*mec* have incomplete inverted repeats and these sequences are recognized by the cassette chromosome recombinase proteins (Katayama *et al.*, 2000).

The *ccrA* and *ccrB* genes are required for excision and integration of SCC*mec*. Spontaneous excision of SCC*mec* occurs at low frequency during bacterial culture *in vitro* or long term frozen storage (Hiramatsu *et al.*, 1990; Katayama *et al.*, 2000). Moreover, loss of the *mec* element from the chromosome has been observed *in vivo* (Deplano *et al.*, 2000a) and evidence for *in vivo* transfer of SCC*mec* from *S. epidermidis* to *S. aureus* has been presented (Wielders *et al.*, 2001). Both precise excision and orientation-specific integration of a recombinant SCC*mec* plasmid into the *orfX* gene of a *recA*⁺ strain N315 have been demonstrated (Katayama *et al.*, 2000). Evidence strongly suggests that the SCC*mec* elements are mobile (see below), prehaps through phage-mediated transduction, but mobility has not yet been demonstrated in a *recA*⁻ *S. aureus* strain.

The *mec* Gene Complex

The *mec* gene complex is a variable region contained within SCC*mec* consisting of *mecA* and neighbouring genes (*mecI* and *mecR1*) involved in regulation of methicillin-resistance (Petinaki *et al.*, 2001). The *mecI* gene encodes a transcriptional repressor and *mecR1* encodes a signal-transduction protein. MecR1 senses the presence of β-lactam antibiotics in the extracellular environment and activates its cytoplasmic metalloprotease domain by autocatalytic cleavage (Zhang *et al.*, 2001). It then cleaves MecI repressor proteins bound to the operator region of the *mecA* gene, releasing the repression of *mecA* transcription (Sharma *et al.*, 1998). Four classes of *mec* gene complex are found among *Staphylococcus* species (Katayama *et al.*, 2001). The class A *mec* gene complex contains intact copies of *mecA*, *mecI*, and *mecR1*. *S. aureus* strains containing a class A *mec* complex are not resistant to methicillin and have been termed pre-MRSA. This lack of resistance is due to MecI-mediated repression (Kuwahara-Arai *et al.*, 1996). In MRSA strains containing a class A *mec* complex, the repressive function of the *mecI* gene has been inactivated by mutations in *mecI* or in the *mecA* promoter region (Kuwahara-Arai *et al.*, 1996; Katayama *et al.*, 2001). In the class B complex, the entire *mecI* gene has been deleted and a fragment of the insertion sequence IS*1272* exists in its place. Class C and D *mec* complexes also contain *mecI* deletions along with different truncations of the *mecR1* gene (Katayama *et al.*, 2001). The recently described type IV SCC*mec* contains a unique combination of a class B *mec* gene complex with a type 2 *ccr* gene complex (Ma *et al.*, 2002). Previously, in SCC*mec* type I elements, the class B *mec* gene complex was found to be linked to the type 1 *ccr* gene complex (Ito et al., 2001).

Apart from the *mecA* complex itself, other chromosomal loci have been identified that have an effect on the methicillin resistance phenotype (De Lencastre *et al.*, 1999). Many of these genes are involved in cell wall synthesis or have regulatory functions and may contribute to the optimal expression of methicillin-resistance in strains containing the *mecA* gene.

SCC-related Elements

A methicillin-susceptible isolate of *S. aureus* recovered in 1945 has a 5.9 kb DNA fragment (IE25923) containing 15-bp direct repeats and incomplete inverted repeats similar to SCC*mec*, located at precisely the same chromosomal location in *orfX* (Ito *et al.*, 2001). However, this fragment lacks the *mecA* and *ccr* genes, suggesting that it is a progenitor of SCC*mec* that existed before the introduction of methicillin. This observation suggested that SCC*mec* has a more general biological function than simply carriage of the *mecA* gene. This hypothesis is supported by the fact that SCC*mec* also contains several integrated plasmids and transposons, which have genes encoding resistance to other antibiotics such as erythromycin, spectinomycin, kanamycin, tobramycin and bleomycin (Ito *et al.*, 1999). Recently, Katayama *et al.* (2003) reported the discovery of an SCC element without a *mecA* gene in *Staphylococcus hominis*. This type I SCC variant is the first to be found which contains a functional *ccrB1* recombinase gene. The authors suggest that the *S. hominis* SCC element may represent a primordial version of SCC prior to accumulation of antibiotic resistance genes.

In addition to antibiotic resistance genes, a gene (*pls*), which encodes a protein (Pls) belonging to the SD-repeat family of staphylococcal surface proteins, is present in the SCC*mec* type II element. Pls has been implicated in preventing adhesion of *S. aureus* to host components and may contribute to dissemination of infection (Vaudaux *et al.*, 1998; Savolainen *et al.*, 2001). Very recently, the genetic element encoding the genes necessary for production of type 1 capsular polysaccharide (SCC*cap1*) was sequenced (Luong *et al.*, 2002). The 27.4 kb SCC*cap1* element has an identical insertion site to SCC*mec* and shares similar conserved repeats in the *att* sites. In addition, SCC*cap1* contains several ORFs homologous to genes in SCC*mec* including a site-specific recombinase (*ccrB*) homologue, which contains a nonsense mutation. Precise excision of SCC*cap1* could only be induced by introduction of a plasmid containing the *ccrA* and *ccrB* genes of type II SCC*mec*. Sequence degeneration appears to have contributed to a large number of pseudogenes in the element and in the left flanking sequence. This suggests that SCC*cap1* was acquired a long time ago and has since lost the ability for self-mobilization. Interestingly a novel enterotoxin gene was identified adjacent to SCC*cap1*, which was presumably acquired independently. Hence SCC*mec* is a very complex genetic element that has the ability to acquire smaller mobile elements such as transposons and plasmids, many of which encode resistance to antimicrobial agents, and which also can contain genes, which contribute to pathogenesis.

The Origin of the Gene Encoding Methicillin Resistance, *mecA*

As noted previously, the overall G+C content of SCC*mec* is within the range of variation for the species *S. aureus*, suggesting that it may have been acquired long ago in evolutionary terms. However, evidence suggests that the *mecA* gene itself is not native to *S. aureus* but was acquired from another organism by an unknown mechanism. The evolutionary origin of *mecA* has been the subject of much speculation and investigation. The *mecA* gene has been identified in at least 13 staphylococcal species (Couto *et al.*, 1996) and a homologue of *mecA* has been identified in *Enterococcus hirae* (Archer *et al.*, 1996). This species also contains an insertion sequence related to IS*1272* from *Staphylococcus haemolyticus*, an element also found inserted near the *mecA* gene in MRSA strains. This discovery led Archer and colleagues to speculate that the *mecA* gene originated in *E. hirae*, was transferred to *S. haemolyticus*, and subsequently spread to *S. aureus* (Archer *et al.*, 1996).

In an effort to index the phylogenetic distribution of *mecA*, Tomasz and colleagues screened 13 different species of staphylococci for the presence of a homologous gene (Couto *et al.*, 1996). All isolates of *Staphylococcus sciuri* examined contained *mecA* homologues, yet virtually all organisms were methicillin-susceptible. A further study revealed the presence of a second *mecA* homologue, which was identical to *mecA* of *S. aureus* in 46% of *S. sciuri* strains examined (Couto *et al.*, 2000). These strains showed resistance to β-lactams. *S. sciuri* is considered to be one of the more phylogenetically old staphylococcal species. The organism is found mainly in animals although it occasionally causes human infection (Wallet *et al.*, 2000). The *S. sciuri mecA* homologue encodes a protein that has 88% similarity with PBP2′. This high level of relatedness suggests that the *S. sciuri mecA* homologue represents the evolutionary origin of the *mecA* gene. Importantly, when *S. sciuri* was grown in the presence of increasing concentrations of methicillin, this 'silent' gene was activated resulting in methicillin resistance (Wu *et al.*, 2001). Activation was caused by a point mutation located at the −10 promoter sequence which resulted in greatly increased gene transcription and subsequent production of PBP2′. Transfer of the gene to a methicillin-susceptible *S. aureus* strain resulted in increased levels of resistance to this drug. These results are consistent with a previous study showing that *mecA*-positive methicillin-susceptible strains of staphylococci can become drug resistant if exposed to β-lactams (Sakoulas *et al.*, 2001). Clearly, *S. sciuri* may represent the source of the *mecA* gene transferred to *S. aureus* and then selected by exposure to methicillin. However, definitive evidence may be difficult to produce.

The Origin of MRSA Strains

There has been considerable speculation and controversy for decades concerning the origin of MRSA strains. As discussed, the SCC*mec* elements have structural features, which indicate the potential for mobility, although horizontal transfer into a *recA*⁻ strain has yet to be demonstrated.

The Single Clone Theory

In 1973, Lacey and Grinsted proposed the single clone origin theory for MRSA strains based on the phenotypic similarities of the strains they examined (Lacey and Grinsted, 1973). The monoclonal theory postulates that all MRSA strains have evolved from a single methicillin-susceptible progenitor cell, which acquired the *mecA* determinant. Under this idea, the newly formed MRSA cell subsequently expanded by clonal dissemination, thereby giving rise to all contemporary MRSA strains. Kreiswirth and colleagues classified MRSA isolates according to their restriction fragment length (RFLP) profiles generated by Southern blot analysis with 2 DNA probes (Kreiswirth *et al.*, 1993). One probe identified polymorphisms of the *mecA* gene region and the other was specific for transposon Tn*554*. Each of the 29 Tn*554* profiles occurred in association with one and only one *mecA* pattern. The investigators reasoned that if acquisition of *mecA* and Tn*554* were independent events, then association of patterns could only occur by clonal micro-evolution, and not by horizontal gene transfer. The data were interpreted to mean that *mecA* was acquired only once by *S. aureus* and that the observed genetic variation in MRSA strains was due to subsequent acquisition and rearrangement of DNA within the *mecA* locus. Although elegant in its simplicity, the interpretation was inconsistent with data published previously (Musser and Kapur, 1992).

The Multi-Clone Theory

Musser and Kapur used multilocus enzyme electrophoresis to estimate overall chromosomal relationships among MRSA isolates recovered from 4 continents between 1961 and 1992, including many of the same isolates examined by Kreiswirth and colleagues (Kreiswirth *et al.*, 1993). Among 254 MRSA isolates, 15 distinct electrophoretic types were identified, indicating the existence of multiple highly divergent clonal types of MRSA strains. A critical observation was the occurrence of the *mecA* gene in association with chromosomal genotypes representing the entire breadth of the species *S. aureus*. Given the recent emergence of MRSA, the only reasonable interpretation is that horizontal gene transfer has contributed to the evolution of MRSA. Consistent with a multiclonal origin, multilocus sequence typing also found that multiple genotypes were present among MRSA isolates recovered in the UK (Enright *et al.*, 2000). Similarly, ribotyping was used to discriminate among MRSA strains, and 3 different ribotypes (A-C) and 3 different SCC types (I-III) were identified, resulting in 5 different clonotype combinations (I-A, II-A, III-A, II-B, and II-C) (Hiramatsu, 1995).

In addition, Roberts and others (Roberts *et al.*, 1998) used PFGE typing to identify 5 major MRSA clonal groups among 261 MRSA isolates from 12 New York City hospitals. These included both the 'Iberian' clone and the 'New York' clone. Shopsin and Kreiswirth then used *spa* typing of the same isolates and identified the same 5 major clonal groups using this technique (Shopsin and Kreiswirth, 2001). To definitively address the controversy, a whole genome DNA microarray formulated on the basis of the genome sequence of *S. aureus* strain COL was used to investigate evolutionary relationships among MRSA and MSSA strains (Fitzgerald *et al.*, 2001). A phylogenetic tree was constructed based on the variation in gene content of the strains examined, including many strains analyzed previously by Kreiswirth *et al.* (1993) and Musser and Kapur (1992). We found that MRSA strains were assigned to at least five distinct and highly divergent chromosomal genotypic groups that differ in some cases by several hundred genes. These genotypic groups have not shared a recent common ancestor. Hence, MRSA strains have evolved multiple times through horizontal transfer of the *mecA* gene to MSSA strains that represent the breadth of diversity in the species. Taken together, there now can be no doubt that multiple clonal lineages of MRSA exist as a consequence of successful horizontal gene transfer of the *mecA* gene and related genetic material in natural populations.

Although it is now definitively established that multiple lateral gene transfer episodes have contributed to the origin and dissemination of MRSA, it is somewhat surprising that SCC*mec* has not been transferred to a more extensive array of chromosomal backgrounds, considering the very large number of methicillin-susceptible genotypes (Musser and Selander, 1990). The relatively restricted number of MRSA genotypes documented thus far could be due to several contributing factors. One theory is that the integration of large numbers of plasmid and insertion sequences into SCC*mec* elements has rendered them incapable of transfer by phage transduction because of their large size (Novick *et al.*, 2001). However, this theory does not take into account the existence of intermediate variants of SCC*mec* which are presumably small enough to be packaged but are nevertheless poorly represented among *S. aureus* genotypes. A second possibility is that lateral gene transfer of *mecA* is frequent but transfer to successful pathogenic strains has been rare. Hence, these genotypes typically are not identified among clinical strain collections because they are uncommon. It seems possible that a relatively small number of MRSA clones has been inordinately successful, as defined by widespread dissemination and ability to cause disease (pathogenicity). If larger collections of strains cultured from infected patients and asymptomatic carriers were studied, perhaps a larger number of MRSA genotypes would be identified.

Community-Acquired MRSA Strains

The Center for Disease Control and Prevention (CDC) in the USA defines a nosocomial infection as 'a localized or systemic condition that 1) results from adverse reaction to the presence of an infectious agent(s) or its toxin(s) and 2) was not present or incubating at the time of admission to the hospital'. For most bacterial nosocomial infections, this means that the infection usually becomes evident 48 h or more after admission. If symptoms occur outside the hospital or within 48 h of admission to hospital, then the infection is defined as 'community-acquired'. In recent years, the prevalence of infections due to MRSA acquired in the community has risen dramatically (Chambers, 2001; Daum and Seal, 2001; Hussain *et al.*, 2001). This observation is important because until recently, infections due to MRSA were confined largely to the hospital setting and could be identified rapidly and treated accordingly. Infections due to *S. aureus* in the community have been routinely treated with oral β-lactam antibiotics. The emergence of MRSA strains in the community means that orally-administered drugs will not be effective against all *S. aureus* infections. Hence, treatment of *S. aureus* becomes a far more complicated and expensive task. In addition, the emergence of community strains, refractory to vancomycin treatment, becomes a greater possibility.

Although relatively little information is available, community-acquired MRSA strains have been reported to be phenotypically and genotypically distinct from hospital-acquired strains. Unlike hospital strains, the community-acquired strains are largely susceptible to other antibiotic classes and often are only resistant to β-lactam antibiotics (Naimi *et al.*, 2001). In addition, PFGE typing has been used to show that the vast majority of strains examined belong to a single clonal group and this clone is unrelated to MRSA strains causing infections in hospitals (Groom *et al.*, 2001; Naimi *et al.*, 2001). Moreover, the identification of a novel type of SCC*mec* element (type IV) associated with community acquired MRSA strains in Chicago (Hiramatsu *et al.*, 2001; Ma *et al.*, 2002) suggested that community-acquired MRSA have arisen by independent clonal acquisition of *mecA* by a community strain and not by dissemination of extant MRSA clones circulating in hospitals. However, these findings are in contrast to a recent report of community-acquired MRSA in the San Francisco region (Charlebois *et al.*, 2002). In common with previous studies these community-acquired isolates showed greater sensitivity to other antibiotic classes than hospital-acquired strains but the same genotypes were identified among concurrently collected community- and hospital-acquired MRSA isolates (Charlebois *et al.*, 2002). This suggests that in San Francisco the same MRSA clonal types are causing both community- and hospital-acquired infections. Importantly, Daum *et al.* identified type IV SCC*mec* elements in 11 of 12 community-acquired MRSA strains of diverse genetic backgrounds in Chicago indicating that recent horizontal transfer of type IV SCC*mec* into multiple clones has taken place (Daum *et al.*, 2002).

It is important to note that epidemiological studies of community-acquired MRSA have been very limited to date. Clearly, further phylogenetic analysis by sensitive techniques such as whole-genome DNA microarray hybridization is required to gain further understanding of the origin and epidemiology of community-acquired MRSA.

Identification of the Major Extant MRSA Clones

Many laboratories have used bacterial typing techniques to investigate the epidemiology and clonal associations of MRSA from specific outbreaks or geographic locations, resulting in the identification of multiple genotypes of MRSA. However, until recently the inter-relatedness of MRSA strains isolated from worldwide sources has not been clear. In addition, the specific MSSA progenitors of contemporary MRSA had not been established. An important publication (Crisostomo *et al.*, 2001) compared MRSA isolates from the UK and Denmark with MSSA isolates from the same time period, and contemporary MRSA isolates. The antibiotic susceptibility profiles, phage type, PFGE type, MLST type, and *spa* type of the isolates were compared. The authors found that there was restricted genetic diversity among MSSA and MRSA isolates. All MRSA strains and the majority of selected MSSA strains from the same era had a similar PFGE pattern. This pattern was very similar to one of the most widely spread contemporary MRSA clones, the 'Iberian clone'. In addition, 3 other distinct MSSA genotypes from the pre-MRSA era had very similar genetic backgrounds to EMRSA-16, which is widely distributed in UK hospitals (Johnson *et al.*, 2001), the 'pediatric' clone, and the 'New York' clone, respectively. Importantly, this study traces the origin of well-characterized contemporary MRSA clones to their original MSSA ancestors. A very recent publication from the same group describes MLS, *spa*, and SCC*mec* typing of 28 MRSA isolates selected to represent the major clonal types among 3,067 MRSA strains from Southern and Eastern Europe, USA and Latin America. The investigators identified 2 major genetic backgrounds among the strains examined. SCC*mec* types I and III were found in association with strains of genetic background A, and SCC*mec* II was found in strains of genetic background B (Oliveira *et al.*, 2001). MRSA strains of the major UK clones EMRSA-15 or EMRSA-16 were not identified among the strains examined and they represent genetic backgrounds distinct from genetic backgrounds A and B, as indicated by MLS typing. A representative strain of clonal type EMRSA-16 is currently being sequenced at the Sanger centre, UK, and it contains an SCC*mec* type II element. The SCC*mec* type present in EMRSA-15 strains has not been established. A model for acqusition of SCC*mec* by *S. aureus* clones is presented in Figure 3.

Recently, Enright and colleagues used MLS typing to examine diversity among 394 MRSA isolates from 22 different countries (Enright *et al.*, 2002).

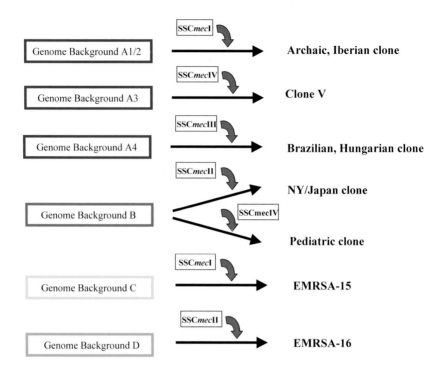

Figure 3. Model of SCC*mec* acquisition by *S. aureus* clones. Genetic backgrounds are characterized on the basis of MLS typing. The commonly reported MRSA clonal types resulting from acquisition of SCC*mec* elements by the different genetic backgrounds are indicated. Adapted from Oliveira *et al.*, 2001.

Clonal complexes are defined as groups of strains in which each isolate is identical to at least one other isolate at five or more of the seven loci. Epidemic MRSA isolates were associated with only seven main sequence types, four of which are members of one clonal complex. This complex (clonal complex I) contains the earliest MRSA isolated from the UK, Switzerland, Egypt and Copenhagen, and members of the Iberian clone. This grouping correlates with data (Crisostomo *et al.*, 2001), which indicate that the predominant 'Iberian clone' is the direct descendant of the original MRSA strain. Interestingly, MLS typing indicates that clonal complex I also includes a recent glycopeptide-intermediately resistant strain (GISA). Enright identifies 3 additional major clones, previously named EMRSA-15 and EMRSA-16 in the UK and an epidemic clone found in both Europe and the UK (Enright *et al.*, 2002).

Taken together, these independently-generated findings have allowed the identification of the limited number of MRSA clones that exist worldwide and traced their probable origins to MSSA progenitor cells. Accordingly, it appears that rapid global dissemination has been fundamental to the spread of MRSA.

All of these findings and interpretations are consistent with the initial population genetic studies (Musser and Selander, 1990; Musser and Kapur, 1992) conducted a decade ago.

Whole-Genome Sequencing of MRSA Strains

Recently, the complete genome sequences were published for an MRSA strain (N315), and an MRSA strain (Mu50) refractory to treatment with vancomycin (Kuroda, 2001). Both of these strains belonged to the clonotype II-A prevalent in Japan and the USA (Ito *et al.*, 1999). The same research group also reported the complete genome sequence of a community-acquired MRSA strain MW2 (Baba et al., 2002). The completed genome sequences of *S. aureus* strains provides an invaluable resource for the staphylococcal scientific community in the future investigations into *S. aureus* pathogenesis and biology. However, the insights provided into methicillin-resistance are restricted to sequence information for the SCC*mec* elements. Strains N315 and Mu50 contained similar type II SCC*mec* elements encoding resistance to β-lactams, bleomycin, macrolide-lincosamide-streptogramin B, aminoglycosides, and spectinomycin whereas MW2 contained a type IV SCC element encoding only resistance to β-lactams. Each sequenced strain also contained plasmids; pN315 in strain N315 was 24.7 kb in size and contained genes encoding cadmium and arsenate resistance, and a Tn*552*-related transposon that harbours the penicillin resistance gene *blaZ*. The Mu50 plasmid pMu50 was 25 kb in size and contains a transposon (*Tn4001*) that encodes aminoglycoside resistance, and genes encoding resistance to quaternary ammonium compounds. Strain MW2 contained a plasmid of 20.7 kb which contains the *blaZ* gene encoding penicillinase. The first sextant of the *S. aureus* chromosome is a site for repeated integration and deletion of genetic elements including SCC*mec* and much of the DNA in this region appears to have been relatively recently acquired from foreign sources. Apart from SCC*mec*, the genomes of strains N315 and strain Mu50 have 26 and 28 putatively mobile elements, respectively. Strain MW2 has markedly less insertion sequences in comparison to N315 and Mu50. The majority of these elements contain genes encoding antibiotic resistance or virulence factors. The potential for lateral gene transfer demonstrates the adaptability of *S. aureus* and probably enhances its ability to adapt to different environmental niches.

Concluding Comments

The widespread existence of *S. aureus* strains resistant to methicillin is one of the biggest problems in nosocomial infections today (Shopsin and Kreiswirth, 2001). The combination of lateral transfer of the *mecA* gene and rapid dissemination of MRSA strains has resulted in the virtual ubiquity of this problem worldwide. In recent years, major breakthroughs have been made in our understanding of MRSA populations, and the genetic elements encoding methicillin resistance. However, there are several large gaps in our understanding of the molecular processes which have led to MRSA strains. These include the mechanism of horizontal transfer of SCC*mec*. This may be a phage-mediated process, but demonstration of this *in vitro* has proved elusive. In addition, the variation and origin of SCC elements of staphylococci is not well understood. Based on the recent identification of SCC*cap1* encoding type 1 capsule production, further investigation may reveal additional SCC elements carrying virulence factor genes.

The recent increase in community-acquired MRSA is a worrisome development. Population genetic analysis of such strains has been limited and more extensive analysis is warranted to improve our understanding of the evolution and epidemiology of MRSA in the community.

Recent studies of many MRSA isolates from different international sources have contributed important information regarding MRSA evolution and dissemination. However, a truly global analysis of MRSA strains would include many isolates from such populous countries as India and China where MRSA is a very serious problem (Verma *et al.*, 2000; Benquan *et al.*, 2002).

The recent identification of MRSA strains which are also completely resistant to vancomycin (Sievert *et al*, 2002; Chang *et al.*, 2003) raises the possibility of staphylococcal infections for which there is no effective antibiotic treatment. The development of alternative therapeutics for such infections is critical. Although enhanced understanding of the emergence and evolution of MRSA has been achieved in recent years, there are still many unanswered questions, which need to be addressed urgently. Improved understanding of the evolution and molecular processes of resistance to methicillin in *S. aureus* may lead to better control strategies or alternative therapeutic measures against this major public health problem.

References

Archer, G.L., Thanassi, J.A., Niemeyer, D.M., and Pucci, M.J. 1996. Characterization of IS*1272*, an insertion sequence-like element from *Staphylococcus haemolyticus*. Antimicrob. Agents. Chemother. 40: 924-929.

Ayliffe, G.A. 1997. The progressive intercontinental spread of methicillin-resistant *Staphylococcus aureus*. Clin. Infect. Dis. 24 Suppl. 1, S74-S79.

Baba, T., Takeuchi, F., Kuroda, M., Yuzawa, H., Aoki, K., Oguchi, A., Nagai, Y., Iwama, N., Asano, K., Naimi, T., Kuroda, H., Cui, L., Yamamoto, K., Hiramatsu, K. 2002. Genome and virulence determinants of high virulence community-acquired MRSA. Lancet 359: 1819-1827.

Barrett, F., McGehee, R., and Finland, M. 1968. Methicillin-resistant *Staphylococcus aureus* at Boston city hospital. Bacteriologic and epidemiologic observations. N. Engl. J. Med. 279: 441-448.

Benquan, W., Yingchun, T., Kouxing, Z., Tiantuo, Z., Jiaxing, Z., and Shuqing, T. 2002. *Staphylococcus* heterogeneously resistant to vancomycin in China and antimicrobial activities of imipenem and vancomycin in combination against it. J. Clin. Microbiol. 40: 1109-1112.

Carbon, C. 1999. Costs of treating infections caused by methicillin-resistant Staphylococci and vancomycin-resistant Enterococci. J. Antimicrob. Chemother. 44: Suppl A, 31-36.

Chambers, H.F. 2001. The changing epidemiology of *Staphylococcus aureus*? Emerg. Infect. Dis. 7: 178-82.

Chang, S., Sievert, D.M., Hageman, J.C., Boulton, M.L., Tenover, F.C., Downes, F.P, Shah S., Rudrik J. T., Pupp, G. R., Brown, W. J., Cardo, D., Fridkin, S. K. 2003. Infection with vancomycin-resistant *Staphylococcus aureus* containing the *vanA* resistance gene. N. Engl. Med. 348: 1342-1347.

Charlebois, E.D., Bangsberg, D.R., Moss, N.J., Moore, M.R., Moss, A.R., Chambers, H. F., and Perdreau-Remington, F. 2002. Population-based community prevalence of methicillin-resistant *Staphylococcus aureus* in the urban poor of San Francisco. Clin. Infect. Dis. 34: 425-433.

Couto, I., de Lencastre, H., Severina, E., Kloos, W., Webster, J.A., Hubner, R.J., Sanches, I.S., and Tomasz, A. 1996. Ubiquitous presence of a *mecA* homologue in natural isolates of *Staphylococcus sciuri*. Microb. Drug. Resist. 2: 377-391.

Couto, I., Sanches, I.S., Sa-Leao, R., and de Lencastre, H. 2000. Molecular characterization of *Staphylococcus sciuri* strains isolated from humans. J. Clin. Microbiol. 38: 1136-1143.

Crisostomo, M.I., Westh, H., Tomasz, A., Chung, M., Oliveira, D.C., and de Lencastre, H. 2001. The evolution of methicillin resistance in *Staphylococcus aureus*: similarity of genetic backgrounds in historically early methicillin-susceptible and -resistant isolates and contemporary epidemic clones. Proc. Natl. Acad. Sci. USA 98: 9865-9870.

Daum, R.S., and Seal R.S. 2001. Evolving antimicrobial chemotherapy for *Staphylococcus aureus* infections: Our backs to the wall. Crit. Care Med. 29: Suppl. 4 N92-96.

Daum, R.S., Ito, T., Hiramatsu, K., Hussain, F., Mongkolrattanothai, K., Jamklang, M., and Boyle-Vavra, S. 2002. A novel methicillin-resistance cassette in community-acvquired methicillin-resistant *Staphylococcus aureus* isolates of diverse genetic backgrounds. J. Infect Dis. 186: 1344-1347.

De Lencastre, H., Wu, S.W., Pinho, M.G., Ludovice, A.M., Filipe, S., Gardete, S., Sobral, R., Gill, S., Chung, M., and Tomasz, A. 1999. Antibiotic resistance as a stress response: complete sequencing of a large number of chromosomal loci in *Staphylococcus aureus* strain COL that impact on the expression of resistance to methicillin. Microb. Drug. Resist. 5: 163-175.

Deplano, A., Tassios, P.T., Glupczynski, Y., Godfroid, E., and Struelens, M.J. 2000. *In vivo* deletion of the methicillin resistance *mec* region from the chromosome of *Staphylococcus aureus* strains. J. Antimicrob. Chemother. 46: 617-620.

Deplano, A., Witte, W., van Leeuwen, W.J., Brun, Y., and Struelens, M.J. 2000. Clonal dissemination of epidemic methicillin-resistant *Staphylococcus aureus* in Belgium and neighboring countries. Clin. Microbiol. Infect. 6: 239-245.

Enright, M.C., Day, N.P., Davies, C.E., Peacock, S.J., and Spratt, B.G. 2000. Multilocus sequence typing for characterization of methicillin-resistant and methicillin-susceptible clones of *Staphylococcus aureus*. J. Clin. Microbiol. 38: 1008-1015.

Enright, M.C., Robinson, D.A., Randle, G., Feil, E.J., Grundman, H., and Spratt, B.G. 2002. The evolutionary history of methicillin-resistant *Staphylococcus aureus* (MRSA). Proc. Natl. acad. Sci. USA 99: 7687-7692.

Fitzgerald, J.R., Sturdevant, D.E., Mackie, S.M., Gill, S.R., and Musser, J.M. 2001. Evolutionary genomics of *Staphylococcus aureus*: Insights into the origin of methicillin-resistant strains and the toxic shock syndrome epidemic. Proc. Natl. Acad. Sci. USA. 98: 8821-8826.

Groom, A.V., Wolsey, D.H., Naimi, T.S., Smith, K., Johnson, S., Boxrud, D., Moore, K.A., and Cheek, J.E. 2001. Community-acquired methicillin-*resistant Staphylococcus aureus* in a rural American Indian community. JAMA. 286: 1201-1205.

Hiramatsu, K. 1995. Molecular evolution of MRSA. Microbiol. Immunol. 39: 531-543.

Hiramatsu, K., Cui, L., Kuroda, M., and Ito, T. 2001. The emergence and evolution of methicillin-resistant *Staphylococcus aureus*. Trends Microbiol. 9: 486-493.

Hiramatsu, K., Suzuki, E., Takayama, H., Katayama, Y., and Yokota, T. 1990. Role of penicillinase plasmids in the stability of the *mecA* gene in methicillin-resistant *Staphylococcus aureus*. Antimicrob. Agents Chemother. 34: 600-604.

Hussain, F.M., Boyle-Vavra, S., and Daum, R.S. 2001. Community-acquired methicillin-resistant *Staphylococcus aureus* colonization in healthy children attending an outpatient pediatric clinic. Pediatr. Infect. Dis. J. 20: 763-7.

Ito, T., Katayama, Y., Asada, K., Mori, N., Tsutsumimoto, K., Tiensasitorn, C., and Hiramatsu, K. 2001. Structural comparison of three types of staphylococcal cassette chromosome *mec* integrated in the chromosome in methicillin-resistant *Staphylococcus aureus*. Antimicrob. Agents Chemother. 45: 1323-1336.

Ito, T., Katayama, Y., and Hiramatsu, K. 1999. Cloning and nucleotide sequence determination of the entire *mec* DNA of pre-methicillin-resistant *Staphylococcus aureus* N315. Antimicrob. Agents Chemother. 43: 1449-1458.

Jevons, M. 1961. Celebenin-resistant Staphylococci. BMJ. 1: 124-125.

Johnson, A.P., Aucken, H.M., Cavendish, S., Ganner, M., Wale, M.C., Warner, M., Livermore, D.M., and Cookson, B.D. 2001. Dominance of EMRSA-15 and -16 among MRSA causing nosocomial bacteraemia in the UK: analysis of isolates from the European Antimicrobial Resistance Surveillance System (EARSS). J. Antimicrob. Chemother. 48: 143-144.

Katayama, Y., Ito, T., and Hiramatsu, K. 2000. A new class of genetic element, *Staphylococcus* cassette chromosome *mec*, encodes methicillin resistance in *Staphylococcus aureus*. Antimicrob. Agents Chemother. 44: 1549-1555.

Katayama, Y., Ito, T., and Hiramatsu, K. 2001. Genetic organization of the chromosome region surrounding *mecA* in clinical staphylococcal strains: role of IS*431*-mediated *mecI* deletion in expression of resistance in *mecA*-carrying, low-level methicillin- resistant *Staphylococcus haemolyticus*. Antimicrob. Agents Chemother. 45: 1955-1963.

Katayama, Y, Takeuchi, F., Ito, T, Ma, X-X. Ui-Mizutani, Y., Kobayashi, I., and Hiramatsu, K. 2003. Identification in methicillin-susceptible *Staphylococcus hominis* of an active primordial mobile genetic element for the staphylococcal cassette chromosome *mec* of methicillin-resistant *Staphylococcus aureus*. 185: 2711-2722.

Kreiswirth, B., Kornblum, J., Arbeit, R.D., Eisner, W., Maslow, J.N., McGeer, A., Low, D.E., and Novick, R.P. 1993. Evidence for a clonal origin of methicillin resistance in *Staphylococcus aureus*. Science. 259: 227-230.

Kuroda, M., Ohta, T., Uchiyama, I., Baba, T., Yuzawa, H., Kobayashi, I., Cui, L., Oguchi, A., Aoki, K., Nagai, Y., Lian, J., Ito, T., Kanamori, M., Matsumaru, H., Maruyama, A., Murakami, H., Hosoyama, A., Mizutani-Ui, Y., Takahashi, N.K., Sawano, T., Inoue, R., Kaito, C., Sekimizu, K., Hirakawa, H., Kuhara, S., Goto, S., Yabuzaki, J., Kanehisa, M., Yamashita, A., Oshima, K., Furuya, K., Yoshino, C., Shiba, T., Hattori, M., Ogasawara, N., Hayashi, H., Hiramatsu, K. 2001. Whole genome sequencing of meticillin-resistant *Staphylococcus aureus*. Lancet. 357: 1225-1240.

Kuwahara-Arai, K., Kondo, N., Hori, S., Tateda-Suzuki, E., and Hiramatsu, K. 1996. Suppression of methicillin resistance in a *mecA*-containing pre-methicillin-resistant *Staphylococcus aureus* strain is caused by the *mecI*-mediated repression of PBP 2' production. Antimicrob. Agents Chemother. 40: 2680-2685.

Lacey, R.W., and Grinsted, J. 1973. Genetic analysis of methicillin-resistant strains of *Staphylococcus aureus*: evidence for their evolution from a single clone. J. Med. Microbiol. 6: 511-526.

Lim, D., and Strynadka, C.J. 2002. Structural basis for the β-lactam resistance of PBP2a from methicillin-resistant *Staphylococcus aureus*. Nature Struct. Biol. 9: 870-876.

Luong, T.T., Ouyang, S., Bush, K., and Lee , C.Y. 2002. Type 1 capsule genes of *Staphylococcus aureus* are carried in a staphylococcal cassette chromosome genetic element. J. Bacteriol. 184: 3623-3629.

Ma, X.X., Ito, T., Tiensasitorn, C., Jamklang, M., Chongtrakool, P., Boyle-Vavra, S., Daum, R.S., and Hiramatsu, K. 2002. Novel type of Staphylococcal cassette chromosome *mec* identified in community-acquired methicillin-resistant *Staphylococcus aureus* strains. Antimicrob. Agents Chemother. 46: 1147-1152.

Matsuhashi, M., Song, M.D., Ishino, F., Wachi, M., Doi, M., Inoue, M., Ubukata, K., Yamashita, N., and Konno, M. 1986. Molecular cloning of the gene of a penicillin-binding protein supposed to cause high resistance to beta-lactam antibiotics in *Staphylococcus aureus*. J. Bacteriol. 167: 975-980.

Musser, J.M., and Kapur, V. 1992. Clonal analysis of methicillin-resistant *Staphylococcus aureus* strains from intercontinental sources: association of the *mec* gene with divergent phylogenetic lineages implies dissemination by horizontal transfer and recombination. J. Clin. Microbiol. 30: 2058-2063.

Musser, J.M., and Selander, R.K. 1990. Genetic analysis of natural populations of *Staphylococcus aureus*. In: Molecular Biology of the Staphylocci. R.P. Novick, ed. VCH publishers, Inc. New York. p. 59-67.

Naimi, T.S., LeDell, K.H., Boxrud, D.J., Groom, A.V., Steward, C.D., Johnson, S.K., Besser, J.M., O'Boyle, C., Danila, R.N., Cheek, J.E., Osterholm, M.T., Moore, K.A., and Smith, K.E. 2001. Epidemiology and clonality of community-acquired methicillin-resistant *Staphylococcus aureus* in Minnesota, 1996-1998. Clin. Infect. Dis. 33: 990-996.

Novick, R.P., Schlievert, P., and Ruzin, A. 2001. Pathogenicity and resistance islands of Staphylococci. Microbes Infect. 3: 585-594.

Oliveira, D.C., Tomasz, A., and de Lencastre, H. 2001. The evolution of pandemic clones of methicillin-resistant *Staphylococcus aureus*: identification of two ancestral genetic backgrounds and the associated *mec* elements. Microb. Drug Resist. 7: 349-361.

Petinaki, E., Arvaniti, A., Dimitracopoulos, G., and Spiliopoulou, I. 2001. Detection of *mecA*, *mecR1* and *mecI* genes among clinical isolates of methicillin-resistant staphylococci by combined polymerase chain reactions. J. Antimicrob. Chemother. 47: 297-304.

Roberts, R.B., de Lencastre, A., Eisner, W., Severina, E.P., Shopsin, B., Kreiswirth, B.N., and Tomasz, A. 1998. Molecular epidemiology of methicillin-resistant *Staphylococcus aureus* in 12 New York hospitals. MRSA Collaborative Study Group. J. Infect. Dis. 178: 164-171.

Roman, R.S., Smith, J., Walker, M., Byrne, S., Ramotar, K., Dyck, B., Kabani, A., and Nicolle, L.E. 1997. Rapid geographic spread of a methicillin-resistant *Staphylococcus aureus* strain. Clin. Infect. Dis. 25: 698-705.

Sakoulas, G., Gold, H.S., Venkataraman, L., DeGirolami, P.C., Eliopoulos, G.M., and Qian, Q. 2001. Methicillin-Resistant *Staphylococcus aureus*: Comparison of susceptibility testing methods and analysis of *mecA*-positive susceptible strains. J. Clin. Microbiol. 39: 3946-3951.

Savolainen, K., Paulin, L., Westerlund-Wikstrom, B., Foster, T.J., Korhonen, T.K., and Kuusela, P. 2001. Expression of *pls*, a gene closely associated with the *mecA* gene of methicillin-resistant *Staphylococcus aureus*, prevents bacterial adhesion *in vitro*. Infect. Immun. 69: 3013-3020.

Sharma, V.K., Hackbarth, C.J., Dickinson, T.M., and Archer, G.L. 1998. Interaction of native and mutant MecI repressors with sequences that regulate *mecA*, the gene encoding penicillin binding protein 2a in methicillin-resistant staphylococci. J. Bacteriol. 180: 2160-2166.

Shopsin, B., and Kreiswirth, B.N. 2001. Molecular epidemiology of methicillin-resistant *Staphylococcus aureus*. Emerg. Infect. Dis. 7: 323-326.

Sievert, D.M, Boulton, M.L., Stoltman, G., Johnson, D., Stobierski, M.G., Downes, F.P., Somsel, P.A., Rudrik, J.T., Brown, W., Hafeez, H., Lundstrom, T., Flanagan, E., Johnson, R., Mitchell, J., Chang, S. 2002. *Staphylococcus aureus* resistant to vancomycin- United States, 2002. MMWR 51: 565-567.

Vaudaux, P.E., Monzillo, V., Francois, P., Lew, D.P., Foster, T.J., and Berger-Bachi, B. 1998. Introduction of the *mec* element (methicillin resistance) into *Staphylococcus aureus* alters *in vitro* functional activities of fibrinogen and fibronectin adhesins. Antimicrob. Agents Chemother. 42: 564-570.

Verma, S., Joshi, S., Chitnis, V., Hemwani, N., and Chitnis, D. 2000. Growing problem of methicillin resistant staphylococci-Indian scenario. Indian J. Med. Sci. 54: 535-540.

Wallet, F., Stuit, L., Boulanger, E., Roussel-Delvallez, M., Dequiedt, P., and Courcol, R. J. 2000. Peritonitis due to *Staphylococcus sciuri* in a patient on continuous ambulatory peritoneal dialysis. Scand. J. Infect. Dis. 32: 697-698.

Wielders, C.L., Vriens, M.R., Brisse, S., de Graaf-Miltenburg, L.A., Troelstra, A., Fleer, A., Schmitz, F.J., Verhoef, J., and Fluit, A.C. 2001. Evidence for *in-vivo* transfer of *mecA* DNA between strains of *Staphylococcus aureus*. Lancet 357: 1674-1675.

Witte, W., Kresken, M., Braulke, C., and Cuny, C. 1997. Increasing incidence and widespread dissemination of methicillin-resistant *Staphylococcus aureus* (MRSA) in hospitals in central Europe, with special reference to German hospitals. Clin. Microbiol. Infect. 3: 414-422.

Wu, S.W., de Lencastre, H., and Tomasz, A. 2001. Recruitment of the *mecA* gene homologue of *Staphylococcus sciuri* into a resistance determinant and expression of the resistant phenotype in *Staphylococcus aureus*. J. Bacteriol. 183: 2417-2424.

Zhang, H.Z., Hackbarth, C.J., Chansky, K.M., and Chambers, H.F. 2001. A proteolytic transmembrane signaling pathway and resistance to beta-lactams in staphylococci. Science. 291: 1962-1965.

From: *MRSA: Current Perspectives*
Edited by: A.C. Fluit and F.-J. Schmitz

Chapter 7

Population Structure of MRSA

Ad C. Fluit and Franz-Josef Schmitz

Abstract

This chapter describes the spread of clonal lineages, and the distribution of different clonal lineages of MRSA strains in hospitals, cities and countries. One or two clonal lineages often dominate in a hospital or sometimes even in individual cities, or countries, however many lineages are found only infrequently. Some clones have a pandemic spread. Directly related to the description of clonal spread is presence of different *mecA* Staphylococcal Cassette Chromosomes (SCC*mec*) types. Four different types, that vary in composition, have been described. The data obtained from the evolutionary appearance in time of different SCC*mec* types, demonstrate a link between the SCC*mec* type and the clonal lineages present. This demonstrates that horizontal transfer of SCC*mec* between staphylococci took place on least several occasions and may be more common than previously believed. The mechanism of this transfer is still unknown, but the genes necessary for the excision and integration of SCC*mec* into the staphylococcal chromosome have been identified.

Introduction

The battle against *Staphylococcus aureus* was thought to be definitively won with the introduction of the earliest antibiotics, sulfonamide and penicillin. However, the widespread use of these antibiotics in the 1950s selected for the predominance of β-lactamase producing resistant strains. At that time approximately 80% of all nosocomial *S. aureus* infections were caused by these resistant organisms (Wilson and Hamburger, 1957; Hassall and Rountree, 1959; Hausmann and Karlish, 1996). The introduction of a semi-synthetic penicillin, celbenin, which later became known as methicillin, was again believed to be the final blow for *S. aureus*. Upon its introduction a major reduction in morbidity and mortality due to staphylococcal infections was seen. Unfortunately, this reduction was short-lived, because methicillin-resistant staphylococci were described soon thereafter (Jevons, 1961).

Methicillin resistance is dependent on the presence of the *mecA* gene which encodes Penicillin Binding Protein 2A (PBP2A). The *mecA* gene did not originate in *S. aureus*, but was acquired from another species by an unknown mechanism. A close homologue of the *mecA* gene was recently identified in *Staphylococcus sciuri*, a taxonomically primitive staphylococcal species (Wu *et al.*, 1996). However, *S. sciuri* isolates are susceptible to β-lactam antibiotics despite the presence of this *mecA* homologue. Stepwise exposure of a *S. sciuri* strain to increasing concentrations of methicillin led to the selection of a resistant derivative. This derivative showed a dramatic increase in the transcription of the *mecA* homologue, which was due to the presence of a point-mutation in the promotor region of the gene. In addition, a new protein appeared, which cross-reacted with an antibody preparation against PBP2A. Transduction of the cloned gene into a methicillin-susceptible *S. aureus* strain resulted in a significant increase of methicillin resistance. Together these data support the hypothesis that the *mecA* homologue, ubiquitous in the antibiotic susceptible species *S. sciuri,* may be an evolutionary precursor of the methicillin resistance gene *mecA* present in MRSA (Wu *et al.*, 2001).

Understanding the mechanism of acquisition of the *mecA* gene and the subsequent spread of MRSA has proved to be challenge during the last 30 years. Establishing the population structure of MRSA and its relationship with the population structure of MSSA may contribute to the elucidation of this problem.

MRSA Clones

Early problems with MRSA appeared to be local, but soon it was appreciated that MRSA posed a greater challenge. Several outbreaks within countries, cross-border, intercontinental and even pandemic spread have been described.

A major reason for the rapid spread of MRSA is the transport of patients as exemplified by the next case. In Western Canada an outbreak of MRSA was detected in a tertiary care hospital. The index patient had recently been hospitalized for 3 months in the Punjab, India and was admitted to the Canadian hospital shortly after arrival. Transfer of the patient to Vancouver and of a subsequently colonized patient to another hospital led to outbreaks at both of these hospitals. This all happened within six weeks of the arrival of the patient in Canada (Roman *et al.*, 1997). This case not only documents the ability of some clones to spread rapidly within an hospital, but also that movement of patients between institutions greatly facilitates this rapid spread. Although this example clearly demonstrates the role of international travel in the spread of MRSA throughout the world, spread within the boundaries of nations are best documented.

Nationwide spread of MRSA has, for example, been documented in Poland. A total of 158 MRSA strains obtained from 18 hospitals and collected from 1990 to 1996: DNA obtained from the strains was seperated by PFGE and *mec* typed. This typing was performed by restriction digestion of the MRSA DNA by *Cla*I followed by hybridization with *mecA* and Tn*554*-specific probes, respectively. The isolates were further subdivided in a group with heterogeneous and a group with homogeneous expression of methicillin resistance (Leski *et al.*, 1998). Seventy five of the 97 isolates showing homogeneous expression belonged to a single PFGE-*mec* type. Ten isolates belonged to the second largest PFGE-*mec* type while the other isolates belonged to a number of different smaller PFGE-*mec* types. The majority of the 61 isolates that showed heterogeneous expression of methicillin resistance exhibited greater variability, with only one clone with more than 10 isolates. Nevertheless nearly half of all the isolates collected belonged to a single clone demonstrating that on a national level a single clone may predominate. Studies with similar results were reported from several countries including Hungary (de Lencastre *et al.*, 1997), Germany (Witte *et al.*, 1997), Denmark (Rosdahl *et al.*, 1994), and the UK (Johnson *et al.*, 2001).

Cross-border spread has been documented in a study centered on Belgium (Deplano *et al.*, 2000). MRSA isolates obtained from Belgium, France, Germany and The Netherlands and collected between 1981 and 1994 were PFGE typed. A total of 171 MRSA isolates were collected from Belgium and 102 isolates were from neighboring countries. In total 32 different PFGE types were found, but 82% of the Belgian and 51% of the MRSA isolates from the other countries belonged to a single PFGE type, which was first observed in 1984. Four other PFGE types were also observed in more than one country, but none in all four. These types were first observed in the period 1991-1994. The spread of two of these international PFGE types was linked to the transfer of two patients to Dutch hospitals. So, predominant types can also be present in broader geographic areas even crossing borders: again the transfer of patients appears to play an important role in the spread of these epidemic clones.

Further evidence of this type of spread was presented by the typing of isolates collected during a seven month period in 1992 and 1993 in a hospital in Lisbon, Portugal (Sanches *et al.*, 1995). The typing methods were similar to those used for the Polish isolates described above. Twenty-three of 43 MRSA isolates collected belonged to one PFGE type and these isolates shared the same *mec* type. This clone was also identified in two outbreaks of MRSA in hospitals in Madrid and Barcelona, Spain, and has been dubbed the Iberian clone. In addition to being genotypically identical, these isolates also possessed identical multidrug resistance patterns and heterogenous methicillin expression. One isolate with an unique clonal type was identical to the dominant clone in another Lisbon hospital. The other isolates belonged to a variety of different clones. This work illustrates once more that frequently one clonal type is predominant in a hospital. In addition, it demonstrates that cross-border spread may take place over large distances. Unfortunately, the route of transmission was not established, although it was suggested that healthy carriers may play a role. This so-called Iberian clone has also been reported in Italy, the UK, Germany, Belgium, Switzerland, France, Czech Republic, Poland, and the USA (Oliveira *et al.*, 2002).

A remarkable example of spread of MRSA is provided by the New York/ Japan clone. The PFGE typing of 55 MRSA at a New York City hospital in 1989 revealed that nearly 70% of the isolates belonged to the same PFGE type and shared an identical *mec* fingerprint. An additional 5 isolates shared the PFGE type, but showed a different *mec* fingerprint. The remaining 12 isolates belonged to 5 different PFGE types and showed different *mec* fingerprints (de Lencastre *et al.*, 1996). The data also show that, although at hospital level one clone may dominate, several other PFGE types can also be found. Extension of this research to 12 New York City hospitals revealed a similar picture (Roberts *et al.*, 1998). A total of 113 of 270 MRSA isolates (42%) collected belonged to a single PFGE-*mec* type. This type was present in all hospitals and was the dominant clone in nine. Thirteen of 15 isolates belonging to a different PFGE-*mec* type came from one hospital. One type was predominant (79%) among MRSA isolates from 28 AIDS patients. One subtype was recovered from 9 of the AIDS patients including 5 that shared a floor in a nursing home. In addition, 67 isolates yielded 37 minor clones with less than 10 isolates. These data suggest that clones may not only be predominant in single hospitals, but clones may also be predominant locally. Despite the presence of a predominant clone a large number of smaller clones may exist as well. However, the predominant clone was not only found to be widely distributed in New York City and some neighboring states, but it made up three quarters of the MRSA isolates in a hospital in Tokyo, Japan: these Japanese showed the same PFGE-*mec* type, although there were slight differences in antibiotic resistance pattern (Aries de Sousa *et al.*, 2000).

Intercontinental spread by the Iberian and Bazilian clones of MRSA has also been documented. The Brazilian clone was predominant across a wide geographic area in Brazil (Teixeira *et al.*, 1995). Two thirds of the MRSA isolates collected in the largest teaching hospital in Portugal during 1992 and 1993 belonged to a single PFGE-*mec* type (established as described above) (Oliveira *et al.*, 1998). This Iberian clone spread through large portions of Southern Europe. However, the prevalence of this clone decreased in subsequent years. In 1994-1995 42% of the isolates belonged to this clone, but in 1996 this had dropped to only 20%. During the same period, the prevalence of the Brazilian clone increased in importance. Its prevalence increased from 5% in 1992-1993 to 36% in 1994-1995 and slightly decreased to 29% in 1996. Together these clones were responsible for nearly 90% of all MRSA infections in this Portugese hospital. The Brazilian clone has also been isolated Argentina, Uruguay, and the Czech Republic (Oliveira *et al.*, 2002). Another major pandemic clone is the pediatric clone, which was first reported in a hospital in Portugal and thereafter also in Poland, the USA, Argentina, and Colombia (Oliveira *et al.*, 2002).

The first vancomycin-resistant (methicillin-resistant) *S. aureus* (VRSA) strain was described in Japan (Hiramatsu *et al.*, 1997): in 1997 this was the only VRSA isolate known. In this study, Hiramatsu *et al.* (1997) reported that the prevalence of MRSA heterogeneously resistant to vancomycin varied from 9.3% to 20% among eight Japanese university hospitals. For non-university hospitals or clinics, the rate varied between 1 and 3%. PFGE typing revealed that the isolates belonged to a MRSA type which is dominant in Japan. This implies that hetero-VRSA strains mimic MRSA strains in that frequently a dominant PFGE type, which is able to spread efficiently, is found. Alternatively, heterogenous vancomycin resistance may develop rather easily from only one or a limited number of PFGE types. The spread of hetero-VRSA was also reported from Brazil (dos Santos Soares *et al.*, 2000). In this study, 65 isolates were analyzed and shown to belong to the same PFGE type and 45 of these had an identical pattern to the so-called Brazilian MRSA clone.

In the past MRSA has nearly always been reported to be hospital-acquired, but recently several reports describe a high rate of MRSA among community-acquired *S. aureus* infections. A study performed by Naimi *et al.* (2001) in Minnesota, reported 354 patients with community-acquired MRSA during 1996-1998. PFGE typing of 174 of the MRSA isolates revealed the presence of 9 PFGE types and a total of 34 subtypes. A total of 150 of these isolates (86%) belonged to one PFGE type and 66% of all isolates typed belonged to 3 subtypes within the dominant PFGE type. In a study of 500 healthy children from Chicago, Illinois, Hussain *et al.* found that a quarter carried *S. aureus* strains and that of the isolates found, 2.5% were MRSA (Hussain *et al.*, 2001). Additional data, obtained from a different study, suggest that the prevalence of MRSA in the community may vary considerably depending on the population or geographic area sampled. A study among an American Indian Community showed that 62% of a total of 112 cases of community-acquired *S. aureus* infection were due to MRSA (Groom *et al.*, 2001). PFGE-typing of 38 of the MRSA isolates showed that 34 isolates were clonally related and distinct from the nosocomial isolates found in the region. The typing data indicate that in community-acquired *S. aureus* infection one PFGE-type is dominant at least at a local level. This is similar to the situation with hospital-acquired MRSA. MRSA isolates were also found among the urban poor in San Francisco (Charlebois *et al.*, 2002). Almost 23% of a sample of 833 homeless and marginally housed adults carried *S. aureus*, 12% of which were MRSA. However, 22 of 23 isolates that were PFGE typed matched with PFGE types found among local clinical isolates. In this population the three major PFGE types accounted for 43.5, 21.7, and 13% of the isolates, respectively. In total 6 distinct PFGE type and 14 subtypes were identified. The data found in this study contrast with the data presented by Groom *et al.* (2001), who reported clear differences between the genotypes in the local hospitals and suggest that care should be taken when isolates are declared to be community-acquired, i.e. isolates may appear to be community-acquired based on clinical criteria, further data may show that this is not the case.

From the description of a selection of epidemics with MRSA it is clear that some PFGE types are capable of massive spread across hospitals, cities, countries, and continents. Some types even appear to be pandemic. This spread is no longer limited to the hospitals, but has reached the community as well. Nevertheless in all environments less dominant PFGE types are found and even sporadic types.

SCC*mec* Structure

Although the distribution of MRSA genotypes gives valuable information about the population structure of MRSA, this information is not complete without the characterization of the genetic environments of the *mecA* gene. Especially as the *mecA* gene in MRSA is part of a larger structure called the Staphylococcal Cassette Chromosome *mec* (SCC*mec*). The composition of this structure appears to be highly variable and a large number of different genetic determinants have been described, which may or may not be present in SCC*mec* (Table 1) (Hiramatsu *et al.*, 2001; Oliveira *et al.*, 2000).

SCC*mec* has been characterized in a number of different ways, most commonly using methods based on Restriction Fragment Length Polymorphisms (RFLP) analysis following prior *Cla*I restriction digestion and, hybridization with a *mecA*- and Tn*554*-specific probes respectively. In addition the *mecI-mecR1* region has been characterized by *Hind*III and *Eco*R1 digests (Hiramatsu, 1995), DNA fingerprinting with southern hybridization using a probe derived from a complete SCC*mec* region, and (for a limited number of isolates) the structure was determined by DNA sequencing (Ito *et al.*, 1999).

Table 1. Most important genetic determinants, which may express functional proteins, present in SCC*mec* (Oliveira *et al.*, 2000; Hiramatsu *et al.*, 2001)

Determinant	Functions
mecA	methicillin resistance
mecI	repressor for *mecA* promotor
mecR1	receptor for β-lactam antibiotics in *mecA* regulation
ccrA	cassette chromosome recombinase A, involved in SCC*mec* excision from and integration into chromosome
ccrB	cassette chromosome recombinase B, involved in SCC*mec* excision from and integration into chromosome
Tn*554*	encodes erythromycin and spectinomycin resistance and transposase genes for Tn*554*
pseudo Tn*554*	cadmium resistance and transposases genes for pseudo Tn*554*
IS*431*	transposase genes for IS*431*
pUB110	integrated plasmid encodes tobramycin and bleomycin resistance and contains *pre* gene encoding pUB110 recombinase
pT181	encodes tetracycline resistance
pI258	resistance against penicillins and heavy metals
IS*256*	transposase
IS257	transposase
pseudoIS1272	transposase

The first published report of a completely nucleotide sequenced SCC*mec* was from the group of Hiramatsu (Ito *et al.*, 1999). This group found that the SCC*mec* from *S. aureus* N315 was 51,669 bp long, and has 27 bp inverted repeats. In addition it was found that both SCC*mec* and the integration site have a copy of a characteristic 15 bp direct repeat. One copy is at the right end of SCC*mec*, whereas the other copy is located outside SCC*mec* and abuts the left end of SCC*mec*. The *mecA* gene is preceded by the regulatory *mecI* and *mecR1* genes. The *mecA* gene is flanked by copies of IS*431*. In addition to resistance to methicillin, the N315 SCC*mec* also encodes resistance to spectinomycin, macrolides, bleomycin and aminoglycosides. The first two resistances are due to the presence of Tn*554*, whereas the latter two are encoded by pUB110 (an integrated plasmid). An interesting set of genes is *ccrA/ccrB*, which encode a site-specific recombinases (Ito *et al.*, 1999). These recombinases will be discussed in more detail in the section about transfer of SCC*mec*. The rest of SCC*mec* is composed of many open reading frames. However, these open reading frames are considered non-functional, because of acquired mutations and deletions (Ito *et al.*, 1999).

Analysis of MRSA collected around the world with ten probes derived from SCC*mec* from strain N315 demonstrated that most of these isolates belonged to three different hybridization patterns (Ito *et al.*, 2001). The organization of the integration site including the repeats was similar for both new types when compared to the structures in strain N315. DNA sequencing confirmed the presence of two new SCC*mec* types. The sizes of the SCC*mec* elements from strain NCTC 10442 and 85/2082 were 34,364 and 66,896 bp, respectively. The integration site into the staphylococcal chromosome for all three types of SCC*mec* was orfX, which was 99% homologous between the strains (Ito *et al.*, 2001).

Two different genetic organizations were found in the region around the *mecA* gene; these are called *mec* complex class A and B. Class A *mec* is composed of *mecI-mecR1-mecA*-IS*431*, whereas Class B *mec* lacks *MecI* and part of *MecR1*, but contains an insertion sequence and is composed of *IS1272-ΔmecR1-mecA*-IS*431*. The organization of *ccrA* and *crrB* differed somewhat between the three types of SCC*mec* complexes, and together with the adjacent open reading frames, these are called *ccr* complex 1-3. The SCC*mec* structures are called type I-III SCC*mec* and were present in MRSA strains NCTC 10442, N315, and 85/2082, respectively (Figure 1) (Ito *et al.*, 2001).

Recently, the Hiramatsu group reported a fourth type of SCC*mec* element (Ma *et al.*, 2002). This is called type IV SCC*mec,* and it was present in community-acquired MRSA. In fact two slightly different variants were detected and designated type IVa and IVb, which were isolated from strain Ca05 and 8/6-3P, respectively. Type IVa is 24,248 bp in length and type IVb is slightly smaller in size with 20,920 bp; both are therefore, much smaller than the three other types of SCC*mec*. The elements possessed the 15 bp direct repeats sequences and degenerate 27 bp inverted repeats. Both elements had type 2 *ccr* complexes, and class B *mec* regions (Figure 1). No integrated plasmids or transposons are present. These elements appear to come close the minimum amount of DNA necessary to confer methicillin resistance and encode functions necessary for the excision and integration of the SCC*mec* element (Ma *et al.*, 2002).

As suggested by the linkage of type IV SCC*mec* with community-acquired MRSA and the recent massive emergence of community-acquired MRSA all types of SCC*mec* were not around when MRSA emerged as a world-wide problem. Several studies performed both in Europe and the USA demonstrated

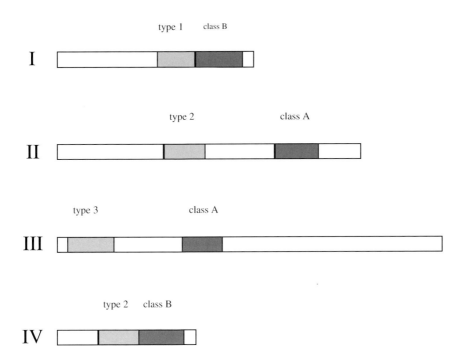

Figure 1. Schematic representation of the different SCC*mec* types (I-IV). The *ccr* complex is indicated in light gray, the position of the sequences representing the *mec* classes is indicated in dark gray.

the absence of *mecI* and *mecR1* in isolates before the 1970s. A study comprising 149 MRSA isolates collected between 1965 and 1990 in the Zürich area of Switzerland showed that all isolates collected before 1972 lacked the regulatory genes. Remarkably, there were two exceptions, an isolate collected during 1966 and one collected during 1972 (Hürlimann-Dalel *et al.*, 1992). Characterization of 25 MRSA isolates from countries around the world showed that only isolates collected since 1981 contained the complete *mecI* and *mecR1* sequences, whereas all isolates collected during the 1960s lacked *mecI* and part of *mecR1* (Suzuki *et al.*, 1993). These data from the distribution of *mecI* and *MecR1* genes suggest that type I SCC*mec* was present in early MRSA and SSC*mec* type II appeared much later in the *S. aureus* population.

The SCC*mec* type for the most important clones has been determined (Table 2). However, between individual clones differences in the details of the SCC*mec* type may be present. For example the Archaic and Iberian clone have a type I element, but the Iberian clone contains pUB110 in the SCC*mec* element and the Brazilian clone lacks pT181, which is present in the SCC*mec* element of the Hungarian clone (Oliveira *et al.*, 2001, 2002).

Table 2. SCC*mec* types present in pandemic MRSA clones (Oliveira *et al.*, 2001, 2002)

Clone	SCC*mec* type	MLST profile
Archaic	I	3-3-1-1/12-4-4-16
Iberian	I	3-3-1-1/12-4-4-16
Clone V	IV	3-3-1-1-4-4-3
Brazilian	III	2-3-1-1-4-4-3
Hungarian	III	2-3-1-1-4-4-3
New York	II	1-4-1-4-12-1-10
Pediatric	IV	1-4-1-4-12-1-10

Also DNA fingerprinting with a probe derived from a SCC*mec* element showed that four major types of SCC*mec* exists (Figure 1) (C.L.C. Wielders, A.T.A. Box, and A.C. Fluit unpublished). These types are identical to types I-IV described by Hiramatsu. Polymorphisms within the fingerprints are in agreement with data from the group of de Lencastre who demonstrated variation within the major types of SCC*mec* (Oliveira *et al.*, 2000, 2001).

Analysis of the emergence of the *mec* types I-IV shows that the different types appear sequentially. Type I is already present in isolates from the 1960s, but type II only appears at the end of the 1970s. Type II was predominant among isolates collected in the USA during the 1980s and is now found worldwide. Isolates containing type III were first collected during the 1980s. MRSA with *mec* type IV were all isolated during the 1990s with only one exception. *Mec* type I-IV were also present in contemporary coagulase-negative staphylococci (C. L.C. Wielders, A.T.A. Box, and A.C. Fluit, unpublished) Characterization of coagulase-negative staphylococci for the structures surrounding the *mecA* gene also demonstrated SCC*mec* structures similar to those observed in MRSA (Katayama *et al.*, 2001).

From these data it can be concluded that four basic types of SCC*mec* are present in the *S. aureus* population, but within these four types large variation is possible. This suggests that the SCC*mec* is a genetic element that can easily evolve by the insertion of new genetic elements depending on the (antibiotic) challenges posed to the MRSA isolate. The different SCC*mec* types seem to emerge at particular points in time and be responsible for subsequent waves of MRSA.

Population Structure

Soon after the introduction of methicillin the first reports of the appearance of MRSA in Africa and Europe appeared (Jevons, 1961; Kreiswirth *et al.*, 1993). Over the next 10 years, an increasing number of outbreaks occurred mainly; in European countries, including the United Kingdom, Denmark, France, Poland, and Switzerland (Borowski *et al.*, 1964; Dyke *et al.*, 1966; Brenner and Kayser, 1968; Parker and Hewitt, 1970). In addition, there were occasional reports of outbreaks in other countries such as Australia, India and Turkey (Cetin and Ang, 1962; Rountree and Beard, 1968). The first major outbreak of MRSA in the USA was reported in 1968 (Barrett *et al.*, 1968), but major inter-hospital spread apparently did not take place for another 5 to 10 years (Jorgensen, 1986 Ayliffe, 1997). Many of the initial MRSA were resistant to multiple drugs including tetracycline and sometimes to streptomycin, erythromycin, lincomycin, neomycin, tobramycin, and novobicin. However some initial isolates were only resistant to β-lactams (Ayliffe, 1997, Hiramatsu *et al.*, 2001).

During the 1970s a decline in the prevalence of MRSA was noted in the European countries. This decrease is poorly understood, but changes in antibiotic policy, especially a reduction in tetracycline use, and the introduction of hygiene measures were offered as explanations (Ayliffe *et al.*, 1979; Jepsen, 1986; Ayliffe, 1997). In the late 1970s a new wave of MRSA emerged with large outbreaks in North America and Australia (Boyce and Causey, 1982; Pavillard *et al.*, 1982). The prevalence of MRSA rose dramatically around the world during the 1980s. In the USA the frequency of isolation of MRSA increased from 2.4% in 1975 to 29% in 1991 (Panlilio *et al.*, 1992) and similar figures were reported from Italy, France, and Japan (Schito and Varaldo, 1988; Reverdy *et al.*, 1993).

Nowadays about 30% of the *S. aureus* isolates in the USA, Latin America, Australia, and Europe are resistant to methicillin (Voss *et al.*, 1994; Verhoef *et al.*, 1999; Diekema *et al.*, 2001). In Asia this figure rises to around 50%. However, large geographic differences may exist. For example in Europe prevalence ranges from more than 50% in Portugal to less than 2% in the Netherlands and some Scandinavian countries (Verhoef *et al.,* 1999). In Asia extremely high rates were reported from Hong Kong (75%) and Japan (72%) (Diekema *et al.*, 2001). In the past MRSA were almost without exception hospital-acquired (Thompson *et al.*, 1982). However, in recent years an increasing number of community-acquired MRSA have been reported. This is particularly true for the USA, where 28% of the community-acquired *S. aureus* isolates may be resistant to methicillin (Herold *et al.*, 1998; Diekema *et al.*, 2001). The community-acquired MRSA are rarely multiresistant in contrast to hospital-acquired MRSA (Adcock *et al.*, 1998; Herold *et al.*, 1998).

Although its clear from the data presented that clonal expansions witnessed in the past came in different waves, the relationship between these waves, MSSA and therefore clonal lineages is still a matter of debate. A long held theory was that the transfer of SCC*mec* to *S. aureus* occurred only once followed by clonal expansion throughout the world. Initial evidence for this theory was provided by Lacey and Grinsted in early 1970s (Lacey and Grindsted, 1973). This evidence is compatible with the appearance of the first wave of MRSA during the 1960s, which contained type I SCC*mec*. Further evidence for the clonal origin of MRSA was provided during the early 1990s by Kreiswirth *et al.* (1993). Based on polymorphisms in the *mecA* gene and Tn*554,* they proposed an evolutionary tree with a single progenitor, although they suggested that the possibility that transfer of SCC*mec* took place in one or two other occasions could not be excluded. However, this view of a single transfer of SCC*mec* was challenged by several studies. Multilocus enzyme electrophoresis (MLEE), which is based on differences in electrophoretic mobility of enzymes as a consequence of amino acid changes due to mutations in their encoding genes, performed on 254 MRSA isolates from 4 continents showed that these isolates belonged to six phylogenetic lineages. The divergent phylogenetic lineages represented a large portion of the breadth of the *S. aureus* population (Musser and Kapur, 1992). It is noteworthy that all isolates from the 1960s belonged to a single electrophoretic type in accordance with the theory that type I SCC*mec* was responsible for the first wave of MRSA.

Analysis of the time of appearance of type I and II SCC*mec* also raised questions about the hypothesis of clonal origin of MRSA. The SCC*mec* from isolates from the 1960s to the 1980s obtained from various geographic locations were analyzed. Isolates from the 1960s lacked *mecI* and *mecR* sequences, whereas these sequences are present in MRSA isolated later. Also the isolate that was the proposed progenitor for MRSA in the clonal expansion hypothesis lacked *mecI* and *mecR* sequences. In the clonal expansion theory one would expect that the structure of SCC*mec* could be explained by evolution of earlier forms to later ones. However, it is not easy to explain how SCC*mec* from MRSA that lack at least part of the *mecI* and *mecR* sequences evolved into a SCC*mec* that contains these sequences. Furthermore, the two SCC*mec* types were also detected in coagulase-negative staphylococci. Taken together these data suggest that transfer of SCC*mec* from coagulase-negative staphylococci to *S. aureus* took place on at least two occasions (Archer and Niemeyer, 1994 Archer *et al.*, 1994).

Compatible data were presented by Hiramatsu (1995). Analysis of the *mecI* and *mecR* genes showed that isolates from the sixties lacked *mecI* and part of *mecR*. Whereas these genes are contained by isolates from the 1980s. In addition to isolates with complete *mecI* and *mecR* genes, isolates with a deleted *mecI* and a partial *mecR* were obtained during this time period, demonstrating that both types co-existed during the 1980s. In addition, a number of different ribotypes were present. Since evolution of rRNA encoding genes and thus the alteration of restriction sites in these genes is believed to be much slower than the mutation of *Sma*I sites, used in PFGE, ribotyping provides a better measure for the long term clonal relationships between isolates. The data presented in this study therefore also indicate that SCC*mec* was acquired at least twice by *S. aureus* form another species. Similar data were also obtained for isolates from the Zürich, Switzerland (Hürlimann-Dalel *et al.*, 1992).

Multilocus sequence typing (MLST), (in which internal fragments of house keeping genes are sequenced and compared) of DNA from the Archaic, Iberian, Brazilian, Hungarian, and Pediatric clones showed that they share identical or closely related MLST profiles. The New York and Pediatric clones also share identical MLST profiles, but these are clearly distinct from the MLST profiles of the first set of clones (Table 2) (Oliveira *et al.*, 2001, 2002). However, the clones represented in the clonal lineages, as defined by these two MLST profiles, contain different SCC*mec* types (Oliveira *et al.*, 2000, 2001, 2002). These data are in agreement with the hypothesis that SCC*mec* was acquired on multiple occasions from coagulase-negative staphylococci.

Compatible data were obtained with MLST of *S. aureus* isolates from invasive disease in Oxford, UK. The DNA sequences of fragments of seven house keeping genes, obtained from 155 isolates, were determined. A total of 53 sequence profiles were obtained. Twenty nine MRSA isolates belonged to 5 sequence profiles and three of these appear related. Twenty-five MRSA isolates belonged to these sequence profiles, but 22 belonged to only one of these three. For two sequence profiles with MRSA isolates, MSSA counterparts were found (Enright *et al.*, 2000). These suggest that the counterparts from which new MRSA lineages arise by acquisition of SCC*mec* are still present as expected for recent horizontal transfer to multiple lineages.

In other studies MSSA counterparts of ancestral MSSA were also found. A study by the groups of de Lencastre and Tomasz went back to the oldest MRSA isolates known. These isolates from the UK and Denmark were compared with MSSA isolates from the same time period and contemporary MRSA isolates. The early MRSA and the MSSA shared a number of properties including antibiogram (except methicillin), phage type, PFGE type, MLST type and an identical gene for protein A. These data strongly suggest that some of the MSSA represented the clonal lineage which was one of the first recipients of SCC*mec* in Europe. Moreover, the genetic background of these MSSA is very similar to that of the Iberian clone, which is now widely disseminated. But also the MLST types of the New York and Pediatric clone, which are identical, were represented among the MSSA, as well as a potential counterpart for EMRSA-16, which is widespread in the UK (Crisostomo *et al.*, 2001).

DNA micro array analysis covering more than 90% the genome of 36 *S. aureus* isolates including 11 MRSA showed that these MRSA belonged to five distinct lineages that are highly differentiated in overall chromosomal gene content (Fitzgerald *et al.*, 2001).

Data form our group obtained by riboprinting, PFGE, and southern hybridization fingerprinting of SCC*mec* are in agreement with those of other studies. SCC*mec* was present in 8 of 10 *S. aureus* ribotypes found among a total of approximately 1000 isolates of MSSA and MRSA. In fact potential

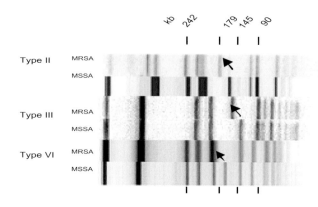

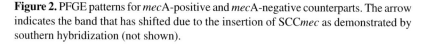

Figure 2. PFGE patterns for *mec*A-positive and *mec*A-negative counterparts. The arrow indicates the band that has shifted due to the insertion of SCC*mec* as demonstrated by southern hybridization (not shown).

MSSA counterparts of MRSA strains could be identified (Wielders *et al.*, 2002) (Figure 2). But the data also revealed that some ribotypes harbored different SCC*mec* types, indicative of not only multiple occasions of horizontal transfer of SCC*mec* to different lineages, but also multiple occasions of horizontal transfer to the same lineage (unpublished observations). Large scale transfer of SCC*mec* between different *S. aureus* strains was invoked to explain the diversity of PFGE isolates observed at some geographic locations (Schneider *et al.*, 1996; Witte *et al.*, 1997).

Additional evidence for multiple occasions of horizontal transfer of SCC*mec* was provided by the riboprinting of 325 unique patients blood stream infection isolates with multidrug-resistant MRSA from 5 continents. A total of 48 ribotypes were obtained, but more than 80% of the isolates belonged to only 10 ribotypes. In fact 30% of the isolates belonged to a single ribotype, which was found on all continents except South America. However, the spread of some other ribotypes was limited to one continent. PFGE typing revealed many clusters in single hospitals or hospitals that were in close proximity, but four PFGE types were found on more than one continent (Diekema *et al.*, 2000). These data strongly support previous data that MRSA cover the the breadth of genetic diversity among the *S. aureus* population. As a consequence the transfer of SCC*mec* between staphylococci must be more common than previously believed.

Also the reports of community-acquired MRSA at geographically widespread locations (Groom *et al.*, 2001; Naimi *et al.*, 2001; Charlebois *et al.*, 2002), which might be associated with SCC*mec* type IV, gives further credence to the theory that transfer of SCC*mec*, probably from coagulase-negative staphylococci is not exceptional.

Recently, a large collection of MRSA and MSSA (n=912) was typed by MLST. A total of 162 sequence types was found and the 359 MRSA isolates belonged to 38 sequence types, but 25 of these types included only a single isolate. Twelve sequence types included isolates that were obtained from more than one country. Several of the major sequence types contained isolates with different SCC*mec* types (Enright *et al.*, 2002). This is compatible with data showing that major ribotypes among MRSA as determined by automated ribotyping contained different SCC*mec* types (C.L.C. Wielders, A.T.A. Box, and A.C. Fluit, unpublished). Several previously identified clones belonged in fact to the same sequence type despite different PFGE types, although they may differ in SCC*mec* type. All 4 known SCC*mec* types were present in isolates belonging to 2 different sequence types. In accordance with results reported by others and unpublished data from our group nearly all isolates from the 1960s belong to a single sequence type and contain SCC*mec* type I (Enright *et al.*, 2002). In accordance with our results (Wielders *et al.*, 2002) MSSA counterparts could be identified for several sequence/SCC*mec* types (Enright *et al.*, 2002). These data also strongly indicate that horizontal transfer of SCC*mec* took place on several occasions.

It is strongly suggested by combining the data on outbreaks during the last forty years and the typing of SCC*mec,* that waves of outbreaks were associated with the appearance of a new type of SCC*mec* in *S. aureus.* Nowadays there appears to be a reasonable number of MRSA lineages and for many of these MRSA lineages potential MSSA counterparts were found. It seems likely that by typing isolates from a larger variety of different geographic locations the number of MRSA lineages will increase. In that respect a whole new reservoir is formed by the community. There is no apparent reason why the majority of MSSA lineages may not acquire a SSC*mec* element. Although horizontal transfer of SCC*mec* between *S. aureus* lineages may play a substantial role, regular transfer of SCC*mec* from coagulase-negative staphylococci should not be discarded. All four types of SCC*mec* described until now were also present in different clinical isolates of coagulase-negative staphylococcal species collected during the late 1990s. However, nothing is known about when and how these SCC*mec* types appeared in coagulase-negative staphylococci.

SCC*mec* Transfer

The population structure of *S. aureus* indicates that SCC*mec* was introduced into a large number of clonal lineages. However, it is almost impossible to reconstruct the history of this introduction. Although *S. sciuri* is believed to be the source of the *mecA* (Wu *et al.*, 1996; Wu *et al.*, 2001), the evolution of SCC*mec* yielding four major types is unclear. Did this evolution exclusively take place in coagulase-negative staphylococci? At least all four types of SCC*mec* are also present among contemporary coagulase-negative species

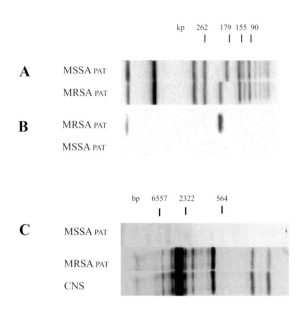

Figure 3. A. PFGE patterns of an MSSA and an MRSA which were subsequently isolated from the same patient. B. The band shift observed for the MRSA corresponds with the insertion of *mecA* DNA as demonstrated by southern hybridization. C. SSC*mec* hybridization fingerprints of the MRSA and a CNS isolated from the patient, the MSSA served as negative control.

and recently evidence was presented that SCC*mec* was transferred from a coagulase-negative staphylococcus to a methicillin-susceptible *S. aureus* isolate in a patient (Wielders *et al.*, 2001). The patient was a male infant with Pierre Robinson syndrome born after 40 weeks of pregnancy. From day 4 after birth he was incubated and mechanically ventilated because of respiratory problems. To treat a suspected respiratory tract infection he was given amoxicillin/ clavulanic acid. He became bacteremic 3 days later and treatment was changed to amoxicillin and cefotaxime. When his blood cultures revealed a *mecA*-negative methicillin-susceptible *S. aureus* treatment was changed to flucloxacillin. He seemed to recover, but at day 32 treatment with amoxicillin/ clavulanic acid was restarted, because of recurrence of the respiratory tract infection. On day 56, routine cultures of nasal swabs unexpectedly showed *mecA*-positive *S. aureus*. This isolate was only resistant to β-lactam antibiotics. Besides this MRSA several *mecA*-positive coagulase-negative staphylococci were recovered from the patient. The patient received no further antibiotic treatment because his situation had improved. PFGE analysis of the *mecA*-positive and negative *S. aureus* isolates revealed identical patterns, except for one fragment, which differed by approximately 40 kb. The fragment from the *mecA*-positive isolate hybridized with a *mecA*-specific probe, whereas its counterpart from the *mecA*-negative isolate did not hybridize with the probe. Fingerprinting of the SCC*mec* from the MRSA and the *mecA*-positive coagulase-negative staphylococci revealed an identical pattern for the MRSA and a *S. epidermidis* isolate (Figure 3).

However, there is no apparent reason why SCC*mec* evolution only should have taken place in coagulase-negative staphylococci before transfer to *S. aureus* occured. Evolution of SCC*mec* could also have taken place in *S. aureus* and subsequently been transferred back into coagulase-negative staphylococci. At least it is plausible that a SCC*mec* element acquired by a clone of *S. areus* may be transferred to other *S. aureus* clones. There is no apparent reason why each *S. aureus* clone should have acquired its SCC*mec* from a methicillin-resistant coagulase-negative staphylococcus. In fact it was proposed that sporadic MRSA strains acquired SCC*mec* from epidemic strains which were in close proximity (Schneider *et al.*, 1996). This suggests that transfer should have happened on a large scale. This proposal was based on the PFGE analysis of 378 MRSA isolates from 180 patients. Fourteen epidemic strains were present in 155 patients, the remaining 25 patients all had unique strains.

As elusive as the frequency of transfer is the mode of transfer of SCC*mec* between different staphylococci. However, the enzymes involved in the excision from and insertion into the chromosome of SCC*mec* are known (Katayama *et al.*, 2000). The genes required for the excision of SCC*mec*, *ccrA* and *ccrB*, are located on SCC*mec* and the predicted lengths for the encoded proteins are 449 and 542 amino acids, respectively. Most likely the two genes are transcribed from the same promotor as a single mRNA. The two predicted protein sequences not only show strong homology with each other, but also have high homology to an integrase from lactococcal bacteriophage TP901-1. After the excision of SCC*mec*, it is circularized by ligation of both ends of the element. The resulting structure at this point was called *att*SCC. Both *att*SCC and the CcrA and CcrB proteins are required for the integration of the circular DNA into the chromosome. This is in contrast to the integration of bacteriophages and conjugative transposons, where only one enzyme, an integrase is required (however, the integrase and a second protein, excisionase, are required for their excision). In addition, little homology is found between CcrA or CcrB and members of the excisionase/integrase family of enzymes from bacteriophages, however the amino-terminal parts of these proteins show homology with the invertase/recombinase family of enzymes.

Although *mecA* carriage *in vivo* seems rather stable (Figueiredo *et al.*, 1991), the loss of *mecA* under certain storage or culture conditions has long been observed (Hiramatsu *et al.*, 1990, Inglis *et al.*, 1990). Loss of part of SCC*mec* starts at the left boundary and a variable part past the *mecA* gene. This makes this process different from the precise excision of SCC*mec* by CcrA and CcrB (Katayama *et al.*, 2000). However, the precise excision at the left boundary makes it tempting to speculate that one or both enzymes are involved in the process. Control of excision may be achieved through control of transciption of the *ccrA* and *ccrB* genes (Katayama *et al.*, 2000).

Although the mechanism by which SCC*mec* is excised from and integrated into the chromosome of staphylococci is known, the mode of transfer of SCC*mec* to different bacterial cells is still unidentified. Both transduction and conjugation are candidates for the process and both types of gene transfer have been described in staphylococci.

Mupirocin resistance is caused by the presence of an alternative isoleucyl tRNA synthetase. Mupirocin resistance is believed to have originated in coagulase-negative staphylococci, but it is also found in increasing frequency in *S. aureus* isolates (Schmitz *et al.*, 1998). The transfer of mupirocin resistance from coagulase-negative staphylococci to *S. aureus* has been well-documented (Connolly *et al.*, 1993). Not only the transfer of muporicin resistance from coagulase-negative staphylococci has been observed, but also conjugative transfer of gentamicin, erythromycin, chloramphenicol, and tetracycline resistance could be observed *in vitro* at frequencies of 10^{-6} to 10^{-8} (Forbes and Schaberg, 1983). Conjugative transfer of gentamicin (Archer and Johnston, 1983; Muhammad *et al.*, 1993) and streptomycin (Muhammad *et al.*, 1993), from coagulase-negative staphylococci to *S. aureus* was observed in other studies. In fact conjugative transfer between different *S. aureus* has been well-documented. Examples include the conjugative transfer of gentamicin (Archer *et al.*, 1983; Goering and Ruff, 1983; Schaberg *et al.*, 1985), and of mupirocin (Morton *et al.*, 1995) between strains of *S. aureus*. In addition conjugative transfer of mupirocin to *Staphylococcus epidermidis* was also be demonstrated (Janssen *et al.*, 1993). In all cases of conjugative transfer the resistance determinant was present on a plasmid.

Some resistance determinants including aminoglycoside resistance and tetracycline resistance could be transferred both by conjugation and transduction, albeit with less efficiency in transduction experiments than in conjugation experiments (El Solh, *et al.*, 1986). Some resistance determinants, e.g., cadmium resistance, could be transferred only by transduction, whereas others could only be transferred by conjugation. However, since these are *in vitro* experiments the mode of transfer and the transfer frequencies need not to reflect the mode and chances of transfer *in vivo*.

None of the conjugation experiments showed transfer of SCC*mec* (or *mecA*), but in experimental setting transduction of *mecA* has been demonstrated by several groups. However, transduction of *mecA* in these experiments seemed to require special properties for the recipient. The requirements vary from the presence of certain integrated phages like phi11 or the presence of a β-lactamase carrying plasmid like pI254, or only β-lactamase, or several of these factors were necessary (Cohen and Sweeney, 1973; Stewart and Rosenblum, 1980; Hiramatsu *et al.*, 1990). Besides these factors several other factors may influence the transfer of *mecA* by transduction (Lacey, 1980). Most transduction experiments were carried out with phages derived from MRSA and used *S. aureus* strains as recipients. But transduction of *mecA* between *S. epidermidis* strain also has been demonstrated (Blanchard *et al.*, 1986).

Although phages often have a narrow host range and only infect a small number strains, exceptions exist. In staphylococci phage phi812 has a broad-host range infecting nearly all *S. aureus* isolates (N=782) tested and nearly 50% of the coagulase-negative staphylococci (N=89) tested (Pantucek *et al.*, 1998). The existence of broad-host range phages suggests that transfer of genetic material via transduction between coagulase-negative staphylococci and *S. aureus* is possible in both directions. But the number of phages potentially able to transduct SCC*mec* is limited especially for the large type I-III SCC*mec*, because most *S. aureus*-specific phages can only harbor 40-45 kb of DNA.

Although DNA sequencing revealed no sequences which would hint at transduction as the transfer mechanism for SCC*mec* (Ito *et al.*, 1999), the existence of phages with a broad host range combined with the fact that transduction of *mecA* has been demonstrated *in vitro*, whereas no experiment has been described that showed *mecA* transfer by conjugation, may suggest that transduction is the mechanism of transfer for SCC*mec*. In accordance with this proposition is the fact that the smaller type IV SCC*mec* appears to disseminate easily in the community and in a relatively short time it has entered a large number of clonal lineages. The smaller type IV DNA fits in a larger number of phages enhancing its chances of transfer, but smaller DNA molecules also may be easier to transfer by conjugation. The fact that sequences associated with transduction are missing from SCC*mec* may be explained by assuming that upon entrance in the bacterial cell the *ccrA*/*ccrB* genes are expressed and SCC*mec* is excised from the phage DNA and integrated in the bacterial chromosome at its usual location.

In conclusion it can be said, that although the enzymes for excision and integration of SCC*mec* are known, its mechanism of transfer is still unknown. Although *in vitro* experiments seem to favor transduction, conjugation can not be ruled out. In addition to conjugation and transduction other mechanisms of transfer can not be excluded (Lacey, 1980). Transfer of resistance determinants between *S. aureus* strains, and also to and from coagulase-negative staphylococci has been described several times suggesting that also the evolution of SCC*mec* is more dynamic than we suspect.

Conclusion

The data from the spread of clones, the distribution of SCC*mec* types both in lineages and time, and sequence data show that some clones have a widespread distribution. The most clear examples of this are the New York/Japan clone, which is found in New York and nearby states and in Tokyo, Japan (Aries de Sousa *et al.*, 2000), the Iberian clone, which has a wide intercontinental distribution (Oliveira *et al.*, 2002) and the Brazilian clone, which found in Latin America and Europe (Oliveira *et al.*, 2002). But besides these very successful clonal lineages, many minor clonal lineages exist (Diekema *et al.*, 2000). Sometimes these minor clonal lineages are distinguished by a different type of SCC*mec* but the genetic background of the strains is identical. Frequently, a MSSA counterpart for the MRSA can be observed (Crisostomo *et al.*, 2001; Oliveira *et al.*, 2002; Wielders *et al.*, 2002). The four SCC*mec* types entered *S. aureus* at different points time. Some clonal lineages (ribotypes) harbored more than one type of SCC*mec*. It is highly likely that transfer of SCC*mec* from coagulase-negative staphylococci took place on multiple occasions although this process is not understood and is not as frequent as with some other resistance determinants. After initial transfer of SCC*mec* from coagulase-negative staphylococci to *S. aureus* transfer to other *S. aureus* lineages may have occurred, but these lineages may have acquired the SCC*mec* element independently from coagulase-negative staphylococci.

References

Adcock, P.M.., Pastor, P., Medley, F., Patterson, J.E., and Murphy, T.V. 1998. Methicillin-resistant *Staphylococcus aureus* in two child care centers. J. Infect. Dis. 178: 577-580.

Archer, G.L., and Johnston, J.L. 1983. Self-transmissible plasmids in staphylococci that encode resistance to aminoglycosides. Antimicrob. Agents Chemother. 24: 70-77.

Archer, G.L., and Niemeyer, D.B. 1994. Origin and evolution of DNA associated with resistance to methicillin in staphylococci. Trends Microbiol. 2: 343-347.

Archer, G.L., Niemeyer, D.M., Thanassi, J.A., and Puccci, M.J. 1994. Dissemination among staphylococci of DNA sequences associated with methicillin resistance. Antimicrob. Agents Chemother. 38: 447-454.

Aries de Sousa, M., de Lencastre, H., Santos Sanches, I., Kikuchi, K., Totsuka, K., and Tomasz, A. 2000. Similarity of antibiotic resistance patterns and molecular typing properties of methicillin-resistant *Staphylococcus aureus* isolates widely spread in hospitals in New York City and in a hospital in Tokyo, Japan. Microb. Drug Resist. 6: 253-258.

Ayliffe, G.A. 1997. The progressive intercontinental spread of methicillin-resistant *Staphylococcus aureus*. Clin. Infect. Dis. 24 Suppl. 1: S74-S79.

Ayliffe, G.A., Lilly, H.A., and Lowbury, E.J.L. 1979. Decline of the hospital staphylococcus? Incidence of multiresistant *Staph. aureus* in three Birmingham hospitals. Lancet I: 539.

Barrett, F.F., McGehee, R.F. Jr, and Finland, M. 1968. Methicillin-resistant *Staphylococcus aureus* at Boston City Hospital. Bacteriologic and epidemiologic observations. N. Engl. J. Med. 279: 441-448.

Benner, E.J., and Kayser, F.H. 1968. Growing clinical significance of methicillin-resistant *Staphylococcus aureus*. Lancet. 5: 741-744.

Blanchard, T.J., Poston, S.M., and Reynolds, P.J. 1986. Recipient characteristics in the transduction of methicillin resistance in *Staphylococcus epidermidis*. Antimicrob. Agents Chemother. 29: 539-541.

Borowski, J., Kamienska, K., and Rutecka. I. 1964. Methicillin-resistnat staphylococci. Brit. Med. J.: 983.

Boyce, J.M., and Causey, W.A. 1982. Increasing occurrence of methicillin-resistant *Staphylococcus aureus* in the United States. Infect. Control. 3: 377-383.

Cetin, E.T., and Ang, O. 1962. Staphylococci resistant to methicillin ("Celbenin"). Brit. Med. J.: 51-52.

Chambers, H.F. 2001. The changing epidemiology of *Staphylococcus aureus*? Emerg. Infect. Dis. 7: 178-182.

Charlebois, E.D., Bangsberg, D.R., Moss, N.J., Moore, M.R., Moss, A.R., Chambers, H. F. and Perdreau-Remington, F. 2002. Population-based community prevalence of methicillin-resistant *Staphylococcus aureus* in the urban poor of San Francisco. Clin. Infect. Dis. 34: 425-433.

Cohen, S., and Sweeney, H.M. 1973. Effect of the prophage and penicillinase plasmid of the recipient strain upon the transduction and stability of methicillin resistance in *Staphylococcus aureus*. J. Bacteriol. 116: 803-811.

Connolly, S., Noble, W.C., and Phillips, I. 1993. Mupirocin resistance in coagulase-negative staphylococci. J. Med. Microbiol. 39: 450-453.

Crisostomo, M.I., Westh, H., Tomasz, A., Chung, M., Oliveira, D.C., and de Lencastre, H. 2001. The evolution of methicillin resistance in *Staphylococcus aureus*: similarity of genetic backgrounds in historically early methicillin-susceptible and -resistant isolates and contemporary epidemic clones. Proc. Natl. Acad. Sci. USA. 98: 9865-9870.

de Lencastre, H., de Lencastre, A., and Tomasz, A. 1996. Methicillin-resistant *Staphylococcus aureus* isolates recovered from a New York City hospital: analysis by molecular fingerprinting techniques. J. Clin. Microbiol. 34: 2121-2124.

de Lencastre, H., Severina, E.P., Milch, H., Thege, M.K., and Tomasz, A. 1997. Wide geographic distribution of a unique methicillin-resistant *Staphylococcus aureus* clone in Hungarian hospitals. Clin. Microbiol. Infect. 3: 289-296.

Deplano, A., Witte, W., van Leeuwen, W.J., Brun, Y., and Struelens, M.J. 2000. Clonal dissemination of epidemic methicillin-resistant *Staphylococcus aureus* in Belgium and neighboring countries. Clin. Microbiol. Infect. 6: 239-245.

Diekema, D.J., Pfaller, M.A., Schmitz, F.J., Smayevsky, J., Bell, J., Jones, R.N., and Beach, M. 2001. Survey of infections due to *Staphylococcus* species: frequency of occurrence and antimicrobial susceptibility of isolates collected in the United States, Canada, Latin America, Europe, and the Western Pacific region for the SENTRY Antimicrobial Surveillance Program, 1997-1999. Clin. Infect. Dis. 32 Suppl. 2: S114-S132.

Diekema, D.J., Pfaller, M.A., Turnidge, J., Verhoef, J., Bell, J., Fluit, A.C., Doern, G.V., and Jones, R.N. 2000. Genetic relatedness of multidrug-resistant, methicillin (oxacillin)-resistant *Staphylococcus aureus* bloodstream isolates from SENTRY Antimicrobial Resistance Surveillance Centers worldwide. 1998. Microb. Drug Resist. 6: 213-221.

dos Santos Soares, M.J., da Silva-Carvalho, M.C., Ferreira-Carvalho, B.T., and Figueiredo, A.M.S. Spread of methicillin-resistant *Staphylococcus aureus* belonging to the Brazilian epidemic clone in a general hospital and emergence of heterogenous resistance to glycopeptide antibiotics among these isolates. J. Hosp. Infect. 44: 301-308.

Dyke, K.G., Jevons, M.P., and Parker, M.T. 1966. Penicillinase production and intrinsic resistance to penicillins in *Staphylococcus aureus*. Lancet. 1: 835-838.

El Solh, N., Allignet, J., Bismuth, R., Buret, B., and Fouace, J.-M. 1986. Conjugative transfer of staphylococcal antibiotic resistance markers in the absence of detectable plasmid DNA. Antimicrob. Agents Chemother. 30: 161-169.

Enright, M.C., Day, N.P., Davies, C.E., Peacock, S.J., and Spratt, B.G. 2000. Multilocus sequence typing for characterization of methicillin-resistant and methicillin-susceptible clones of *Staphylococcus aureus*. J. Clin. Microbiol. 38: 1008-1015.

Enright, M.C., Robinson, D.A., Randie, G., Feil, E.J., Grundmann, H., and Spratt, B.G. 2002. The evolutionary history of methicillin-resistant *Staphylococcus aureus* (MRSA). Proc. Natl. Acad. Sci. USA 99: 7687-7692.

Figueiredo, A.M., Ha, E., Kreiswirth, B.N., de Lencastre, H., Noel, G.J., Senterfit, L., and Tomasz, A. 1991. *In vivo* stability of heterogeneous expression classes in clinical isolates of methicillin-resistant staphylococci. J. Infect. Dis. 164: 883-887.

Fitzgerald, J.R., Sturdevant, D.E., Mackie, S.M., Gill, S.R., and Musser, J.M. 2001. Evolutionary genomics of *Staphylococcus aureus*: insights into the origin of methicillin-resistant strains and the toxic shock syndrome epidemic. Proc. Natl. Acad. Sci. USA. 98: 8821-8826.

Forbes, B.A., and Schaberg, D.R. 1983. Transfer of resistance plasmids from *Staphylococcus epidermidis* to *Staphylococcus aureus*: Evidence for conjugative exchange of resistance. J. Bacteriol. 153: 627-634.

Goering, R.V., and Ruff, E.A. 1983. Comparative analysis of conjugative plasmids mediating gentamicin resistance in *Staphylococcus aureus*. Antimicrob. Agents Chemother. 24: 450-452.

Groom, A.V., Wolsey, D.H., Naimi, T.S., Smith, K., Johnson, S., Boxrud, D., Moore, K.A. and Cheek, J.E. 2001. Community-acquired methicillin-resistant *Staphylococcus aureus* in a rural American Indian community. JAMA. 286: 1201-1205.

Gürtler, V., and Mayall, B.C. 2001. Genetic transfer and evolution in MRSA. Microbiol. 147, 2001: 3195-3197.

Hassall, J.E., Rountree, P.M. 1959. Staphylococcal septicemia. Lancet i: 213-217.

Hausmann, W., and Karlish, A.J. 1996. Staphylococcal pneumonia in adults. Br. Med. J. 2: 845-847.

Herold, B.C., Immergluck, L.C., Maranan, M.C., Lauderdale, D.S., Gaskin, R.E., Boyle-Vavra, S., Leitch, C.D., and Daum, R.S. 1998. Community-acquired methicillin-resistant *Staphylococcus aureus* in children with no identified predisposing risk. JAMA. 279: 593-598.

Hiramatsu, K. 1995. Molecular evolution of MRSA. Microbiol. Immunol. 39: 531-543.

Hiramatsu, K., Aritaka, N., Hanaki, H., Kawasaki, S., Hosoda, Y., Hori, S., Fukuchi, Y., and Kobayashi, I. 1997. Dissemination in Japanese hospitals of strains of *Staphylococcus aureus* heterogeneously resistant to vancomycin. Lancet 350:1670-1673.

Hiramatsu, K., Cui, L., Kuroda, M., and Ito. T. 2001. The emergence and evolution of methicillin-resistant *Staphylococcus aureus*. Trends Microbiol. 9: 486-493.

Hiramatsu, K., Hanaki, H., Ito, T., Yabuta, K., Oguri, T., and Tenover, F.C. Methicillin-resistant *Staphylococcus aureus* clinical strain with reduced vancomycin susceptibility. J. Antimicrob. Chemother. 40: 135-146.

Hiramatsu, K., Suzuki, E., Takayama, H., Katayama, Y. and Yokota, T. 1990. Role of penicillinase plasmids in the stability of the *mecA* gene in methicillin-resistant *Staphylococcus aureus*. Antimicrob. Agents Chemother. 34: 600-604.

Hookey, J.V., Richardson, J.F., and Cookson, B.D. 1998. Molecular typing of *Staphylococcus aureus* based on PCR restriction fragment length polymorphism and DNA sequence analysis of the coagulase gene. J. Clin. Microbiol. 36: 1083-1089.

Hürliman-Dalel, R.L., Ruffel, C., Kayser, F.H., and Berger-Bächi, B. 1992. Survey of methicillin-resistance-associated genes *mecA, mecR1-mecI*, and *femA-femB* in clinical isolates of methicillin-resistant *Staphylococcus aureus*. Antimicrob. Agents Chemother. 36: 2617-2621.

Hussain, F.M., Boyle-Vavra, S., and Daum, R.S. 2001. Community-acquired methicillin-resistant *Staphylococcus aureus* colonization in healthy children attending an outpatient pediatric clinic. Pediatr. Infect. Dis. J. 20: 763-767.

Inglis, B., Matthews, P.R., and Stewart, P.R. 1990. Induced deletions within a cluster of resistance genes in the *mec* region of the chromosome of *Staphylococcus aureus*. J. Gen. Microbiol. 136: 2231-2239.

Ito, T., Katayama, Y., Asada, K., Mori, N., Tsutsumimoto, K., Tiensasitorn, C., and Hiramatsu, K. 2001. Structural comparison of three types of staphylococcal cassette chromosome *mec* integrated in the chromosome in methicillin-resistant *Staphylococcus aureus*. Antimicrob. Agents Chemother. 45: 1323-1336.

Ito, T., Katayama, Y., and Hiramatsu, K. 1999. Cloning and nucleotide sequence determination of the entire *mec* DNA of pre-methicillin-resistant *Staphylococcus aureus* N315. Antimicrob. Agents Chemother. 43: 1449-1458.

Janssen, D.A., Zarins, L.T., Schaberg, D.R., Bradley, S.F., Terpenning, M.S., and Kauffman, C.A. 1993. Detection and characterization of mupirocin resistance in *Staphylococcus aureus*. Antimicrob. Agents Chemother. 37: 2003-2006.

Jepsen, O.B. 1986. The demise of the 'old' methicillin-resistant *Staphylococcus aureus*. J. Hosp. Infect. 7 Suppl. A: 13-17.

Jevons, M.P. 1961. 'Celbenin'-resistant staphylococci. Br. Med. J. 1: 124-125.

Johnson, A.P., Aucken, H.M., Cavendish, S., Ganner, M., Wale, M.C., Warner, M., Livermore, D.M. and Cookson, B.D. 2001. Dominance of EMRSA-15 and -16 among MRSA causing nosocomial bacteraemia in the UK: analysis of isolates from the European Antimicrobial Resistance Surveillance System (EARSS). J. Antimicrob. Chemother. 48: 143-144.

Jorgensen, J.H. 1986. Laboratory and epidemiologic experience with methicillin-resistant *Staphylococcus aureus* in the USA. Eur. J. Clin. Microbiol. 5: 693-696.

Katayama, Y., Ito, T., and Hiramatsu, K. 2000. A new class of genetic element, staphylococcus cassette chromosome *mec*, encodes methicillin resistance in *Staphylococcus aureus*. Antimicrob. Agents Chemother. 44: 1549-1555.

Katayama, Y., Ito, T., and Hiramatsu, K. 2001. Genetic organization of the chromosome region surrounding *mecA* in clinical staphylococcal strains: role of IS*431*-mediated mecI deletion in expression of resistance in *mecA*-carrying, low-level methicillin-resistant *Staphylococcus haemolyticus*. Antimicrob. Agents Chemother. 45: 1955-1963.

Kreiswirth, B., Kornblum, J., Arbeit, R.D., Eisner, W., Maslow, J.N., McGeer, A., Low, D.E., and Novick, R.P. 1993. Evidence for a clonal origin of methicillin resistance in *Staphylococcus aureus*. Science. 259: 227-230.

Lacey, R.W. 1980. Bacteriophages and spread of resistance in *Staphylococcus aureus*. J. Antimicrob. Chemother. 6: 567-575.

Lacey, R.W., and Grinsted, J. 1973. Genetic analysis of methicillin-resistant strains of *Staphylococcus aureus*: evidence for their evolution from a single clone. J. Med. Microbiol. 6: 511-526.

Leski, T., Oliveira, D., Trzcinski, K., Sanches, I.S., de Sousa, M.A., Hryniewicz, W., and de Lencastre H. 1998. Clonal distribution of methicillin-resistant *Staphylococcus aureus* in Poland. J. Clin. Microbiol. 36: 3532-3539.

Ma, X.X., Ito, T., Tiensasitorn, C., Jamklang, M., Chongtrakool, P., Boyle-

Vavra, S., Daum, R.S., and Hiramatsu, K. 2002. Novel type of Staphylococcal Cassette Chromosome *mec* identified in community-acquired methicillin-resistant *Staphylococcus aureus* strains. Antimicrob. Agents Chemother. 46:1147-1152.

Marshall, S.A., Wilke, W.W., Pfaller, M.A., and Jones, R.N. 1998. *Staphylococcus aureus* and coagulase-negative staphylococci from blood stream infections: frequency of occurrence, antimicrobial susceptibility, and molecular (*mecA*) characterization of oxacillin resistance in the SCOPE program. Diagn. Microbiol. Infect. Dis. 30: 205-214.

Morton, T.M., Johnston, J.L., Patterson, J., and Archer, G.L. 1995. Characerization of a conjugative staphylococcal mupirocin resistance plasmid. Antimicrob. Agents Chemother. 39: 1272-1280.

Muhammad, G., Hoblet, K.H., Jackwood, D.J., Bech-Nielsen, S., and Smith, K.L. 1993. Interspecific conjugal transfer of antibiotic resistance among staphylococci isolated from bovine mammary gland. Am. J. Vet. Res. 54: 1432-1440.

Musser, J.M., and Kapur, V. 1992. Clonal analysis of methicillin-resistant *Staphylococcus aureus* strains from intercontinental sources: association the *mec* gene with divergent phylogenetic lineages implies dissemination by horizontal transfer and recombination. J. Clin. Microbiol. 30: 2058-2063.

Musser, J.M., and Selander, R.K. 1990. Genetic analysis of natural populations of *Staphylococcus aureus*. In: Molecular biology of the Staphylococci. R.P. Novick, ed. VCH publishers, Inc., New York. p. 59-67.

Naimi, T.S., LeDell, K.H., Boxrud, D.J., Groom, A.V., Steward, C.D., Johnson, S.K., Besser, J.M., O'Boyle, C., Danila, R.N., Cheek, J.E., Osterholm, M.T., Moore, K.A., and Smith, K.E. 2001. Epidemiology and clonality of community-acquired methicillin-resistant *Staphylococcus aureus* in Minnesota, 1996-1998. Clin. Infect. Dis. 33: 990-996.

Oliveira, D., Santos-Sanches, I., Mato, R., Tamayo, M., Ribeiro, G., Costa, D., and de Lencastre, H. 1998. Virtually all methicillin-resistant *Staphylococcus aureus* (MRSA) infections in the largest Portuguese teaching hospital are caused by two internationally spread multiresistant strains: the 'Iberian' and the 'Brazilian' clones MRSA. Clin. Microbiol. Infect. 4: 373-384.

Oliveira, D.C., Tomasz, A., and de Lencastre, H. 2001. The evolution of pandemic clones of methicillin-resistant *Staphylococcus aureus*: identification of two ancestral genetic backgrounds and the associated *mec* elements. Microb. Drug Resist. 7: 349-361.

Oliveira, D.C., Tomasz, A., and de Lencastre, H. 2002. Secrets of success of a human pathogen: molecular evolution of pandemic clones of methicillin-resistant *Staphylococcus aureus*. Lancet Infect. Dis. 2:180-189.

Oliveira, D.C., Wu, S.W., and de Lencastre, H. 2000. Genetic organization of the downstream region of the *mecA* element in methicillin-resistant *Staphylococcus aureus* isolates carrying different polymorphisms of this region. Antimicrob. Agents Chemother. 44:1906-1910.

Panlilio, A.L., Culver, D.H., Gaynes, R.P., Banerjee, S., Henderson, T.S., Tolson, J.S., and Martone, W.J. 1992. Methicillin-resistant *Staphylococcus aureus* in U.S. hospitals, 1975-1991. Infect. Control Hosp. Epidemiol. 13: 582-586.

Pantucek, R., Rosypalova, A., Doskar, J., Kailerova, J., Ruzickova, V., Borecka, P., Snopkova, S., Horvath, R., Götz, F., and Rosypal, S. 1998. The polyvalent staphylococcal phage phi812: its host-range mutants and related phages. Virol. 246: 241-252.

Parker, M.T., and Hewitt, J.H. 1970. Methicillin resistance in *Staphylococcus aureus*. Lancet 1: 800-804.

Pavillard, R., Harvey, K., Douglas, D., Hewstone, A., Andrew, J., Collopy, B., Asche, V., Carson, P., Davidson, A., Gilbert, G., Spicer, J., and Tosolini, F. 1982. Epidemic of hospital-acquired infection due to methicillin-resistant *Staphylococcus aureus* in major Victorian hospitals. Med. J. Aust. 29: 451-454.

Reverdy, M.E., Bes, M., Brun, Y., and Fleurette. J. 1993. [Evolution of resistance to antibiotics and antiseptics of hospital *Staphylococcus aureus* strains isolated from 1980 to 1991.] Pathol Biol (Paris). 41: 897-904.

Roberts, R.B., de Lencastre, A., Eisner, W., Severina, E.P., Shopsin, B., Kreiswirth, B.N., and Tomasz, A. 1998. Molecular epidemiology of methicillin-resistant *Staphylococcus aureus* in 12 New York hospitals. MRSA J. Infect. Dis. 178:164-171.

Roman, R.S., Smith, J., Walker, M., Byrne, S., Ramotar, K., Dyck, B., Kabani, A. and Nicolle, L.E. 1997. Rapid geographic spread of a methicillin-resistant *Staphylococcus aureus* strain. Clin. Infect. Dis. 25: 698-705.

Rosdahl, V.T., Witte, W., Musser, M., and Jarlov, J.O. 1994. *Staphylococcus aureus* strains of type 95. Spread of a single clone. Epidemiol. Infect. 113: 463-470.

Rountree, P.M., and Beard, M.A. 1968. Hospital strains of *Staphylococcus aureus*, with particular reference to methicillin-resistant strains. Med. J. Aust. 28: 1163-1168.

Sanches, I.S., Ramirez, M., Troni, H., Abecassis, M., Padua, M., Tomasz, A., and de Lencastre, H. 1995. Evidence for the geographic spread of a methicillin-resistant *Staphylococcus aureus* clone between Portugal and Spain. J. Clin. Microbiol. 33: 1243-1246.

Schaberg, D.R., Power, G., Betzold, J., and Forbes, B.A. 1985. Conjugative R plasmids in antimicrobial resistance of *Staphylococcus aureus* causing nosocomial infections. J. Infect. Dis. 152: 43-49.

Schito, G.C., and Varaldo, P.E. 1988. Trends in the epidemiology and antibiotic resistance of clinical *Staphylococcus* strains in Italy. J. Antimicrob. Chemother. 21 Suppl C: 67-81

Schmitz, F.-J., Lindenlauf, E., Hofmann, B., Fluit, A.C., Verhoef, J., Heinz, H.P., and Jones, M.E. 1998. The prevalence of low and high-level mupirocin-resistance in staphylococcal clinical isolates from 19 European hospitals. J. Antimicrob. Chemother. 42: 489-495.

Schneider, C., Wendel, M., and Brade, V. 1996. Frequency, clonal heterogeneity and antibiotic resistance of methicillin-resistant *Staphylococcus aureus* (MRSA) isolated in 1992-1994. Zbl. Bakt. 283: 529-542.

Stewart, G.C., and Rosenblum, E.D. 1980. Transduction of methicillin resistance in *Staphylococcus aureus*: Recipient effectiveness and beta-lactamase production. 1980. Antimicrob. Agents Chemother. 18: 424-432.

Suzuki,E., Kuwahara-Arai, K., Richardson, J.F., and Hiramatsu, K. 1993. Distribution of *mec* regulator genes in methicillin-resistant *Staphylococcus* clinical strains. Antimicrob. Agents Chemother. 37: 1219-1226.

Thompson, R.L., Cabezudo, I., and Wenzel, R.P. 1982. Epidemiology of nosocomial infections caused by methicillin-resistant *Staphylococcus aureus*. Ann. Intern. Med. 97: 309-317.

Teixeira, L.A., Resende, C.A., Ormonde, L.R., Rosenbaum, R., Figueiredo, A.M., de Lencastre, H., and Tomasz, A. 1995. Geographic spread of epidemic multiresistant *Staphylococcus aureus* clone in Brazil. J. Clin. Microbiol. 33: 2400-2404.

Verhoef, J., Beaujean, D., Blok, H., Baars, A., Meyler, A., van der Werken, C., and Weersink, A. 1999. A Dutch approach to methicillin-resistant *Staphylococcus aureus*. Eur. J. Clin. Microbiol. Infect. Dis. 18: 461-466

Voss, A., Milatovic, D., Wallrauch-Schwarz, C., Rosdahl, V.T., and Braveny, I. 1994. Methicillin-resistant *Staphylococcus aureus* in Europe. Eur. J. Clin. Microbiol. Infect. Dis. 13: 50-55.

Wielders, C.L.C., Fluit, A.C., Brisse, S., Verhoef, J., and Schmitz, F.J. 2002. *mecA* gene is widely disseminated in *Staphylococcus aureus* population. J. Clin. Microbiol. 40: 3970-3975.

Wielders, C.L.C., Vriens, M.R., Brisse, S., de Graaf-Miltenburg, L.A., Troelstra, A., Fleer, A., Schmitz, F.-J., Verhoef, J., and Fluit, A.C. 2001. *In-vivo* transfer of *mecA* DNA to *Staphylococcus aureus* [corrected]. Lancet 357:1674-1675.

Wilson, R., and Hamburger, M. 1957. Fifteen years' experience with *Staphylococcus* septicemia in a large city hospital. Am. J. Med. 22: 437-457.

Witte, W., Kresken, M., Braulke, C., and Cuny, C. 1997. Increasing incidence and widespread dissemination of methicillin-resistant *Staphylococcus aureus* (MRSA) in hospitals in central Europe, with special reference to German hospitals. Clin. Microbiol. Infect. 3: 414-422.

Wu, S.W., de Lencastre, H., and Tomasz, A. 2001. Recruitment of the *mecA* gene homologue of *Staphylococcus sciuri* into a resistance determinant and expression of the resistant phenotype in *Staphylococcus aureus*. J. Bacteriol. 183: 2417-2424.

Wu, S., Piscitelli, C., de Lencastre, H., and Tomasz, A. 1996. Tracking the evolutionary origin of the methicillin resistance gene: cloning and sequencing of a homologue of *mecA* from a methicillin susceptible strain of *Staphylococcus sciuri*. Microb. Drug Resist. 2:435-441.

From: *MRSA: Current Perspectives*
Edited by: A.C. Fluit and F.-J. Schmitz

Chapter 8

Vancomycin-Resistant *Staphylococcus aureus*

Longzhu Cui and Keiichi Hiramatsu

Abstract

The glycopeptide antibiotics, vancomycin and teicoplanin, were the only licensed antibacterial compounds, until recently, to which methicillin-resistant *Staphylococcus aureus* clinical isolates have remained uniformly susceptible. However, the worldwide increase in the incidence of *S. aureus* clinical isolates with reduced susceptibility to vancomycin and teicoplanin raises the possibility that glycopeptide resistance in *S. aureus* is becoming an important clinical problem. *S. aureus* has evolved genetic and biochemical ways of resisting these antimicrobial actions. Genetic mechanisms include multistep mutations in bacteria chromosome without acquisition of new DNA. A review of available biochemical studies suggests that *S. aureus* does not resist glycopeptides by inactivating the drugs or altering the drug target, but reducing drug access to bacterial cell membrane by thickened cell wall and changed cell wall components. This article outlines the background to this developing issue with a focus on the mechanisms of resistance. The geographic prevalence, potential for continued spread, and proposed strategies for prevention and control are also discussed.

Introduction

Strains of *Staphylococcus aureus*, that exhibited reduced susceptibility to vancomycin, the remaining drug of choice for treating life-threatening infections due to *S. aureus*, were found in Japan in 1996 (Hiramatsu *et al.*, 1997b) and the United States in 1997 (Smith *et al.*, 1999). As strains with reduced susceptibility to vancomycin (vancomycin-resistant *S. aureus*, VRSA) are expected to emerge continuously and evolve, perhaps to full resistance, there is an urgent need to fully characterize them and conduct well-designed basic research on the mechanism of resistance and epidemiological studies.

    S. aureus is an aggressive pathogen and is a common cause of infections both in hospitals and in community. *S. aureus* bacteraemia was associated with an 82% mortality and the diseases often occurred in young adults before the introduction of antibiotics (Waldvogel, 1995). The advent of penicillin marked an historic breakthrough in the treatment of serious *S. aureus* infections. However, the benefits only lasted 10 to 15 years because of the widespread penicillin-resistance among *S. aureus* strains. In the late 1950's, during an influenza outbreak, many people died of what was again untreatable multi-resistant *S. aureus* infections. Fortunately, we overcame this problem in the late 50s and early 60s by developing new antibiotics. Vancomycin was developed in 1956 and its release was closely followed by beta-lactamase stable penicillins, represented by methicillin. Cloxacillin and cephalosporins, which have activity against *S. aureus,* also subsequently became available. Unfortunately, the emergence of methicillin-resistant *S. aureus* (MRSA) in the 1960's and 70's and its increase at an alarming rate through the world let us run short of efficient beta-lactam antibiotics to MRSA infection. Vancomycin became the only remaining antibiotic for treating serious *S. aureus* infections that were refractory to multiple antibiotics. Until recently there has been no *S. aureus* strains that did not respond to vancomycin, provided that vancomycin could reach the site of infection in sufficient amount. However, now that the strains with reduced susceptibility to vancomycin have emerged, it is expected that the therapeutic effect of vancomycin will decrease significantly, considering its limited tissue permeability.

Emergence of Vancomycin-resistant *Staphylococcus aureus* (VRSA)

Staphylococcus aureus strains with reduced susceptibility to vancomycin, for which the MIC of vancomycin is 8 µg/mL, have been defined in different terminology, such as VRSA, VISA and GISA, which correspond to different criteria in different countries (Table 1), causing confusion among infectious disease and microbiology specialist. The National Committee for Clinical Laboratory Standard (NCCLS) defines *S. aureus* strains requiring 8-16 µg/mL

Table 1. Glycopeptide breakpoints values for *Staphylococcus aureus*

	Vancomycin			Teicoplanin		
	Susceptible	Intermediate	Resistant	Susceptible	Intermediate	Resistant
NCCLS[a]	≤ 4 mg/L	8-16 mg/L	32 mg/L	8 mg/L	8-16 mg/L	32 mg/L
BSAC[b]	4 mg/L	-	8 mg/L	4 mg/L	-	8 mg/L
SRGA-M[c]	4 mg/L	-	8 mg/L	4 mg/L	-	8 mg/L

a, The National Committee for Clinical Laboratory Standards, USA.
b, British Society for Antimicrobial Chemotherapy.
c, Swedish Reference Group for Antibiotics.

of vancomycin for growth inhibition as VISA (Vancomycin-intermediate *S. aureus*), those requiring a concentration of 4 µg/mL as VSSA (Vancomycin-susceptible *S. aureus*), and those requiring concentrations 32 µg/mL as VRSA (Vancomycin-resistant *S. aureus*) (Anonymous, 2000). However, the British Society for Antimicrobial Chemotherapy (BSAC), and Swedish Reference Group for Antibiotics (SRGA-M) define those requiring concentrations 8 µg/mL as VRSA, and those requiring 4 µg/mL as VSSA (Anonymous 1998) (Olsson-Liljequist *et al.*, 1997). We use the term VRSA for the strains with vancomycin MIC of 8 µg/mL, as represented by Mu50, because Mu50 and majority of the strains subsequently reported to have vancomycin MIC of 8 µg/mL resisted prolonged vancomycin therapy (Hiramatsu *et al.*, 1997b; Ploy *et al.*, 1998; Sieradzki, *et al.*, 1999b; Smith, *et al.*, 1999; Chesneau *et al.*, 2000; Ferraz *et al.*, 2000; Hood, 2000; Kim *et al.*, 2000; Wong *et al.*, 2000; Boyle-Vavra *et al.*, 2001; Hageman *et al.*, 2001; Oliveira *et al.*, 2001; Paton *et al.*, 2001). Thus, they are clearly 'clinically resistant'. Therefore, we propose to use the term VRSA for those *S. aureus* strains with vancomycin MIC of 8 µg/mL .

In July 1996, we isolated the first VRSA strain from the infected incision site of a 4-month old baby patient who underwent heart surgery on pulmonary atresia. This was designated Mu50, and had a vancomycin MIC of 8 µg/mL The patient was treated with vancomycin (45 mg/kg per day for 29 days) with no resolution of the MRSA infection. Arbekacin (30 mg per day) was added to the regimen for 12 days and wound healing was observed. Twelve days after cessation of antibiotic treatment the infection recurred. Complete resolution of the infection required 23 days of sulbactam/ampicillin (300 mg per day) and arbekacin (30 mg per day) therapy followed by surgical debridement of a subcutaneous abscess at the incision site (Hiramatsu *et al.*, 1997b). The clinical course of the patient clearly showed, that it was the loss of susceptibility of the MRSA strain itself that determined therapeutic failure of vancomycin, because other antibiotic regimens were successful. Soon thereafter, a report of 2 additional cases from the United States was published in 1997 (Smith *et al.*, 1999), and other newly identified cases have been subsequently reported. To date, eighteen VRSA infection cases have been reported worldwide including Japan (Hiramatsu *et al.*, 1997b), United States of America (Sieradzki *et al.*, 1999b; Smith *et al.*, 1999; Boyle-Vavra *et al.*, 2001; Hageman *et al.*, 2001), UK (Hood, 2000; Paton *et al.*,2001), France (Ploy *et al.*, 1998; Chesneau *et al.*, 2000), South Korea (Kim *et al.*, 2000), South Africa (Ferraz *et al.*, 2000) and Brazil (Oliveira *et al.*, 2001). According to the literature, the majority of

Figure 1. Vancomycin killing curve for vancomycin-susceptible MRSA strain 87/20 and vancomycin hetero-resistant MRSA strain Mu3. The strains were cultivated in Mueller-Hinton broth at 37°C overnight and inoculated in the same pre-warmed broth containing various concentration of vancomycin. The test tubes were re-incubated 37°C with gentle shaking. The viable cells were counted after exposure to vancomycin at time 0, 24, 48, 72, 96 and 120 hours.

A

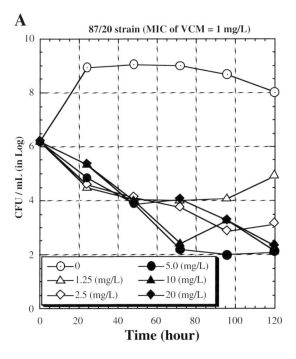

87/20 strain (MIC of VCM = 1 mg/L)

B

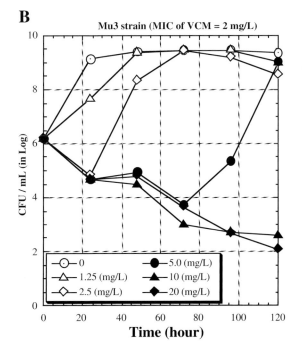

Mu3 strain (MIC of VCM = 2 mg/L)

the patients had received prolonged therapy with vancomycin in the months preceding VRSA infection. The most common clinical picture was the patient having an underlying illness, long-term exposure to vancomycin, and the fact that the vancomycin treatment was apparently unsuccessful. Subsequent worldwide reports of VRSA have confirmed that emergence of vancomycin resistance in *S. aureus* is a global issue.

Heterogeneously Vancomycin-resistant *Staphylococcus aureus* (hetero-VRSA)

In 1997, we reported another category of strains with reduced susceptibility to vancomycin: designated hetero-VRSA (Hiramatsu *et al.*, 1997a). The first hetero-VRSA strain Mu3 was isolated from a 64-year-old male patient who developed MRSA pneumonia 3 days after surgery for lung cancer. The pneumonia responded favorably to vancomycin (2 g per day) at the beginning, but exacerbated after 7 days of vancomycin therapy. The vancomycin therapy was continued for another 4 days with apparent exacerbation of the patient pneumonia. Finally, combination therapy with ampicillin/sulbactam (6 g per day) and arbekacin (0.2 g per day) was initiated, and resulted in complete clearing of the pneumonia and resolution of other signs of infection after 11 days treatment. Mu3 was isolated from sputum taken from patient on the last day of vancomycin therapy, at this point pneumonia was aggravating. The clinical course of vancomycin therapy failure clearly correlated with an *in vitro* time-killing curve of Mu3 strain, as illustrated in Figure 1. The Mu3 strain, despite its MIC of 2 µg/mL, required 10 µg/mL of vancomycin to inhibit completely the growth of 10^6 CFU/ml cell, whereas vancomycin-susceptible strain 87/20 (MIC of 1 µg/mL) was completely inhibited with 1.25 µg/mL of vancomycin. Mu3 showed continued cell death over 3 days with 5 µg/mL vancomycin but then started to re-grow. As illustrated in Figure 2, the population curve of Mu3 shows that it contains resistant sub-populations which are capable of growth in the presence of 5-9 µg/mL of vancomycin, which contrasts to those of the H1 strain that was isolated from a patient with MRSA pneumonia who responded favorably to vancomycin therapy. When the colonies that grew on the plates containing vancomycin of 5-9 µg/mL were picked up and re-analyzed with population analysis, their patterns were similar to that of VRSA strain Mu50 (Hiramatsu *et al.*, 1997a).

Hetero-VRSA has been mistakenly judged to be 'susceptible' to vancomycin on the basis of any of the MIC breakpoint systems in the world (Table 1). However, it spontaneously generates VRSA cells at a high frequency within its cell population, which are considered to be responsible for the clinical vancomycin treatment failure. We defined a hetero-VRSA strain as one that satisfies all of the following criteria: (i) its vancomycin MIC is less than 8 µg/mL when determined by NCCLS-based broth dilution methods (ii) it contains

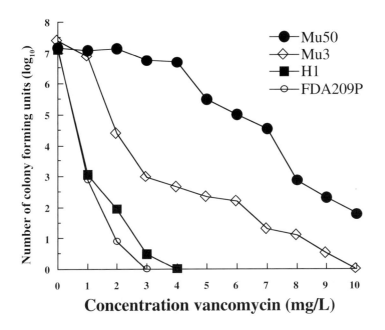

Figure 2. Analysis of vancomycin-resistant subpopulations of *S. aureus* strain Mu50 (vancomycin-resistant MRSA, VRSA), Mu3 (hetero-VRSA), H1 (vancomycin-susceptible MRSA), and FDA209P (vancomycin-susceptible MSSA). The analysis was performed by spreading 0.1 ml of the starting cell suspension and its serial dilutions onto BHI agar plates containing 1 to 10 µg/mL of vancomycin with 1µg/mL increments of vancomycin. The plates were then incubated at 37°C for 48 h before the number of colony forming units was counted. The number of resistant cells contained in 0.1 mL of the starting cell suspension was calculated and plotted semi-logarithmically.

subpopulations of cells resistant to higher concentration of vanomycin, including 4 µg/mL of vancomycin (iii) mutant strains with increased vancomycin resistance (a MIC of 8 µg/mL) can be obtained from the strain by one-step vancomycin-selection procedure with a frequency of 1 in 1,000,000 or greater (Hiramatsu *et al.*, 1997a; Aritaka *et al.*, 2001). The strains tend to be misjudged as vancomycin-susceptible because their MICs are by definition equal to or less than the NCCLS breakpoint (Table 1) for vancomycin susceptibility (4 µg/mL). Despite its low MIC, however, they are clearly distinguishable from ordinary vancomycin-susceptible strains by their high frequency of generating mutants of VRSA. This salient feature of producing resistant mutants makes hetero-resistant strain important not only as the precursor of VRSA, but also as the cause vancomycin-refractory infection in certain clinical settings as have been reported by us and other investigators (Hiramatsu *et al.*, 1997; Rotun *et al.*, 1999; Wong *et al.*, 2000; Trakulsomboon *et al.*, 2001; Ward *et al.*, 2001; Benquan *et al.*, 2002).

Scope of Current Problem of VRSA and h-VRSA

Until recently therapeutic failure or slow response of vancomycin against staphylococcal infection had been attributed to such factors such as its slower killing rate than beta-lactam antibiotics, limited penetration to infected tissues, the presence of foreign bodies, or the underlying compromised health status of the patient (Karchmer, 1991; Houlihan *et al.*, 1997). In some reports, vancomycin treatment failure occurs in as many as 40% of the treated cases (Small and Chambers, 1990; Moise and Schentag, 2000), and yet the *S. aureus* strains are reported to be susceptible to vancomycin. This apparently contradictory argument in vancomycin therapy against MRSA infection has remained untouched until the reports of the emergence and prevalence of VRSA and h-VRSA strains became popular in the last few years. It is particularly worrying that there are h-VRSA strains that appear to be susceptible by conventional testing but can generate resistance at a high frequency. These strains, which are currently difficult to detect in most clinical laboratories, may potentially give rise to strains homogeneously resistant to vancomycin *in vivo* (Aeschlimann *et al.*, 1999; Pfeltz *et al.*, 2000; Sugino *et al.*, 2000).

The unstable nature of vancomycin-resistant phenotype is another important issue for understanding the controversy about the low frequency of isolation of VRSA strains and the high rate of vancomycin therapeutic failure. The vancomycin-resistant phenotype tends to be lost after a while in the absence of selective antibiotic pressure (Boyle-Vavra *et al.*, 2000; Pfeltz *et al.*, 2000). Our experiments for looking at the stability of the vancomycin-resistant phenotype with the 16 clinical VRSA strains in the world showed that the vancomycin MICs for all the strains do return to 'susceptible' level (MIC, 2 μg/mL) after 10 to 84 days' serial passages in the drug-free medium. The hetero-VRSA phenotype of certain strains may also be unstable. The hetero-VRSA strains obtained from a VSSA strain N315 by *in vitro* selection with 1 μg/mL of vancomycin maintained the hetero-resistance phenotype for only one week during the serial daily passage in drug-free media (Cui, L. *et al.*, unpublished). These phenotypic reversions may explain many clinical and laboratory observations made to date about the frequency of isolation of vancomycin-resistant strains. The vancomycin-resistant isolates may not be isolated even from the blood of patients with persistent *S. aureus* bacteremia who are receiving vancomycin therapy, because the bacteria need to be cultured on non-selective media before a susceptibility test is performed. In addition to this problem in handling specimen, there is a difficulty with the detection of the reduced vancomycin susceptibility of staphylococci. Tenover *et al.* (2001) recently described the problem. They tested the ability of 130 laboratories worldwide to identify 7 different resistance phenotypes. Most phenotypes (including TEM-3 extended-spectrum beta-lactamase producing *Klebsiella pneumoniae*, VRE, MRSA, and beta-lactam resistant *Enterobacter cloacae*) were correctly identified by >90% of laboratories. On the other hand, only

64.1% of laboratories correctly identified penicillin-resistance in a *mefE*-positive *Streptococcus pneumoniae* isolate, and only 25.4% correctly identified vancomycin resistance of a *S. epidemidis* strain. Thus detection of vancomycin resistance in staphylococci is a problem.

It has been reported that VRSA is not obtained easily from VSSA strains (Sugino *et al.*, 2000; Bobin-Dubreux *et al.*, 2001). Therefore, recent isolates of clinical VRSA strains and the observed easy selection of VRSA from h-VRSA (see below) may reflect a special genetic background of certain clinical MRSA strains that are suited for the relatively stable expression of hetero-resistance. The VRSA strains of Japan and USA share the same MLST type that is different from predominant MLST types of MRSA (Enright *et al.*, 2002). This indicates that the different capability of *S. aureus* to acquire reduced susceptibility to vancomycin is due to the difference in the genetic backgrounds of MRSA clones. Indeed we observed that the 'susceptible' mutants obtained from 16 VRSA strains still express heterogeneous vancomycin resistance similar to that of Mu3. Exposure of these mutants to 4 μg/mL of vancomycin can select VRSA (8 μg/mL, MIC) strains at very high frequencies of 10^{-4} ~ 10^{-5} (Cui, L. *et al.*, unpublished). This indicates that, although a VRSA strain may not disseminate itself with a stable resistance phenotype and tends to return to hetero-VRSA status, it readily reverts to VRSA again if it is exposed to vancomycin.

Hetero-VRSA is a cause of great public health concern because it is considered to be the main source of emergence of staphylococcal vancomycin resistance. Hetero-VRSA is difficult to identify in clinical microbiological laboratories. Since its MIC is within the susceptible range (1-4 μg/mL), neither MIC nor disk-diffusion tests are diagnostic (Tenover *et al.*, 1998). Population analysis is the gold standard for the detection of hetero-VRSA, but it is too laborious for processing many samples (Hanaki and Hiramatsu, 2001). In addition to the considerable prevalence of hetero-VRSA in Japanese hospitals (Hiramatsu *et al.*, 1997a), a recent survey of MRSA strains from 13 Asian countries using population analysis has shown that hetero-VRSA is widely disseminated among MRSA strains in Korea, India, and Thailand (Suh *et al.*, 2002). It is predictable that VRSA would emerge much more easily on vancomycin therapy in those countries where hetero-VRSA strains are prevalent. Thus, hetero-VRSA constitutes a risk factor for vancomycin therapeutic failure in the hospital, and warrants periodical surveillance trials in those countries where MRSA is prevalent. The clinical and biological significance of hetero-VRSA is well reviewed elsewhere (Hiramatsu, 1998, 2001; Hiramatsu *et al.*, 1998, 1999, 2001).

Recently, a report from Brazil described an outbreak of VRSA in a state-owned hospital (Oliveira *et al.*, 2001). A total of five VRSA infection cases were reported in a short period of time (From December 1998 to January 1999), and four were from the same unit of the hospital. Indeed the PFGE pattern of VRSA strains was completely the same in four of the five strains. In view of the temporal development of clinical symptoms and location of the patients in the hospital, the authors suspected that VRSA strains were transmitted from patient to patient within the hospital. Alternatively, it might have been hetero-VRSA strains that disseminated in the hospital and determined the incidence of vancomycin-refractory infection. In fact, many hetero-VRSA strains were found in the Brazilian hospital in addition to the multiple VRSA isolates (Oliveira *et al.*, 2001).

Resistance Mechanism of VRSA

Vancomycin is a relatively large glycopeptide antibiotic (MW 1,450) derived from *Nocardia orientalis* (formerly known as *Streptomyces orientalis*) (Barna and Williams, 1984). It is active against most Gram-positive bacteria including streptococci, corynebacteria, clostridia, *Listeria*, and *Bacillus* species. Vancomycin does not interact with or block any enzyme involved in the cell-wall synthesis as do the beta-lactam antibiotics; it physically blocks the important substrates for the cell wall synthesizing machinery, the D-Ala-D-Ala termini of lipid II precursor and/or D-ala-D-ala-containing subunits of pre-existing cell wall (Watanakunakorn, 1981; Reynolds, 1989). Thereby it inhibits further processing of the substrate at later stages of cell-wall biosynthesis. As a result of binding to D-Ala-D-Ala groups in wall intermediates, vancomycin inhibits, apparently by steric hindrance, the formation of the backbone glycan chains (catalysed by peptidoglycan polymerase, transglycosylase) from the simple wall subunits as they are extruded through the cytoplasmic membrane. The subsequent transpeptidation reaction that imparts rigidity to the cell wall is also thus inhibited. It has been known that this unique mechanism of action, i.e. involving binding of the bulky inhibitor to the substrate outside the membrane so that the active sites of two enzymes (transglycosylase and transpeptidase) cannot align themselves correctly, renders the acquisition of resistance to the vancomycin more difficult than that to other antibiotic groups (Geisel *et al.*, 2001; Hiramatsu, 2001). Therefore, the mechanism(s) of vancomycin resistance in *S. aureus* has been the subject of intense research after the emergence of vancomycin-resistant *S. aureus* reported (Hiramatsu, 1998; Hiramatsu *et al.*, 1998, 1999, 2001; Geisel *et al.*, 2001; Hiramatsu, 2001).

Current Development on the Resistance Mechanism

In view of the mode of action of vancomycin, studies on the resistance mechanism in *S. aureus* have mostly focused on the machinery involved in the later stages of cell-wall biosynthesis. To date, several factors related with resistance mechanism in *S. aureus* are reported including: thickening of the cell wall (Hanaki *et al.*, 1998; Cui *et al.*, 2000), accumulation of surplus cell wall material (Sieradzki and Tomasz, 1999), reduced peptidoglycan cross-linking (Hanaki *et al.*, 1998; Boyle-Vavra *et al.*, 2001), inactivation of penicillin-binding protein 4 (PBP4) (Sieradzki *et al.*, 1999; Finan *et al.*, 2001), and/or other cell wall alterations, such as increased glycan-chain length (Komatsuzawa *et al.*, 2002). Furthermore, *pbp2* (Shlaes *et al.*, 1993; Moreira *et al.*, 1997), *sigB* (Bischoff *et al.*, 2001; Morikawa *et al.*, 2001), *ddh* (Milewski *et al.*, 1996; Boyle-Vavra *et al.*, 1997), *tcaR-B* operon (Brandenberger *et al.*, 2000), *vraR* (Kuroda *et al.*, 2000) were identified as factors involved in vancomycin or teicoplanin resistance. In view of these reports, it seems that several different mutations can cause vancomycin resistance in *S. aureus*. These alternations however do not affect the structure of vancomycin-binding target (D-Ala-D-Ala residue of peptidoglycan). Although the precise role of these individual factors on the resistance mechanism is unclear, they seem to create an obstacle that prevents vancomycin from reaching its target on the cytoplasmic membrane where nascent peptidoglycan biosynthesis is ongoing.

Thickening Cell Wall is the Common Determinant to Vancomycin Resistance in *S. aureus*

The thickened cell wall seems to play an important role in the vancomycin resistance mechanism by acting as a physical barrier that protects the cytoplasmic membrane from attack by vancomycin. Soon after our report on the isolation of the first VRSA, the Mu50 strain, we described the unusually thickened cell-wall of Mu50, as observed with electron micrography (Hanaki *et al.*, 1998). The same observation was made with the subsequently reported VRSA strains (Kim *et al.*, 2000; Oliveira *et al.*, 2001).

Recently, a study on the 16 clinical VRSA isolates reported from USA (Sieradzki *et al.*, 1999b; Smith *et al.*, 1999; Boyle-Vavra *et al.*, 2001), UK (Hood, 2000; Paton *et al.*, 2001), France (Ploy *et al.*, 1998), South Korea (Kim *et al.*, 2000), South Africa (Ferraz *et al.*, 2000) and Brazil (Oliveira *et al.*, 2001) revealed that all of the isolates showed distinctly thickened cell-walls as compared to vancomycin-susceptible *S. aureus* (Cui, L., *et al.*, unpublished). It seems that the thickening of the cell-wall is the common phenomenon for clinical VRSA strains. In the same study, vancomycin-susceptible mutants were obtained from each of 16 VRSA by serial daily passage of strains in drug-free medium, and vancomycin-resistant double mutants were selected from these vancomycin-susceptible mutants by one-step selection. The comparison of these triple sets of strains, parent VRSA, VS-mutants and VR-double mutants revealed that the degree of cell-wall thickness correlated well with the level of vancomycin resistance as measured by the MIC value (the correlation coefficient, 0.908) ($P < 0.001$). Thus the thickened cell-wall of VRSA strains return to normal with the loss of vancomycin resistance during drug-free passage, and it becomes thicker again following selection by vancomycin (Figure 3). Therefore, we proposed that

Strain Name	Mu50	Mu50-S	Mu50-SR	MI	MI-S	MI-SR
EM Photo						
Cell-Wall Thickness (nm)	35.02±4.01	24.45±7.80	34.19±4.96	32.03±3.60	24.13±5.54	32.03±6.75
VCM MIC (mg/L)	9	2	7	10	2	7

Strain Name	AMC11094	AMC11094-S	AMC11094-SR	BR1	BR1-S	BR1-SR
EM Photo						
Cell-Wall Thickness (nm)	34.57±4.82	24.04±7.43	32.35±4.95	34.92±7.18	20.67±5.69	34.84±5.47
VCM MIC (mg/L)	8	2	7	9	2	9

Figure 3. Transmission electron microscopy for the representative VRSA, its vancomycin-susceptible mutants (with suffix '-S') and vancomycin-resistant double mutants (with suffix '-SR'). The original cultures of VRSA were passaged daily in BHI broth until the MIC of vancomycin dropped to 2 or 3 μg/mL. The culture was then referred to as a vancomycin susceptible mutant (given with '-S') if the decreased MIC-value stayed the same during the subsequent passage of at least 10 days. The vancomycin-resistant double mutants were generated by one-step selection on the plates containing 4 μg/mL of vancomycin by spreading the cultures of susceptible mutants, and given name with suffix '-SR'. Preparation and examination of cells by transmission electron microscopy was performed as described previously (Cui *et al.*, 2000). Morphometric evaluation of the cell-wall thickness was performed using photographic images at 30,000 x final magnification, and the cell-wall thickness was measured as previously described (Cui *et al.*, 2000). Thirty cells of each strain with nearly equatorial cut surfaces were measured for the evaluation of cell-wall thickness, and results were expressed as mean value ± standard deviation (SD). Note that the change of cell-wall thickness was companied by the degrees of MIC of vancomycin. Reprinted with permission from Cui *et al.*, J. Clin. Microbiol. 2003. 41: 5-14.

the cell-wall thickness is the determinant for vancomycin resistance in *S. aureus*. One approach to prove this hypothesis is to prepare a syngeneic set of cells with different cell wall thickness, and analyze the difference in the vancomycin susceptibility. Since their genetic capability for vancomycin resistance is the same for each set of comparators, these experimentally created syngeneic cells provide a powerful analysis tool: since the only difference is in cell-wall thickness, this can be correlated with the difference in the vancomycin susceptibility.

We prepared the cells with different cell wall thickness from a single colony each of the two VRSA strains (Mu50 and MI, a secondly reported VRSA stain from USA), by incubating the colony in resting medium containing different nutrient components (Cui *et al.*, 2000). Five sorts of cells with different cell-wall thickness from each of Mu50 and MI strains were prepared. Then these preparations were compared for vancomycin susceptibility by observing the time required for them to regrowth (TRG) in the BHI media containing 30 μg/mL of vancomycin (Cui *et al.*, 2000). Results showed a close statistical linear-correlation between the cell-wall thickness and the level of vancomycin resistance as measured by TRG. The correlation coefficients were 0.976 (P< 0.01) and 0.993 (P< 0.01) for Mu50 and MI, respectively (Cui, L., unpublished).

Thickened Cell Wall Directly Contributes to Vancomycin Resistance by Retarding the Vancomycin Penetration Through the Cell-wall

In the experiment described above, vancomycin consumption in the media drops below 5-7 μg/mL before the regrowth of the cells is observed. Moreover, vancomycin is gradually consumed by *S. aureus* cells and the rate of consumption is high in cells with thick cell-wall and low in the cells with thin

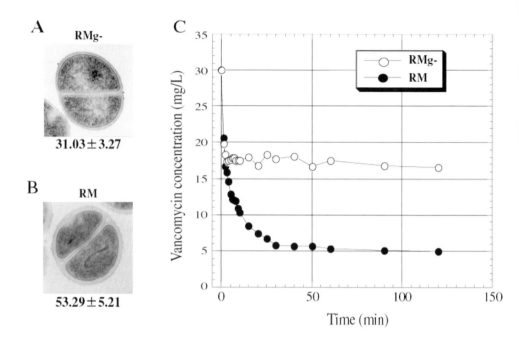

Figure 4. Preparations of cells with different cell-wall thickness and their consumption of vancomycin. The cells with thick cell-wall and thin cell-wall from a single colony of Mu50 was prepared in resting media (RMs) containing different component, which allows cell-wall biosynthesis without cell multiplication, as described previously (Cui *et al.*, 2000). Briefly, the cells of Mu50 from single colony were cultivated in BHI to an OD_{600} of 0.7, washed with RMg- twice, and incubated at 37°C for two hours with shaking in the medium with RM (RMg- plus 30 mM D-glucose) or RMg-, and then the cells with thick cell-wall (A) and thin cell-wall (B) were generated. The cell-wall thickness was shown under each picture in mean and SD values in nm (A and B). The preparations were washed twice with RMg-, and incubated in RMg- containing vancomycin of 30 μg/mL, and the vancomycin consuming by these cells were determined by measuring vancomycin concentration remained in medium using HPLC (C). Note that the cells with thick cell-wall consume more vancomycin than cells with thin cell-wall. The time required for binding of the maximun amount of vancomycin takes longer than 30 min in the cells with thick cell-wall, and only about 4 min in the cells with thin cell-wall.

cell-wall. However, the amount of vancomycin in the media was high enough to saturate all the peptidoglycan layers of the cells with thickened cell-wall, and inhibit their *de novo* cell-wall synthesis. To elucidate the reason for this refractoriness, i.e. the vancomycin-mediated inhibition of cell-wall synthesis of the cells with thickened cell-wall, we performed the following experiment. To study the effect of cell-wall thickness on the rate of vancomycin consumption, syngeneic cells with thick cell-wall and thin cell-wall were prepared from Mu50 in RM and RMg-, respectively (Cui *et al.*, 2000). Then vancomycin consumption by these cells was compared by quantifying the vancomycin concentration in the media by HPLC. Figure 4 shows that vancomycin concentration drops very sharply within the first two minutes for both the cells. However, the consumption by the cells with thin cell-walls nearly stops after 4 min (prepared in RMg-), while for the cells having thick cell walls (prepared in RM), gradual consumption continues for 30 to 60 min, until finally 25 µg/mL is consumed. This indicates that vancomycin molecules gradually perpetrate through the cell-wall until they saturate all the cell-wall layers before they finally reach the membrane to inhibit cell-wall synthesis. In other words, more time is required for vancomycin to stop cell-wall synthesis in the thick cell-wall cells. To prove that, the cells exposed to vancomycin for different time length were harvested, to measure their cell multiplication capability. The cells were inoculated in pre-warmed drug-free BHI medium to monitor the time required for growth (TRG), which reflects the degree of vancomycin-mediated cell-wall synthesis inhibition. As shown in Figure 5, the cells with thick cell walls (Figure 5B) started to grow within 5 hours eventhough the cells had been previously exposed to vancomycin for different lengths of time. The cells with thin cell-walls (Figure 5A) that had been subjected to prior exposure to vancomycin, did not start to grow more than 5 hours after inoculation. Most interestingly, the temporal order of TRG correlated with the length of the time of pre-exposure of the cells to vancomycin prior to transfer to the BHI medium (Figure 4). It is noticed that vancomycin-mediated inhibition of cell multiplication gradually proceeds as seen in Figure 4 and the longer the exposure to vancomycin, the more retarded the resumption of cell multiplication. Therefore, the inhibition of cell-multiplication by vancomycin is clearly reduced by the thickening of cell wall.

A further experiment showed that the observed reduction of the vancomycin effect on the cell multiplication occurs at the cell-wall biosynthesis level. The cells with thick cell walls still continually biosynthesize the cell-wall in the presence of vancomycin, whilst that in the cells with thin cell walls is soon inhibited as observed with the *N*-acetyl-glucosamine and D-glucose incorporation into the cell-wall in the presence of the vancomycin (Cui *et al.*, unpublished). These results indicate that the molecular mechanism of vancomycin resistance provided by the thickened cell wall is due to the increased 'affinity trapping' of vancomycin molecules, retarding the traffic of vancomycin through thickened cell-wall layers.

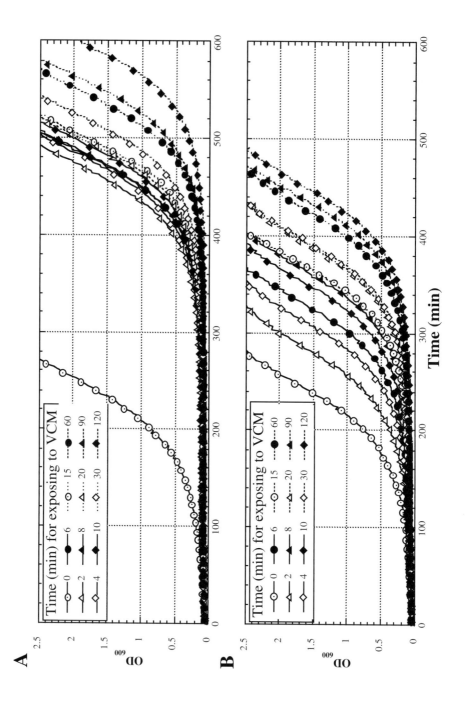

Figure 5. Comparison of the cell multiplication capability of cells with thick cell walls and thin cell walls following -preexposure to vancomycin for different lengths of time. A 0.5 mL portion of the cell preparations with thick cell wall (B) and thin cell wall (A) (see Figure 4) sampled for determination of the vancomycin concentration were inoculated in 10 mL pre-warmed BHI medium, and were further cultivated in the photo-recording incubator (TN-261, ADVANTEC, Tokyo, Japan) at 37°C with shaking to measure the time required for re-growth (TRG) by monitoring OD_{600} value in every two min. Note that the cells with thick cell wall (B) all re-grow within 5 hours regardless of the cells time for pre-exposure to vancomycin, while the cells with thin cell-wall (A) exposed to vancomycin did not start to grow until 5 hours after inoculation. The temporal order of re-growth of both cell types reflects the length of time the cell exposed to vancomycin before being transferred to the BHI broth.

Alteration of Cell-wall Component Contributes to the Vancomycin Resistance

In addition to the cell-wall thickening, the alteration of a cell-wall component is also correlated with the reduced susceptibility to vancomycin in *S. aureus*. An increased non-amidated murein monomer ratio in the cell-wall peptidoglycan (Hanaki *et al.*, 1998; Boyle-Vavra *et al.*, 2001), and an increased glycan-chain length (Komatsuzawa *et al.*, 2002) is correlated with vancomycin resistance. Our previous report showed that the unit weight of purified peptidoglycan with high content of non-amidated muropeptides consumes more vancomycin molecules than that with a low content of non-amidated muropeptides (Cui *et al.*, 2000). The non-amidated muropeptide is considered to directly contribute to the increased consumption of vancomycin due to its greater binding affinity to vancomycin than that of the amidated counterpart, and indirectly by decreasing peptidoglycan cross linkage (non-amidated murein monomer is an inefficient substrate for PBPs to perform cross-bridge formation). It is noteworthy in this regard that a significant decrease in the cell-wall cross-linkage is also observed in an *in vitro* vancomycin-resistant mutant strain (Sieradzki and Tomasz, 1997). However, the under cross-linkage of cell-wall peptidoglycan does not make the cell resistant to vancomycin if it is not accompanied by a thickening of cell wall, as was demonstrated using the *femC* mutant strain BB589. The production of non-amidated muropeptide, a characteristic of the *femC* mutant BB589, is due to the insertion of a Tn*551* in the vicinity of glutamine synthetase (GS) repressor gene (*glnR*), thereby exerting a polar effect on the transcription of glutamine synthetase gene (*glnA*) (Gustafson *et al.*, 1994). Its contribution to vancomycin resistance becomes effective only when the strain has thickened cell wall (Cui *et al.*, 2000). Therefore, cell wall thickening is considered to be the pre-requisite for vancomycin resistance.

Multiple Pathways are Involved in Achieving the Thickening of Cell-wall in VRSA

Mu50 cells have an enhanced rate of incorporation of *N*-acetylglucosamine (GlcNAc) into the cell wall, and an increased pool size of the cytoplasmic murein monomer precursor (UDP-*N*-acetylmuramyl-pentapeptide). Another pathway for cell-wall biosynthesis, by supplying UDP-GlcNAc from Embden-Meyerhof pathway at the Fru-6-P to generate GlcN-6-P, is also enhanced in Mu50 (Hanaki *et al.*, 1998; Hanaki *et al.*, 1998). This enhancement is considered to be due to the increased GlmS enzyme activity as well as GS activities (Hiramatsu *et al.*, 1998; Cui *et al.*, 2000). It also fits the observation that a significantly greater amount of glucose is taken up and incorporated in the cell wall of Mu50 as compared to vancomycin susceptible control strains (Cui *et al.*, 2000). These observations point to the view that Mu50 produces more peptidoglycan than vancomycin susceptible control strains, and explain the characteristic cell-wall thickening of Mu50 observed by transmission electron microscopy (Hiramatsu *et al.*, 1999; Hiramatsu, 2001). As mentioned previously, thickening of the cell wall is a common characteristic of vancomycin resistance in all VRSA strains isolated so far (see above), but the mechanism of cell wall thickening is not uniform. The Michigan (MI) strain does not exhibit an enhanced cell-wall synthesis as Mu50. Instead, the strain has remarkably suppressed autolytic activity, which is absent in the vancomycin-susceptible mutants (Cui, L., unpublished). This indicates that Michigan strain thickens the cell wall not by making it in excess but by retarding cell-wall turnover. Therefore, there are various genetic mechanisms for vancomycin resistance, but all, in their effect, are directed to the common phenotype of thickened cell-wall peptidoglycan layers that serves as the direct obstacle for glycopeptide action.

A complex Relationship Between Glycopeptide and Beta-lactam Resistance

Glycopeptides and beta-lactams are both inhibitors of cell-wall synthesis. Therefore, it is reasonable to suspect that there is an intimate relationship between glycopeptide and beta-lactam resistance expressed by VRSA strains. A 'seesaw' phenomenon between beta-lactam and vancomycin resistance has been proposed based on the observation with *in vitro*-trained VRSA strains (Sieradzki and Tomasz, 1997). In the series of *in vitro* resistant mutants, any decrease in methicillin resistance is accompanied by a rise in vancomycin resistance with a gradual decrease in cross-linkage in the cell-wall peptidoglycan (Sieradzki and Tomasz 1999). Decrease of peptidoglycan cross-linkage leads to decreased durability of the cell wall to high internal cell osmotic pressure. Thus, it certainly increases the vulnerability of the cell wall to beta-lactam action. On the other hand, since a reduction of peptidoglycan cross-linkage results in an increase of D-alanyl-D-alanine residues in the cell-wall, it contributes to vancomycin resistance by enhancing the 'affinity trapping' of vancomycin molecules (Cui *et al.*, 2000). This seems to be the reason why beta-lactam resistance and vancomycin resistance appear to be inversely proportionate in this particular experimental system. However, in the clinical VRSA strains analyzed by us, the relationship between beta-lactam and vancomycin resistance is not that simple. As known from our recent experiment that a typical seesaw phenomenon is seen only with four of 16 pairs of VRSA and its vancomycin-susceptible mutants (Cui *et al.*, unpublished).

Penicillin-binding protein 2 (PBP2) expression is essential for high-level methicillin resistance (Pinho *et al.*, 1997), while its overexpression raises glycopeptide resistance especially against teicoplanin (Hanaki *et al.*, 1998) (see below). PBP2 expression is enhanced in Mu50 but is decreased significantly in some strains such as MI, NJ, and PC (Cui, L., unpublished). We have also shown that certain genetic alterations responsible for high methicillin resistance are associated with an increase in glycopeptide resistance whereas others are not (Hiramatsu *et al.*, 1999). Therefore, the mechanisms relating beta-lactam and glycopeptide resistance are interconnected in *S. aureus* cell physiology, however this is a complicated issue and may well differ from strain to strain: dependent upon the strain genetic background and the history of exposure of the strains to these antibiotics.

Additional Mechanism for Teicoplanin Resistance

Teicoplanin resistance also decreases with the diminution of cell-wall thickness. In some strains, such as MI, teicoplanin resistance is completely lost in the vancomycin susceptible mutants (Cui, L. *et al.*, unpublished). It is reasonable to assume that teicoplanin, by having the same target-binding activity as vancomycin is also affected by the 'affinity-trapping' mechanism described for vancomycin resistance, i.e. its penetration into the cell wall is prevented by thickened peptidoglycan mesh (Cui *et al.*, 2000). However, in addition to the resistance provided by thickened peptidoglycan layers, other mechanisms are likely to be important for teicoplanin resistance. Overproduction of penicillin-binding protein 2 (PBP2) is observed in teicoplanin-resistant *S. aureus* clinical strains (Shlaes *et al.*, 1993). It is observed in both Mu3 (hetero-VRSA) and Mu50. Experimental overproduction of PBP2 in a glycopeptide-susceptible *S. aureus* strain raised teicoplanin MIC significantly from 1 to 8 µg/mL, but that of vancomycin only slightly from 1 to 2 µg/mL (Hanaki *et al.*, 1998). This coincides well with the observation that the susceptible mutant Mu50-S, having a decreased cell-wall thickness, still expresses teicoplanin resistance (MIC, 7 µg/mL). This is also consistent with the observation that the Michigan strain, having very little PBP2 expression, lost teicoplanin resistance completely following the loss of cell-wall thickness.

The explanation for why teicoplanin and vancomycin are differentially affected by PBP2 overproduction has yet to be elucidated, but it probably reflects their preferential mode of inhibition of *S. aureus* peptidoglycan synthesis. Overproduction of PBP2 does not increase cell-wall thickness, but it increases cross-linkage of cell-wall peptidoglycan (Murakami *et al.*, 1999). A more substantial rise in teicoplanin resistance than vancomycin resistance due to PBP2 overproduction can be explained by postulating that teicoplanin, inhibits the cross-linkage formation step of peptidoglycan synthesis more effectively than the polymerization step of nascent peptidoglycan chain. In agreement with this is the observation that beta-lactam antibiotics, such as ampicillin and carbapenems, act highly synergistically with teicoplanin against hetero-VRSA and VRSA strains (Hanaki and Hiramatsu, 1999), whereas the combination of the beta-lactams and vancomycin is antagonistic in some clinical strain (Aritaka *et al.*, 2001).

Prevention and Control of Emergence of VRSA

Although clinical experience of dealing with VRSA infection has been limited, prevention and control measures for VRSA infections has been proposed by some authors (Anonymous, 1997; Hiramatsu, 1998; Wenzel and Edmond, 1998; Hiramatsu, 2001; Oliveira *et al.*, 2001). Based on the experience of MRSA, the major strategies for prevention and controlling of the emergence of VRSA would be the cautious use of vancomycin and infection control practice. Recently, Oliveira *et al.* reported an outbreak of VRSA in a Brazilian hospital. They warn us with their experience that VRSA may be transmissible if adequate hospital infection control measures are not taken (Oliveira *et al.*, 2001). Curiously, we showed that some beta-lactam antibiotics selected vancomycin hetero-resistanct MRSA (Hiramatsu , 2001). Therefore, strict control of both vancomycin and beta-lactams use should be important to prevent emergence of VRSA. In view of the resistance mechanism of VRSA and its high biological cost we should be able to control emergence of VRSA by using the antibiotics sparingly.

References

Aeschlimann, J. R., Hershberger, E., and Rybak, M.J. 1999. Analysis of vancomycin population susceptibility profiles, killing activity, and postantibiotic effect against vancomycin-intermediate *Staphylococcus aureus*. Antimicrob. Agents Chemother. 43: 1914-1918.

Anonymous. 1997. Interim guidelines for prevention and control of staphylococcal infection associated with reduced susceptibility to vancomycin. Morb Mortal Wkly Rep. 46: 626-628, 635.

Anonymous. 1998. Revised guidelines for the control of methicillin-resistant *Staphylococcus aureus* infection in hospitals- British Society for Antimicrobial Chemotherapy. J. Hosp. Infect. 39: 253-290.

Anonymous. 2000. National Committee for Clinical laboratory Standard. Methods for dilution antimicrobial susceptibility tests for bacteria that grow aerobically, 5th edn. Approved standard M7-A5. Villanova, Pa. NCCLS.

Aritaka, N., Hanaki, H., Cui, L., and Hiramatsu, K. 2001. Combination effect of vancomycin and beta-lactams against a *Staphylococcus aureus*, strain Mu3, with heterogeneous resistance to vancomycin. Antimicrob. Agents Chemother. 45: 1292-1294.

Barna, J., and Williams, D. 1984. The structure and mode of action of glycopeptide antibiotics of the vancomycin group. Ann. Rev. Microbiol. 38: 339-357.

Benquan, W., Yingchun, T., Kouxing, Z., Tiantuo, Z., Jiaxing, Z., and Shuqing, T. 2002. *Staphylococcus* heterogeneously resistant to vancomycin in China and antimicrobial activities of imipenem and vancomycin in combination against it. J. Clin. Microbiol. 40: 1109-1112.

Bischoff, M., Roos, M., Putnik, J., Wada, A., Glanzmann, P., Giachino, P., Vaudaux, P., and Berger-Bächi, B. 2001. Involvement of multiple genetic loci in *Staphylococcus aureus* teicoplanin resistance. FEMS Microbiol. Lett. 194: 77-82.

Bobin-Dubreux, S., Reverdy, M.E. Nervi, C., Rougier, M., Bolmstrom, A., Vandenesch, F., and Etienne, J. 2001. Clinical isolate of vancomycin-heterointermediate *Staphylococcus aureus* susceptible to methicillin and *in vitro* selection of a vancomycin- resistant derivative. Antimicrob. Agents Chemother. 45: 349-352.

Boyle-Vavra, S., Berke, S.K., Lee, J.C., and Daum, R.S. 2000. Reversion of the glycopeptide resistance phenotype in *Staphylococcus aureus* clinical isolates. Antimicrob. Agents Chemother. 44: 272-277.

Boyle-Vavra, S., de Jonge, B.L., Ebert, C.C., and Daum, R.S. 1997. Cloning of the *Staphylococcus aureus ddh* gene encoding NAD$^+$-dependent D-lactate dehydrogenase and insertional inactivation in a glycopeptide-resistant isolate. J. Bacteriol. 179: 6756-6763.

Boyle-Vavra, S., Labischinski, H., Ebert, C.C., Ehlert, K., and Daum, R.S. 2001. A spectrum of changes occurs in peptidoglycan composition of glycopeptide-intermediate clinical *Staphylococcus aureus* isolates. Antimicrob. Agents Chemother. 45: 280-287.

Brandenberger, M., Tschierske, M., Giachino, P., Wada, A., and Berger-Bächi, B. 2000. Inactivation of a novel three-cistronic operon *tcaR-tcaA-tcaB* increases teicoplanin resistance in *Staphylococcus aureus*. Biochim. Biophys. Acta. 1523: 135-139.

Chesneau, O., Morvan, A., and Solh, N.E. 2000. Retrospective screening for heterogeneous vancomycin resistance in diverse *Staphylococcus aureus* clones disseminated in French hospitals. J. Antimicrob. Chemother. 45: 887-890.

Cui, L., Murakami, H., Kuwahara-Arai, K., Hanaki, H., and Hiramatsu, K. 2000. Contribution of a thickened cell wall and its glutamine nonamidated component to the vancomycin resistance expressed by *Staphylococcus aureus* Mu50. Antimicrob. Agents Chemother. 44: 2276-2285.

Enright, M.C., Robinson, D.A., Randle, G., Feil, E.J., Grundmann, H., and Spratt, B.G. 2002. The evolutionary history of methicillin-resistant *Staphylococcus aureus* (MRSA). Proc. Natl. Acad. Sci. USA. 99(11):7687-7692.

Ferraz, V., Duse, A., Kassel, M., Black, A., Ito, T., and Hiramatsu, K. 2000. Vancomycin-resistant *Staphylococcus aureus* occurs in South Africa. S. Afr. Med. J. 90: 1108-1009.

Finan, J., Archer, G., Pucci, M., and Climo, M. 2001. Role of penicillin-binding protein 4 in expression of vancomycin resistance among clinical isolates of oxacillin-resistant *Staphylococcus aureus*. Antimicrob. Agents Chemother. 45: 3070-3075.

Geisel, R., Schmitz, F-J., Fluit, A., and Labischinski, H. 2001. Emergence, mechanism, and clinical implications of reduced glycopeptide susceptibility in *Staphylococcus aureus*. Eur. J. Clin. Microbiol. Infect. Dis. 20: 685-697.

Gustafson, J., Strassle, A., Hachler, H., Kayser, F.H., and Berger-Bächi, B. 1994. The *femC* locus of *Staphylococcus aureus* required for methicillin resistance includes the glutamine synthetase operon. J. Bacteriol. 176: 1460-1467.

Hageman, J., Pegues, D., Jepson, C., Bell, R., Guinan, M., Ward, K., Cohen, M., Hindler, J., Tenover, F., McAllister, S., Kellum, M., and Fridkin, S. 2001. Vancomycin-Intermediate *Staphylococcus aureus* in a home health-car patient. Emerg. Infect. Dis. 7: 1023-1025.

Hanaki, H., and Hiramatsu, K. 1999. [Combination effect of teicoplanin and various antibiotics against hetero-VRSA and VRSA]. Kansenshogaku Zasshi 73: 1048-1053.

Hanaki, H., and Hiramatsu, K. 2001. Detection methods of glycopeptide-resistant *Staphylococcus aureus* I: Susceptibility testing. In: Antibiotic Resistance Methods and Protocols. S. H. Gillespie, ed. Humana Press, Totowa, New Jersey. p. 85-92.

Hanaki, H., Kuwahara-Arai, K., Boyle-Vavra, S., Daum, R.S., Labischinski, H., and Hiramatsu, K. 1998. Activated cell-wall synthesis is associated with vancomycin resistance in methicillin-resistant *Staphylococcus aureus* clinical strains Mu3 and Mu50. J. Antimicrob. Chemother. 42: 199-209.

Hanaki, H., Labischinski, H., Inaba, Y., Kondo, N., Murakami, H., and Hiramatsu, K. 1998. Increase in glutamine-non-amidated muropeptides in the peptidoglycan of vancomycin-resistant *Staphylococcus aureus* strain Mu50. J. Antimicrob. Chemother. 42: 315-320.

Hiramatsu, K. 1998. Vancomycin resistance in staplylococci. Drug Resist. 1:135-150.

Hiramatsu, K. 2001. Vancomycin-resistant *Staphylococcus aureus*: a new model of antibiotic resistance. Lancet Infect. Dis. 1: 147-155.

Hiramatsu, K., Aritaka, N., Hanaki, H., Kawasaki, S., Hosoda, Y., Hori, S., Fukuchi, Y., and Kobayashi, I. 1997a. Dissemination in Japanese hospitals of strains of *Staphylococcus aureus* heterogeneously resistant to vancomycin. Lancet. 350: 1670-1673.

Hiramatsu, K., Cui, L., Kuroda, M., and Ito, T. 2001. The emergence and evolution of methicillin-resistant *Staphylococcus aureus*. Trends Microbiol. 9:486-493.

Hiramatsu, K., Hanaki, H., Ino, T., Yabuta, K., Oguri, T., and Tenover, F.C. 1997b. Methicillin-resistant *Staphylococcus aureus* clinical strain with reduced vancomycin susceptibility. J. Antimicrob. Chemother. 40: 135-136.

Hiramatsu, K., Ito, T., and Hanaki, H. 1998. Glycopeptide resistance in staphylococci. Curr. Opin. Infect. Dis. 11: 653-658.

Hiramatsu, K., Ito, T., and Hanaki, H. 1999. Mechanisms of methicillin and vancomycin resistance in *Staphylococcus aureus*. In: Bailliere's Clinical Infectious Diseases. W. R. G. Finch, R. J., ed. vol. 5. Bailliere Tindall, London. p.211-242.

Hood, J., Edwards, G.F.S., Cosgrove, B., Curran, E., Morrison, D. and Gemmell, C.G. 2000. Vancomycin-intermediate *Stapyylococcus aureus* at a Scottish hospital. J. Infect. 40: A11.

Houlihan, H.H., Mercier, R.C., and M.J. Rybak. 1997. Pharmacodynamics of vancomycin alone and in combination with gentamicin at various dosing intervals against methicillin-resistant *Staphylococcus aureus*-infected fibrin-platelet clots in an *in vitro* infection model. Antimicrob. Agents Chemother. 41: 2497-2501.

Karchmer, A.W. 1991. *Staphylococcus aureus* and vancomycin: the sequel. Ann. Intern. Med. 115: 739-741.

Kim, M.N., Pai, C.H., Woo, J.H., Ryu, J.S., and Hiramatsu, K. 2000. Vancomycin-intermediate *Staphylococcus* aureus in Korea. J. Clin. Microbiol. 38: 3879-3881.

Komatsuzawa, H., Ohta, K., Yamada, S., Ehlert, K., Labischinski, H., Kajimura, J., Fujiwara, T., and Suga, I.M. 2002. Increased glycan chain length distribution and decreased susceptibility to moenomycin in a vancomycin-resistant *Staphylococcus aureus* mutant. Antimicrob. Agents Chemother. 46: 75-81.

Kuroda, M., Kuwahara-Arai, K., and Hiramatsu, K. 2000. Identification of the up- and down-regulated genes in vancomycin- resistant *Staphylococcus aureus* strains Mu3 and Mu50 by cDNA differential hybridization method. Biochem. Biophys. Res. Commun. 269: 485-490.

Milewski, W.M., Boyle-Vavra, S., Moreira, B., Ebert, C.C., and Daum, R.S. 1996. Overproduction of a 37-kilodalton cytoplasmic protein homologous to NAD^+-linked D-lactate dehydrogenase associated with vancomycin resistance in *Staphylococcus aureus*. Antimicrob. Agents Chemother. 40: 166-172.

Moise, P., and Schentag, J. 2000. Vancomycin treatment failures in *Staphylococcus aureus* lower respiratory tract infections. Int. J. Antimicrob. Agents 16 Suppl. 1: S31-S34.

Moreira, B., Boyle-Vavra, S., deJonge, B.L., and Daum, R.S. 1997. Increased production of penicillin-binding protein 2, increased detection of other penicillin-binding proteins, and decreased coagulase activity associated with glycopeptide resistance in *Staphylococcus aureus*. Antimicrob. Agents Chemother. 41: 1788-1793.

Morikawa, K., Maruyama, A., Inose, Y., Higashide, M., Hayashi, H., and Ohta, T. 2001. Overexpression of sigma factor, varsigma(B), urges *Staphylococcus aureus* to thicken the cell wall and to resist beta-lactams. Biochem. Biophys. Res. Commun. 288: 385-389.

Murakami, H., Kuwahara-Arai, K., Tsuchihashi, H., Hanaki, H. and Hiramatsu, K. 2002. Overproduction of PBP2 is associated with vancomycin- and teicoplanin-resistance in *Staphylococcus aureus*. J. Antimicrob. Chemother. 44: Suppl. A: 167

Oliveira, G.A., Aquila, A.M., Masiero, R.A., Levy, C., Gomes, S.G., Cui, L., Hiramatsu, K., and Mamizuka, E.M.. 2001. Isolation in Brazil of nosocomial *Staphylococcus aureus* with reduced susceptibility to vancomycin. Infect. Control Hosp. Epidemiol. 22: 443-448.

Olsson-Liljequist, B., Larsson, P., Walder, M., and Miorner, H. 1997. Antimicrobial susceptibility testing in Sweden. III. Methodology for susceptibility testing. Scand. J. Infect. Dis. 105 (Suppl.): 13-23.

Paton, R., Snell, T., Emmanuel, F., and Miles, R.. 2001. Glycopeptide resistance in an epidemic strain of methicillin-resistant *Staphylococcus aureus*. J. Antimicrob. Chemother. 48: 941-942.

Pfeltz, R.F., Singh, V.K., Schmidt, J.L., Batten, M.A., Baranyk, C.S., Nadakavukaren, M.J., Jayaswal, R.K., and Wilkinson, B.J. 2000. Characterization of passage-selected vancomycin-resistant *Staphylococcus aureus* strains of diverse parental backgrounds. Antimicrob. Agents Chemother. 44: 294-303.

Pinho, M., Ludovice, A., Wu, S., and de Lencastre, H. 1997. Massive reduction in methicillin resistance by transposon inactivation of the normal PBP2 in a methicillin-resistant strain of *Staphylococcus aureus*. Microb. Drug Resist. 3: 409-413.

Ploy, M.C., Grelaud, C., Martin, C., de Lumley, L., and Denis, F. 1998. First clinical isolate of vancomycin-intermediate *Staphylococcus aureus* in a French hospital. Lancet. 351: 1212.

Reynolds, P. 1989. Structure, biochemistry and mechanism of action of glycopeptide antibiotics. Eur. J. Clin. Microbiol. Infect. Dis. 8: 943-950.

Rotun, S., McMath, V., Schoonmaker, D., Maupin, P., Tenover, F., Hill, B., and Ackman, D. 1999. *Staphylococcus aureus* with reduced susceptibility to vancomycin isolated from a patient with fatal bacteremia. Emerg. Infect. Dis. 5: 147-149.

Shlaes, D.M., Shlaes, J.H., Vincent, S., Etter, L., Fey, P.D., and Goering, R.V. 1993. Teicoplanin-resistant *Staphylococcus aureus* expresses a novel membrane protein and increases expression of penicillin-binding protein 2 complex. Antimicrob. Agents Chemother. 37: 2432-2437.

Sieradzki, K., Pinho, M.G., and Tomasz, A. 1999a. Inactivated pbp4 in highly glycopeptide-resistant laboratory mutants of *Staphylococcus aureus*. J. Biol. Chem. 274: 18942-18946.

Sieradzki, K., Roberts, R.B., Haber, S.W., and Tomasz, A. 1999b. The development of vancomycin resistance in a patient with methicillin-resistant *Staphylococcus aureus* infection. N. Engl. J. Med. 340: 517-523.

Sieradzki, K., and Tomasz, A. 1997. Inhibition of cell wall turnover and autolysis by vancomycin in a highly vancomycin-resistant mutant of *Staphylococcus aureus*. J. Bacteriol. 179: 2557-2566.

Sieradzki, K., and Tomasz, A. 1999. Gradual alterations in cell wall structure and metabolism in vancomycin- resistant mutants of *Staphylococcus aureus*. J Bacteriol 181:7566-70.

Small, P., and Chambers, H. 1990. Vancomycin for *Staphylococcus aureus* endocarditis in intravenous drug users. Antimicrob. Agents Chemother. 34: 1227-1231.

Smith, T.L., Pearson, M.L., Wilcox, K.R., Cruz, C., Lancaster, M.V., Robinson-Dunn, B., Tenover, F.C., Zervos, M.J., Band, J.D., White, E., and Jarvis, W.R. 1999. Emergence of vancomycin resistance in *Staphylococcus aureus*. Glycopeptide-intermediate *Staphylococcus aureus*. N. Engl. J. Med. 340: 493-501.

Sugino, Y., Iinumab, Y., Ichiyama, S., Ito, Y., Ohkawa, S., Nakashima, N., Shimokata, K., and Hasegawa, Y. 2000. *In vivo* development of decreased susceptibility to vancomycin in clinical isolates of methicillin-resistant *staphylococcus aureus*. Diagn. Microbiol. Infect. Dis. 38: 159-167.

Suh, J.Y., Kapi, M., Ito, T., Jung, S.I., Kim. W.S., Peck, N.Y., Lee, K., Hiramatsu, K. and Song, J.H. 2002. Prevalence of heterogenous vancomycin-resistant *Staphylococcus* aureus (h-VRSA) among methicillin-resistant strains from Asian countries. 10th International Symposium on Staphylococci and Staphylococcal Infections. Tsukuba, Japan, Abstract ISSSI-284-Abs-01, p.143.

Tenover, F.C., Lancaster, M.V., Hill, B.C., Steward, C.D., Stocker, S.A., Hancock, G.A, O'Hara, C.M., McAllister, S.K., Clark, N.C., and Hiramatsu, K. 1998. Characterization of staphylococci with reduced susceptibilities to vancomycin and other glycopeptides. J. Clin. Microbiol. 36: 1020-1027.

Tenover, F.C., Mohammed, M.J., Stelling, J., O'Brien, T., and Williams, R. 2001. Ability of laboratories to detect emerging antimicrobial resistance: proficiency testing and quality control results from the World Health Organization's external quality assurance system for antimicrobial susceptibility testing. J. Clin. Microbiol. 39: 241-250.

Trakulsomboon, S., Danchaivijitr, S., Rongrungruang, Y., Dhiraputra, C., Susaemgra, W., Ito, T., and Hiramatsu, K. 2001. First report of methicillin-resistant *Staphylococcus aureus* with reduced susceptibility to vancomycin in Thailand. J. Clin. Microbiol. 39: 591-595.

Waldvogel, F.A. 1995. *Staphylococcus aureus* (including toxic shock syndrome). In: Principles and Practice of Infectious Diseases., Fourth ed, vol. 2. G. L. Mandell, J. E. Bennett, and R. Dolin, eds. Churchill Livingstone, New York. p. 1757-777.

Ward, P., Johnson, P., Grabsch, E., Mayall, B., and Grayson, M. 2001. Treatment failure due to methicillin-resistant *Staphylococcus aureus* (MRSA) with reduced susceptibility to vancomycin. Med. J. Aust. 175: 480-483.

Watanakunakorn, C. 1981. The antibacterial action of vancomycin. Rev. Infect. Dis. 3 Suppl.: S210-S215.

Wenzel, R., and Edmond, M. 1998. Vancomycin-resistant *Staphylococcus aureus*: infection control considerations. Clin. Infect. Dis. 27:245-249.

Wong, S.S., Ng, T.K., Yam, W.C., Tsang, D.N., Woo, P.C., Fung, S.K., and Yuen, K.Y. 2000. Bacteremia due to *Staphylococcus aureus* with reduced susceptibility to vancomycin. Diagn. Microbiol. Infect. Dis. 36: 261-268.

From: *MRSA: Current Perspectives*
Edited by: A.C. Fluit and F.-J. Schmitz

Chapter 9

Virulence Mechanisms in MRSA Pathogenesis

Jesse S. Wright III and Richard P. Novick

Abstract

The multifactorial virlulence mechanisms of *Staphylococcus aureus* represent the evolution of a well-adapted human pathogen. The ability of *S. aureus* to colonize the host safely and its capacity to acquire and exchange genetic information are elements that contribute to its success as a pathogen. Here we discuss the factors that contribute to *S.aureus* pathogenesis and the potential impact of methicillin resistance on virulence. Virulence factors can be generally separated into two main classes; surface-associated factors, providing the mechanisms for adherence, attachment and immune evasion and secreted factors, which cause tissue destruction and are used to counteract host cell responses. While the roles of several virulence factors in pathogenesis have been elucidated, many more play an unknown role in the infectious process. Elaboration of these virulence determinants occurs in a preprogrammed fashion *in vitro* by an extensive regulatory network. These regulatory factors integrate environmental parameters to coordinate the expression of several, if not all, virulence genes. While the acquisition of antibiotic resistance, and most notably, methicillin resistance, by *S. aureus* has had an alarming impact on the treatment of staphylococcal infections, there is little information that links these elements with increased virulence. Complicating the relationship between resistance and virulence is the high degree of genomic diversity found among both antibiotic sensitive and resistance *S. aureus* isolates. Although gain of antibiotic resistance elements by themselves does not appear to be a major factor in the overall pathogenesis of the organism, three specific areas have been identified in which methicillin resistance may contribute indirectly to virulence.

Introduction

It hardly needs to be reiterated that MRSA, which has developed since the introduction of methicillin around 1960, has spread alarmingly throughout the world and are now being isolated increasingly from community acquired staphylococcal infections as well as, traditionally, from the majority of nosocomial ones. This state of affairs naturally raises the question, addressed in this chapter, of whether MRSA possess enhanced pathogenicity along with their antibiotic resistance. To set the stage for an evaluation of this issue, we briefly review the determinants of staphylococcal pathogenesis and their regulation. This is followed by an equally brief review of the general phenotypic and genotypic features of the MRSA, and we conclude with an attempt to integrate these two lines of information. The presentation is predicated on the view that antibiotic resistance contributes to pathogenesis only in so far as it compromises the treatment of infections. Though the importance of this is obvious enough from the clinical standpoint, it has no heuristic value in terms of an effort to understand pathogenesis. Thus our overall view is that the *mecA* gene, like other resistance genes, does not, per se, enhance staphylococcal pathogenicity, *sensu stricto*. However, there are three features of MRSA that are particularly relevant in the present context, namely the spread of MRSA in relation to colonization, the possible enhancement of virulence by subinhibitory β-lactam antibiotics, and the association of community-acquired methicillin resistance with an alarming constellation of pathogenicity determinants giving rise to lethal pneumonia. It is predicted that the continuing overuse of methicillin and related β-lactams will lead to an increase in the prevalence of infections by such organisms, as well as to the continuing spread of resistance to methicillin and to the other antibiotics whose resistance is associated with methicillin resistance.

Staphylococcal Pathogenesis and its Regulation

The ability of a bacterial pathogen to cause disease involves two distinct types of bacterial activities - those that are directly injurious to the host and those that are not specifically injurious but are simply required for growth or for protection against host defenses. The latter category includes all of the general housekeeping genes that are required for growth, including the acquisition of nutrients. It also includes products that are specifically directed against host defenses, such as peroxidases and superoxide dismutases. These are not discussed in detail here, except insofar as their functions overlap with those that are directly injurious. In this chapter, we first summarize information on the known pathogenicity factors of *S. aureus* and then review the genetics and regulation of these factors, noting any relationships to the MRSA genotype. This presentation is not encyclopedic, but rather emphasizes information that may be relevant to the pathogenicity of MRSA. For more extensive information on staphylococcal virulence and its regulation, readers are referred to the following recent reviews, plus references cited therein: (Novick, 2000; Arvidson and Tegmark, 2001; Cheung *et al.*, 2002a, 2002b; Novick, 2003).

Pathogenicity factors

The pathogencity factors of *S. aureus* that are directly or indirectly injurious are of 3 general types: surface associated factors, extracellular enzymes, and toxins. These are involved in attachment and adherence, tissue degradation and penetration, and cytotoxicity, respectively, and are listed in Table 1.

Surface Proteins: Adhesins and Immune Evasion Factors

Generally expressed early in growth *in vitro*, staphylococcal surface proteins mediate adherence to host and bacterial cells and the extracellular matrix, and to inert surfaces (for a review see Foster and Hook (1998)). *S. aureus* has a wide repertoire of such proteins, which are known as MSCRAMMS (microbial surface components recognizing adhesive matrix molecules); these share many structural and biological features and possess certain domains in common. There appears to be functional redundancy in that certain MSCRAMS can compensate, in part, for others *in vitro*, and most likely, *in vivo*. Most staphylococcal surface proteins are covalently anchored to the cell wall by the action of sortase A (SrtA), a transpeptidase that catalyzes the exchange of the threonine in the conserved LPXTG motif for a glycine residue in the pentaglycine cross-bridge (Mazmanian *et al.*, 1999; Navarre and Schneewind, 1999). Sortase A mutants are defective in surface protein linkage, giving rise to a "bald" phenotype, and have reduced virulence (Mazmanian *et al.*, 2000). A second sortase, SrtB is iron regulated and recognizes a NPQTN motif (Mazmanian *et al.*, 2002). Whether the *srtA-* or *srtB-* genotypes have any impact on methicillin resistance or vice versa is presently unknown.

Table 1. Extracellular proteins, including virulence factors

Gene	Location	Product	Activity/Function	Timing	ACTION OF REGULATORY GENES							References
					agr	saeRS	sarA	sarS	sarT	tst/entB		
Superantigens												
sea	Phage	Enterotoxin A	Food poisoning, TSS	xp**	O					O		Tremaine et al., 1993; Vojtov et al., 2002
seb	SaPI3	Enterotoxin B	Food poisoning, TSS	pxp	+		*			-		Gakill and Khan, 1988; Vojtov et al., 2002
sec	SaPI4	Enterotoxin C	Food poisoning, TSS	pxp	+							Regassa et al., 1991
sed	Plasmid	Enterotoxin D	Food poisoning, TSS	pxp	+							Zhang and Stewart, 2000
eta	ETA phage	Exfoliatin A	Scalded skin syndrome	pxp	+							Sheehan et al., 1992
etb	Plasmid	Exfoliatin B	Scalded skin syndrome	pxp	+							Yamaguchi et al., 2001
tst	SaPI1,2, bov1	Toxic shock toxin-1	Toxic shock syndrome	pxp	+		*			-		Recsei et al., 1986; Vojtov et al., 2002
Cytotoxins												
hla	Chrom	α-hemolysin	Hemolysin, cytotoxin	pxp	+	+	*	-	-	-		Giraudo et al., 1997; Recsei et al., 1986; Schmidt et al., 2001; Tegmark et al., 2000; Vojtov et al., 2002
hlb	Chrom	β-hemolysin	Hemolysin, cytotoxin	pxp	+	+	*					Giraudo et al., 1997; Recsei et al., 1986
hld	Chrom	δ-hemolysin	Hemolysin, cytotoxin	xp	+	O	+	+	O/-			Giraudo et al., 1997; Recsei et al., 1986; Schmidt et al., 2001; Tegmark et al., 2000
hlg	Chrom	γ-hemolysin	Hemolysin, cytotoxin	pxp	+		*					Bronner et al., 2000
lukS/F	PVL Phage	P-V leukocidin	Leukolysin	pxp	+					-		Vojtov et al., 2002

Enzymes and other secreted proteins

Gene	Location	Protein	Function	Expression						Reference
ssp	Chrom	V8 protease	Spreading factor	pxp	+	O		O/-	-	Arvidson and Tegmark, 2001; Chan and foster, 1998; Vojtov et al., 2002
aur		Metalloprotease (aureolysin)	Processing enzyme?	pxp	+			-		Arvidson and Tegmark, 2001; Chan and foster, 1998
sspB		Cysteine protease	Processing enzyme?	pxp	+			-		Arvidson and Tegmark, 2001
scp		Staphopain (protease II)	Spreading, nutrition	pxp	+			-		Arvidson and Tegmark, 2001
geh	Chrom	Glycerol ester hydrolase	Spreading, nutrition	pxp	+	O	*		-	McNamarra and Iandolo, 1998; Smeltzer et al., 1993
lip		Lipase (butyryl esterase)	Spreading, nutrition	pxp	+	O	*			Chamberlain and Imanoel, 1996
fme	Chrom	FAME	Fatty acid esterification	pxp	+		*			Chamberlain and Imanoel, 1996
plc		PI-phospholipase C	Spreading, nutrition	pxp	+					Arvidson and Tegmark, 2001
nuc	Chrom	Nuclease	nutrition	pxp	+	+				Carpenter and Chesbro, 1974
hys	Chrom	Hyaluronidase	Spreading factor	xp	*					Farrell et al., 1995
coa	Chrom	Coagulase	Clotting, clot digestion	exp	+/-		+			Giraudo et al., 1997; Lebeau et al., 1994; Wolz et al., 1996
sak	Phage	Staphylokinase	Plasminogen activator	pxp	+	O				Recsei et al., 1986
	Plasmid, ETD phage	EDIN	ADP-ribosyltransferase							Czech et al., 2001; Yamaguchi et al., 2001; Yamaguchi et al., 2002
	Phage	CHIP	Blocks innate immunity							Veldkamp et al., 2000

Table 1, Continued

Gene	Location	Product	Activity/Function	Timing	agr	saeRS	sarA	sarS	sarT	Tst/ entB	References
					ACTION OF REGULATORY GENES						
Surface proteins											
spa	Chrom	Protein A	Anti-immune, Anti-PMN	exp	-	*	-	+			Chien *et al.*, 1999; Giraudo *et al.*, 1997; Recsei *et al.*, 1986
cna	PT islet	Collagen BP	Collagen binding	pxp	O		-				Blevins *et al.*, 1999
fnbA	Chrom	Fibronectin BPA	Fibronectin binding	exp	-		+				Saravia-otten *et al.*, 1997
fnbB	Chrom	Fibronectin BPB	Fibronectin binding	exp	-		+				Saravia-otten *et al.*, 1997
clfA	Chrom	Clumping factor A	Fibrinogen binding	exp	O						Wolz *et al.*, 1996
clfB	Chrom	Clumping factor B	Fibrinogen binding	exp	O		O				McAleese *et al.*, 2001
		Lactoferrin BP	Lactoferrin binding		-						Arvidson and Tegmark, 2001
Capsular polysaccharides											
cap5	Chrom	Polysacch. cap. type 5	Antiphagocytosis ?	pxp	+		+				Pohlman-Dietze *et al.*, 2000
cap8	Chrom	Polysacch. cap. type 8	Antiphagocytosis ?	pxp	+						Pohlman-Dietze *et al.*, 2000

*Controversial

**xp - throughout exponential phase; exp - early exponential phase only; pxp - postexponential phase; O = No effect of gene on expression; + =upregulated; - = down-regulated.

Protein A

The best known of the staphylococcal surface proteins is protein A (Spa), whose several activities may serve to hide the organism from the innate immune system at the critical early time in infection when bacterial cell numbers are low (Patel *et al.*, 1987; Patel *et al.*, 1992). Protein A contains five repeating homologous α-helical domains (A,B,C,D,E) of about 60 amino acids each, which are individually involved in binding the Fc region of IgG (Starovasnik *et al.*, 2001). The organism is thus coated with IgG in the orientation opposite to that required for IgG function. This inhibits phagocytosis and could also disguise the organism against the innate immune system, preventing opsonization-dependent activation of the complement cascade. Protein A could also act as an antibody sponge, titrating out antibodies that may be crucial for preventing early establishment of the organism. Recently, Protein A has also been implicated in binding to platelets via the gC1qR/p33 receptor (Nguyen *et al.*, 2000) and the von Willebrand factor (Hartleib, *et al.*, 2000). Finally, Protein A can also bind the Fab portions of antibodies (Graille *et al.*, 2000) leading to the clonal expansion and stimulation of B-cells to generate a Th1 response in a similar manner to the action of superantigens on T-cells (Sinha *et al.*, 1999).

Fibronectin Binding Proteins

S. aureus produces two different proteins, FnbA and B, that mediate the binding of *S. aureus* to fibronectin. FnbA also binds fibrinogen (Wann, *et al.*, 2000). The Fnbs, encoded by homologous genes at separate loci, possess extended repeat regions which appear to mediate binding to different regions of the fibronectin molecule. Peptides corresponding to these repeating units bind fibronectin and can block bacterial binding (Joh *et al.*, 1999). Fibronectin and fibrinogen are abundant in the extracellular matrix, are important for wound healing, and are prominently absorbed to foreign surfaces such as catheters and prosthetic devices (Francois *et al.*, 1996). Thus, the Fnbs may have a major role in the establishment of *S. aureus* following entry into the subcutaneous tissues or encounter with implanted foreign bodies. The Fnbs form a tetrameric complex between themselves, fibronectin and the α5β1 integrin, which mediates binding to and invasion of epithelial and endothelial cells as well as fibroblasts (Sinha *et al.*, 1999). Invasiveness is concomitant with host cell tyrosine kinase activity (Dziewanowska, *et al.*, 2000) presumably stimulated in part by the conformation of the α5β1 integrin-fibronectin-FnbAB complex.

Fibrinogen Binding Proteins

S. aureus encodes two distinct fibrinogen binding proteins, known as clumping factors (ClfA and B) because they are responsible for the well-known phenotype of bacterial clumping in the presence of human plasma. ClfA and ClfB are structurally related, bind to different sites on the fibrinogen molecule, and also promote binding to fibrinogen coated biomaterials (McDevitt *et al.*, 1994; McDevitt, *et al.*, 1995; Ni Eidhin *et al.*, 1998). The binding of ClfA to fibrinogen is inhibited by Ca^{2+} and exhibits binding characteristics akin to the platelet αIIbβ3 integrin (O'Connell *et al.*, 1998). ClfA also binds to platelets (Siboo *et al.*, 2001) and is a potent competitor of platelet fibrinogen binding and aggregation (O'Connell *et al.*, 1998). Since platelet aggregation, mediated by fibrinogen, stimulates the release of antimicrobial peptides, inhibition of platelet aggregation by ClfA is a potential means of avoiding these peptides. Indeed, ClfA mutants are attenuated in a rat endocarditis model (Que *et al.*, 2001; Siboo *et al.*, 2001; Stutzman Meier *et al.*, 2001), and ClfA is a protective antigen in a murine arthritis model (Josefsson *et al.*, 2001). Interestingly, while ClfB is regulated in standard temporal fashion (*i.e.*, expressed primarily during the postexponential phase of growth), ClfA appears to be expressed throughout growth *in vitro* (Ni Eidhin *et al.*, 1998).

Collagen Binding Protein

A single collagen binding protein, Cna, has been identified and implicated as a virulence factor in *S. aureus* deep tissue infections such as osteomyelitis (Elasri *et al.*, 2002) keratitis (Rhem *et al.*, 2000), and septic arthritis (Patti *et al.*, 1994). Cna is encoded by a "pathogenicity islet" (Gillaspy *et al.*, 1997) that is not carried by all strains, so that introduction of the gene confers adherence to collagen (Switalsky *et al.*, 1993). Interestingly, the fact that fibronectin can also bind to collagen (Hahn and Yamada, 1979) suggests that *cna-* strains may be able to bind collagen indirectly via a Fnb-fibronectin bridge.

Other MSCRAMMS

S. aureus strains may bind to elastin, laminin, bone sialoprotein, thrombospondin, and vitronectin (Foster and Hook, 1998; Hook and Foster, 2000). With the exception of EbpS, an integral membrane protein lacking an LPXTG motif, which mediates elastin binding (Park *et al.*, 1999; Downer *et al.*, 2002) bacterial receptors for these host proteins have not yet been identified. Candidates are the SdrCDE (for Serine/Aspartate-repeat) proteins, which have homology to ClfA and ClfB, other LPXTG-containing ORFs, and surface associated non-LPXTG containing proteins (Josefsson *et al.*, 1998). One interesting example of the latter is the MHC II analagous protein or Map (also known as Eap or extracellular adherence protein) (McGavin *et al.*, 1993; Johnsson *et al.*, 1995; Chavakis *et al.*, 2002). Map/Eap interacts with a variety of extracellular host proteins and peptides and may inhibit leukocyte recruitment. Its close homology with MHC II could be of considerable interest both in terms of other potential functions in pathogenesis and in evolution, implying possible horizontal gene transfer between prokaryotic pathogens and eukaryotic hosts.

Intercellular Adhesion and Biofilm Formation

Biofilm formation has two general requirements: an adhesin to mediate attachment to a substrate, and an intercellular aggregation substance to enable the formation of a multilayered structure. *S. aureus* is an accomplished biofilm generator, using Fnbs or other surface proteins such as Bap (biofilm associated protein) (Cucarella *et al.*, 2001), to mediate adhesion, and an extracellular polysacchride adhesin, PIA, identical to that produced by slime-forming *S. epidermidis*, for intercellullar attachment and layering. Synthesis of PIA is determined by the *ica* locus (Cramton *et al.*, 1999; Arciola *et al.*, 2001; Fowler *et al.*, 2001), which may function only after the binding of *S. aureus* to specific extracelluar matrix proteins in the presence (or even in the absence) of foreign material. PIA is a linear β-1,6-linked glucosaminylglycan, synthesized from UDP-N-acetylglucosamine. Although the *ica* locus is critical for cell-cell adhesion, Ica has not been shown to be a virulence factor. Bap is a very large surface-anchored adhesin only found to be produced by some bovine mastitis isolates (Cucarella *et al.*, 2001).

Capsule

More than 90% of all clinical isolates of *S. aureus* possess polysaccharide capsules. These are usually very thin (<0.05µm) and are known as microcapsules. Rarely, thick "macrocapsules" are produced, which result in mucoidy and in diffuse growth in serum-soft agar. Capsules are antigenic and eleven serotypes have been defined, of which the most common are 5 and 8, which are microcapsules and together account for ~80% of clinical isolates (Arbeit *et al.*, 1984). Mucoid capsules mask surface proteins such as Cna (Gillaspy *et al.*, 1998) and also mask complement factor C3b bound to the cell wall, blocking phagocytosis and enhancing virulence (Cunnion *et al.*, 2001). The role of microcapsules in pathogenesis, however, is controversial. The microcapsule affects the outcome of an infection in some but not in other animal models and may thus be important for specific types of infections only (Lee *et al.*, 1986; Albus *et al.*, 1991; Baddour *et al.*, 1992; Nilsson *et al.*, 1997; Thakker *et al.*, 1998; Luong and Lee, 2002). Thus, the microcapsule reduces adherence to human endothelial cells (Pohlman-Dietze *et al.*, 2000) and may directly block other surface bound proteins, but appears to be a major determinant of abscess formation in certain animal models, possibly by activating CD4[+] T-cells (Tzianabos *et al.*, 2001). A vaccine candidate based upon a capsular component expressed *in vivo*, poly-N-succinyl β-1-6 glucosamine (PNSG) was protective against *S. aureus* kidney infections and death in a murine model (McKenny *et al.*, 1999) and a cap5/cap8 conjugate vaccine has shown promise in clinical trials (Shopsin *et al.*, 2000).

Secreted Proteins: Enzymes and Toxins

S. aureus strains produce a large variety of extracellular enzymes and toxins, most of which are directly or indirectly involved in pathogenesis.

Enzymes

Coagulase

Among the staphylococci, coagulase is produced almost exclusively by *S. aureus* (with the exception of a few strains of *S. intermedius*) and is generally used to differentiate *S. aureus* from other staphylococci clinically. Coagulase is a secreted protein, but is partially associated with the cell wall though it does not contain the LPXTG or other sorting motif. Coagulase binds prothrombin, forming a complex known as staphylothrombin, which can convert fibrinogen to fibrin (Kawabata *et al.*, 1985). Unlike the physiological activation of thrombin by proteolytic cleavage, coagulase appears to activate prothrombin by a non-proteolytic mechanism. Coagulase has a variable N-terminal region that has been used to define serotypes and is the site of binding to prothrombin; at the C-terminus of the coagulase molecule fibrinogen binding occurs (Phonimdaeng *et al.*, 1990; McDevitt *et al.*, 1992). Conversion of fibrinogen to fibrin may serve a useful purpose for the bacterium, by promoting formation of the abscess pocket and limiting further immune cell recruitment or by generating a protective fibin layer around the bacterium. However, the role of coagulase during pathogenesis is not clear, as in many infection models coagulase-defective mutants do not appear to be diminished in virulence (Phonimdaeng *et al.*, 1990; Baddour *et al.*, 1994; Sawai *et al.*, 1997). Perhaps coagulase is useful only at cell densities too low for analysis in animal models or in certain tissue locations.

Staphylokinase

Staphylokinase (Sak) is a potent activator of plasminogen (Collen, 1998) and is a potent thrombolytic agent. Like Coa, Sak is not an enzyme but forms a 1:1 complex with plasmin that has high plasminogen-activating activity (Kim *et al.*, 2002). Sak also binds to plasminogen but this complex is apparently inactive. In contrast to the streptococcal plasminogen activator, streptokinase, Sak results in localized but not systemic thrombosis (Molkanen *et al.*, 2002). The Sak-plasmin complex is inhibited by α2-antiplasmin, which may explain the localization of the complex (Collen *et al.*, 1993; Silence *et al.*, 1993; Okada *et al.*, 1994;). Interestingly, the Sak-plasmin-fibrin complex is >100-fold less sensitive to α2-antiplasmin than is the Sak-plasmin complex. Coagulase production and staphylokinase production are reciprocally regulated *in vitro*, with coagulase being produced early in growth and staphylokinase postexponentially (Recsei *et al.*, 1986). This reciprocal regulation may reflect a subtle orchestration of the use of these opposing factors by *S. aureus* during the infective process. Interestingly, mice lacking urokinase, a plasminogen activator, incur spontaneous *S. aureus* infections (Shapiro *et al.*, 1997), suggesting that both host and *S. aureus*-derived anti-clotting factors may play a critical role in pathogenesis, even though mutants defective in either *coa* or *sak* are not demonstrably less virulent than their wt parents in animal models (Vaudaux *et al.*, 1998), despite the high prevalence of these genetic elements in clincial isolates. This suggests, parenthetically, that there are subtle features of the pathogenetic process that are not readily revealed by standard models of infection.

Extracellular Evasion Factors

A second protein that, like Protein A, interferes with innate immunity has recently been described by Veldkamp *et al.* (2000). This is a small extracellular protein, known as CHIP (chemotaxis inhibitory protein), that interferes with the mobilization of PMNs and blocks the early activation of the C5a complement cascade. CHIP is encoded by a common temperate phage that also variably encodes enterotoxin A and staphylokinase, and insertionally inactivates *hlb* (Coleman *et al.*, 1989; Smeltzer *et al.*, 1994). Another secreted protein, Eap, has also been found to prevent neutrophil recruitment (Chavakis *et al.*, 2002).

Proteases

S. aureus proteases are believed to play a role in evasion of host defenses, by degrading antibacterial proteins, in invasiveness, by degrading matrix components as well as staphylococcal adhesins, and in nutrition, by converting host proteins to amino acids. The most important of the staphylococcal proteases is a serine protase, SspA (also known as V8 protease), which usually cleaves at glutamic acid and less frequently at aspartic acid. A functionally undescribed putative cysteine protease, SspB may be co-transcribed with SspA (Rogolsky *et al.*, 1974; Massimi *et al.*, 2002). Most proteases are lethal in the intracellular space and are synthesized as inactive pro-enzymes that are activated by cleavage following secretion. SspA, for example, is extracellularly activated by a zinc-dependent endopeptidase, aureolysin (Aur) (Banbula *et al.*, 1998). SspA can cleave the Fbps and other staphylococcal adhesins and this activity is thought to have a role is releasing the organism from its attachment sites and enabling it to spread (McGavin *et al.*, 1997; Karlsson *et al.*, 2001). Further, protease activity on extracelluar matrix components may generate fragments that bind these adhesins and interfere with secondary adhesion to the same substrates. SspA can also cleave the heavy chains of all human Ig classes providing a pathway to inactivate antibodies (Rousseaux *et al.*, 1983). Whether it can cleave Protein A bound antibodies is unknown. It also cleaves human α1-proteinase inhibitor, which is the major inhibitor of elastase released from PMNs upon phagocytosis of invading microorganisms (Prevost *et al.*, 1995). The uncontrolled activity of host elastase may contribute to tissue damage and degradation of proteins involved in host defense. SspA, incidentally, is efficiently inhibited by α2-macroglobulin, a protease inhibitor found in human plasma. *S. aureus* produces a variety of other proteases (see Table 1), whose potential roles in pathogenesis are generally unknown, but at least one of which has strong activity on elastin (Potempa *et al.*, 1988).

Lipases, Esterases, and FAME

S. aureus produces at least 3 lipid esterases with different substrate specificities and a fatty acid esterifying enzyme (FAME). The best known of these, Geh lipase, is a serine esterase which can hydrolze long-chain triaglycerols, shorter water-soluble triaglcyerols and tweens (Arvidson, 2000). The attachment site for phage L54a is in the *geh* structural gene so that L54a lysogens do not produce the enzyme (Lee and Iandolo, 1986). A second serine esterase, Lip, has preference for short-chain triaglcyerols (Simons *et al.*, 1996) and a third lipase has been demonstrated (G. Abramochkin and RPN, unpublished data), which may correspond to a putative lipase gene detected in the *S. aureus* genome. The products of lipase action are long-chain free fatty acids, which impair the host immune system and are also bacteriocidal for *S. aureus*. FAME is a fatty acid monoesterifying enzyme that inactivates the bactericidal fatty acids produced in infected tissues by bacterial lipase action and directly by host cells, esterifying them to cholesterol. FAME producing strains are more virulent in a murine model of infection (Mortensen *et al.*, 1992). Lipases may promote interstitial spreading of the organisms and may also be important for bacterial nutrition. The varied pH optima of lipases (Arvidson, 2000) characterized so far suggest that different lipases may be more or less active in certain host environments. Two phospholipases are also secreted by *S. aureus*, β-hemolysin (Hlb), a sphingomyelinase, and Plc, a phosophatidylinositol-specific phospholipase (PI-PLC). Plc degrades membrane-associated inositol phospholipids and releases glycan-PI-anchored cell surface proteins which can impair host cell functions such as adhesion and cell signaling.

Hyaluronate Lyase and Nuclease

Over 90% of staphylococcal isolates in one study produced hyaluronate lyase (HysA), and virtually all strains produce nuclease (Nuc). Hyaluronic acid is a component of the extracellular matrix of vertebrates and the *hysA* product is secreted by *S. aureus* ostensiblly to degrade this extracelluar polysacharide, possibly enabling the organism to spread (Farrell *et al.*, 1995). A well-characterized calcium-dependent thermostable DNA nuclease, Nuc, is also secreted by *S. aureus* (Anfinsen *et al.*, 1971). Nuc catalyzes the 5' phosphodiester cleaveage of both single and double stranded DNA and RNA. Whether HysA and Nuc are involved in pathogenesis is unknown at present.

Toxins

Alpha Hemolysin

Alpha hemolysin (Hla) is a secreted, cytolytic pore-forming toxin. It is a monomer of 293 amino acids that oligomerizes to form a ring shaped heptameric pore and can also form a hexamer (Song *et al.*, 1996). Hla has low affinity and high affinity binding characteristics. Low affinity binding can cause damage to liposomes at high concentrations. The high affinity binding site is unknown but probably a host membrane protein (Bashfors *et al.*, 1996). Hla producers are generally more virulent than isogenic non-producers (O'Reilly, 1986; Ji *et al.*, 1995; Kielian *et al.*, 2001) yet overproduction of Hla leads to decreased virulence in at least one animal model (Bayer *et al.*, 1997). *In vivo* and *in vitro* Hla is hemolytic, cytotoxic, dermonecrotic, and is lethal for mice at doses in the low microgram range. Several cell types, including erythrocytes, mononuclear immune cells, epithelial and endothelial cells and platelets can be killed by α-toxin. Cell death from membrane damage has been attributed to impaired osmoregulation, cation and small molecule influx/efflux and apoptosis (Fink, 1989; Menzies and Kourteva, 2000). Hla can stimulate prostaglandin and/or leukotriene generation in target cells, resulting in vasoactive effects which may augment lethality (Bhakdi, 1991). Proinflamatory cytokines and procoagulatory compounds released from activated monocytes and platelets, respectively, contribute to the systemic effects of α-toxin activity as well (Onogawa, 2002).

Beta Hemolysin

Beta hemolysin (Hlb) is a sphingomylenase, closely related to enzymes of the same type produced by soil bacilli, which disrupts the membranes of erythrocytes and other mammalian cells (Walev *et al.*, 1996; Marshall *et al.*, 2000). It is a "hot-cold" hemolysin, meaning that its activity is best demonstrated by a period at 4°C after initial exposure of erythrocytes at 37°C (Smyth *et al.*, 1975). It shows the classic synergistic hemolysis phenomenon - enhancement of hemolytic activity synergistic with delta hemolysin (Herbert and Hancock, 1985). Since its frequently deactivated by phage L54a integration its role in human and animal disease remains unclear.

Delta Hemolysin

Delta hemolysin (Hld) is a 26-amino acid amphiphilic peptide (Tappin, 1988) encoded within the *agr* locus (see below) and produced by most non-*aureus* staphylococci as well as by *S. aureus* (Donvito *et al.*, 1997). It is heat stable, has surfactant properties and is lytic for many types of membranes, including those of erythrocytes, organelles, and even bacterial protoplasts, probably disrupting membranes by its surfactant action (Pokorny *et al.*, 2002). It does not appear to have a signal peptide and may be secreted by passive diffusion, since some 50% of the extracellular molecules have retained their N-terminal formyl methionine (Fitton *et al.*, 1980). Hld is inhibited by binding to proteins, cholesterol, and phospholipids in serum. It has been reported to be involved in necrotizing enterocolitis of newborn infants caused by *S. epidermidis* (Schlievert *et al.*, 1982); however, these reports have not been confirmed.

Gamma Hemolysin and Leukocidin

The products of three genes, HlgA (LukS-like), HlgC (LukS-like), and HlgB (LukF), combine to form two distinct synergohymenotrophic bicomponent toxins, gamma hemolysin and Panton-Valentin leukocidin (PVL), each composed of one lukS-like and one LukF subunit (Cooney *et al.*, 1993; Konig *et al.*, 1997). Over 99% of clinical *S. aureus* isolates carry the *hlg* locus, encoding a pair of these subunits that form gamma hemolysin (Prevost *et al.*, 1995). A small subset of these, 2-5%, encode a second LukS-like subunit that combines with the common LukF or a second LukF to form PVL. PVL can stimulate and lyse neutrophils and macrophages (Ferreras *et al.*, 1998; Szmigielski *et al.*, 1998). Other normal cells aren't affected by PVL, nor is PVL hemolytic. Hlg, however, is strongly hemolytic but 90-fold less leukotoxic. For Hlg, the F subunit binds to erythrocytes first and then the S subunit to form a pore. PVL binds similarly to neutrophils, but the S subunit binds first. Both toxins have been implicated in staphylococcal pathogenesis (Foster *et al.*, 1990; Novick and Skurray, 1990). PVL has long been associated with cutaneous infections, and most recently has been implicated in a fulminant necrotizing pneumonia, which is particularly relevant to the subject of this book since all of the 16 isolates thus far implicated in this syndrome are community-acquired MRSA (CMRSA), and incidentally, are all *agr* group III (Gillet *et al.*, 2002). The PVL components are generally encoded by a converting phage known as PVL-phage (Zou *et al.*, 2000; Narita *et al.*, 2001), whereas the Hlg subunits are usually chromosomal.

ADP-ribosyltransferases

At least three secreted enzymes with ADP-ribosyltransferase activity are produced by many *S. aureus* strains (Czech *et al.*, 2001; Yamaguchi *et al.*, 2001, 2002). Interestingly, these C3-like enzymes have homology to ADP-ribosyltransferases from other Gram-positive pathogens, but appear to have wider substrate specificity. The *in vitro* targets for these transferases are several eukaryotic low molecular weight Rho and Ras GTPases, whose modification results in deactivation. This leads to gross changes in cellular architechture due to disaggregation of the actin cytoskeleton. However, unlike typical ADP-ribosylating toxins such as Pertussis toxin and others, there is no domain associated with the staphylococcal toxins for membrane translocation and hence these are ineffective when added to cells extracellularly (Wilde *et al.*, 2001). There may be another factor, yet unidentified, which promotes this translocation or provides membrane-associating activity in trans. Intriguingly, these ADP-ribosyltransferases could reach their cellular targets following the invasion of host cells and escape from the phagosome and could thus play a role in cytoxicity.

Superantigens and Related Exotoxins

S. aureus strains produce a large number of pyrogenic exotoxins known as superantigens (SAgs) owing to their unusual interaction with the immune system. These are responsible for all of the known staphylococcal toxinoses (diseases wholly attributable to a single toxin), including the toxic shock syndrome, staphylococcal scarlet fever, food poisoning, and exfoliative dermatitis (scalded skin syndrome) (McCormick *et al.*, 2001). SAgs are potent T-cell mitogens. They bind to major histocompatibility complex class II (MHC II) proteins which stabilizes the interaction with CD4 T-cell receptor β-chains resulting in Th1 type response stimulating the proliferation and the secretion of various interleukins and cytokines such as TNF and IFN-γ. The typical outcome of this is an acute systemic illness known as the toxic shock syndrome (TSS). The overexpression of these cytokines has an important role in the pathogenesis of TSS, but additionally, TSST-1 binds to macrophages of the reticulo-endothelial system, particularly the Kupffer cells of the liver, and interferes with the ability of these cells to clear foreign substances, particularly gram-negative endotoxin (Krakauer, 1999; Dinges *et al.*, 2000). Thus, the hypotension in TSS can be viewed as largely a consequence of endotoxin shock; indeed, the mitogenic and lethal activities of the toxin can be differentiated mutationally. The commonest form of TSS is associated with tampon use in menstruating women and is caused almost exclusively by toxic shock syndrome toxin-1 (TSST-1) because TSST-1 is the only SAg capable of traversing an intact epithelial membrane (Schlievert, personal communication). The other staphylococcal SAgs, especially enterotoxins A,.B, and D, are frequently responsible for TSS at extra-vaginal sites, and, unlike TSST-1, are common causes of staphylococcal food poisoning. Interestingly, food poisoning is not the result of SAg function but is rather due to a specific interaction of the toxin with parasympathetic nerve endings in the GI tract, and is associated with a specific loop in the toxin structure (the "emesis loop") that is absent in TSST-1 (McCormick *et al.*, 2001).

TSST-1-producing *S. aureus* secrete other staphylococcal exoproteins at only very low levels (Schmidt *et al.*, 2001), and, remarkably, cause local tissue infections that are largely apurulent, often confounding the diagnosis (Fast *et al.*, 1988). Indeed, it has recently been demonstrated that the intracellular precursor of TSST-1, and probably also of enterotoxin B (SEB) sharply down-regulate the transcription of most exoprotein genes (Vojtov *et al.*, 2002). Thus TSST-1 could interfere with the influx of PMNs in two ways - by blocking the production of chemotactic factors elaborated by the organism, and by inducing anti-inflammatory cytokines.

With one exception, the SAgs are produced only by strains of *S. aureus*. The exception is the exfoliative toxins, ETA and ETB, which are cysteine proteases that specifically attack the dermal-epidermal junctions and are also produced by certain strains of *S. hyicus* (Sato *et al.*, 1999; Plano *et al.*, 2000); however, the status of these toxins as SAgs has been questioned.

Invasion and Intracellular Survival

S. aureus appears to be surprisingly efficient at invasion and intracellular survival (Sinha *et al.*, 1999).Moreover, in epithelial cell lines, *S. aureus* escapes the phagosome and induces apoptosis (Bayles *et al.*, 1998; Wesson *et al.*, 1998; Kahl *et al.*, 2000), possibly by producing proteases. Overall, little is known about the molecular requirements for *S. aureus* intracellular survival and its contribution during infection. One clear trend is that enhanced surface-protein expression increases the frequency of uptake into host cells (Dziewanowska *et al.*, 1999; Joh *et al.*, 1999; Dziewanowska *et al.*, 2000; Jett and Gilmore, 2002). This is intriguing in light of the fact that *S. aureus* elaborates several antiphagocytic factors (Protein A, capsule, ect.) which are thought to help it to avoid uptake by phagocytes. *S. aureus* may be found intracellularly as a result of host-immune clearance or by invasion of cells as a virulence mechanism. Intracellular *S. aureus* may remain viable in PMNs (Lundqvist-Gustafsson *et al.*, 2001); adoptive transfer of infected PMNs to healthy mice indicates that this intracellular bacterial population can initiate infection (Gresham *et al.*, 2000). Intracellular persistence frequently results in mutations affecting respiratory electron transport, causing very slow growth and aminoglycoside resistance. These mutants, known as "small colony variants", are inherently antibiotic resistant and are defective in the production of exotoxins and other virulence factors, which may be responsible for their long-term intracellular persistence in conditions such as chronic osteomyelitis (Proctor *et al.*, 1995).

Regulation of Staphylococcal Virulence

Except for the toxinoses, staphylococcal diseases are multifactorial, with a variety of factors contributing in small and often subtle ways to the overall disease process. Consistent with this concept is a complex regulatory network by which the expression of genes encoding these various factors is controlled, perhaps with attention to the vicissitudes of the environmental niche corresponding to any given infection site. Thus far, most of the information on the regulation of virulence genes has come from *in vitro* studies; it is widely assumed that the regulatory environment *in vivo* is more or less similar to that *in vitro* so that *in vitro* results are meaningful. Virulence factor production in conventional flask cultures follows a well-established overall temporal pattern in which many of the surface protein genes are expressed shortly after the initiation of exponential growth and are down-regulated shortly later, during the mid-exponential phase, whereas many of the genes encoding secreted proteins are expressed primarily during the postexponential phase (Novick, 2002). This pattern of gene expression is generally thought to represent a pathogenic strategy of the organism, in which surface proteins are needed during the establishment phase of the infection, to aid in adherence to host tissues and to protect the organism from host defenses such as opsonization-phagocytosis, complement-mediated killing, etc., whereas secreted enzymes and cytotoxins are needed later, to aid in spreading, acquisition of nutrients, and killing of phagocytes.

The primary regulatory function by which this temporal program is effected is the *agr* locus, which encodes a quorum-sensing signal transduction pathway activated by an *agr*-encoded autoinducing peptide (AIP), the ligand for the system's signal receptor (Ji *et al.*, 1995). *Agr* regulates the transcription of most of the virulence factor genes and independently regulates the translation of at least two of the proteins, Alpha toxin and Protein A (Morfeld *et al.*, 1995, Patel *et al.*, 1992). In general, *agr* downregulates surface protein genes during

Table 2. Known accessory gene regulators and transcription factors in *S. aureus*

Regulatory unit	Description	Role	Reference
AgrACDB/rnaIII	TCS, autoinduced by peptide	Regulates many extracellular and cytoplasmic protein accessory genes	Novick et al., 1993
SaePQRS	TCS, autoinduced	Regulates many extracellular protein genes	Giraudo et al., 1999
ArlRS	TCS	Regulates autolysis and certain accessory genes	Fournier et al., 2001
SrrAB/srhSR	TCS	Regulates certain accessory genes at low P_{O2} (*resDE* homolog)	Throup et al., 2001; Yarwood et al., 2001
SvrA	Predicted membrane associated factor	Required for transcription of the *agr* locus	Garvis et al., 2002
σ^B	Rpo sigma factor	Active in late exponential phase; regulates many accessory genes	Kullik et al., 1998
SarA	Transcription factor	Assists in *agr* autoinduction under certain conditions; pleiotropic repressor	Cheung et al., 1996
SarS	Transcription factor	Activates transcription of *spa* and possibly other surface protein genes	Tegmark et al., 2000
SarT	Transcription factor	Represses transcription of *hla* and possibly other exoprotein genes	Schmidt et al., 2001
SarR	Transcription factor	Minor transcription factor for *sarA* and possibly *sarS*	Manna and Cheung, 2001
Rot	Transcription factor	Minor transcription factor for *hla* and possibly other exoprotein genes	McNamarra et al., 2000
Rlp	Transcription factor	Similar to Rot	McNamarra et al., 2001
Tst/entB	Superantigens	Autorepressors; Down-regulate transcription of most exoprotein genes	Vojtov et al., 2002

mid-exponential phase and up-regulates secreted protein genes postexponentially. The putative signal(s) that up-regulate the surface protein genes early in exponential phase are unknown. However, there is an extensive collection of transcription factors and other regulatory mediators, including several other TCS, that interact with *agr* and with the virulence factor genes in complex ways that are not presently clear. A summary of the known regulatory genes is presented in Table 2 (readers are referred to a recent review by Novick (2003) for a full update on current knowledge of these regulatory interactions). In addition to the intracellular regulatory modalities set in motion by the *agr* autoinduction system, environmental variables such as ionic strength, pH, temperature, O_2 and CO_2 levels, etc. also feed into the overall system, presumably helping the organism to recognize and respond appropriately to its local environment.

Genetics of the Staphylococcal Virulon

The staphylococcal virulon is composed of two general types of genetic units - those that belong to the standard chromosomal gene set, and those that are encoded by variable genetic elements, many of which are demonstrably mobile. Information on these is included in Table 1 and can be summarized as follows: i) *Converting phages:* at least 90% of clinical *S. aureus* isolates carry temperate phages of serogroup F that have a common attachment site within the structural gene for β-hemolysin, and are therefore responsible for the low frequency of β-hemolysin production by clinical isolates (Coleman *et al.*, 1989). These phages generally encode one or more of the following: enterotoxin A, staphylokinase, and CHIP, and it has been suggested that this complex of converting genes constitutes a pathogenicity island within the phage genome (Lindsay *et al.*, 1998; Novick *et al.*, 2001; Wagner and Waldor, 2002). A different phage, ETA phage, encodes exfoliative toxin A, and another, PVL phage, encodes PVL. Finally, phage 54 insertionally inactivates the lipase gene, *geh,* but is not known to encode any virulence factor. ii) *Plasmids:* Only two of the known staphylococcal virulence factors are known to be naturally plasmid-coded, exfoliative toxin B (Rogolsky *et al.*, 1974) and enterotoxin D (Bayles and Iandolo, 1989) iii) *Pathogenicity islands:* most of the known staphylococcal SAgs are encoded by a newly described family of phage-related pathogenicity islands (Lindsay *et al.*, 1998; Novick *et al.*, 2001), several of which are known to encode 3 or more individual SAgs. One of these does not encode any SAg, but rather encodes a protein known as BAP, a very large (240 kDa) surface-linked adhesin that is found in bovine isolates and required for biofilm formation (Cucarella *et al.*, 2001). Sequencing of *S. aureus* genomes has revealed several additional genetic units that resemble these pathogenicity islands, and encode Hlg, restriction systems, families of enterotoxins or serine-protease-like products (Jarraud, 2001; Kuroda *et al.*, 2001; Baba *et al.*, 2002). Remarkably, the plasmids and other mobile genetic elements that carry antibiotic resistance determinants rarely, if ever carry virulence genes. This is particularly interesting with regard to the SCC*mec* elements (covered extensively elsewhere in this volume), which seem to be magnets for the insertion of transposons, IS elements and plasmids - all of which carry only resistance genes.

The MRSA Biotype

It is well-established that methicillin resistance is based on the presence of a novel penicillin binding protein, PBP2a (or PBP2'), that has been added to the standard set of PBPs required for the synthesis of the staphylococcal cell wall (Hartman and Tomasz, 1984). β-lactam antibiotics inactivate the standard PBPs leaving peptidoglycan assembly largely or entirely to PBP2a. Though the peptidoglycan synthesized under these conditions provides only moderate resistance to methicillin and is not entirely normal, it adequately fulfills the role of the normal peptidoglycan - has the usual multilayered structure and provides mechanical protection, remodeling during division, etc. Secondary mutations in a variety of genes, the *fem* genes, greatly enhance methicillin resistance, though they do not confer resistance on their own. The *fem* genes affect various aspects of cell wall biosynthesis. The best characterized, *femA, B, C,* and *X,* encode peptidyl transferases that are responsible for assembling the pentaglycine bridge between adjacent peptidoglycan subunits (Berger-Bächi, 1999). The mutations affecting them generally increase resistance to the lytic enzyme lysostaphin as well as to methicillin, but, remarkably, do not significantly compromise the integrity of the cell wall. Nor do they cause any detectable increase - or, for that matter, decrease, in pathogenicity or in the expression of known virulence genes.

It is particularly relevant in the present context that in *S. aureus* , *mecA* is always carried by a mobile genetic element, the SCC*mec* element (Katayama *et al.*, 2000), that is anywhere from 20 to 60 kb in length, and carries a corresponding number of additional genes depending on its size. There is no doubt that the SCC*mec* element has been acquired by horizontal genetic transfer to *S. aureus,* presumably from some related species, and that it serves as a preferred site for the insertion of other mobile elements, including plasmids, transposons, ISs, and others less well defined. In consequence, the SCC*mecs* are a repository of genes for resistance to other antibiotics as well as to methicillin. They are also, however, a prime example of the general rule that mobile genetic elements in bacteria carry genes either for resistance or for pathogenesis, but rarely for both. Thus, with the exception of a recently described gene that affects the function of certain adhesins (Vaudaux *et al.*, 1998; Huesco *et al.*, 2002), the SCC*mecs* have not been directly implicated in pathogenesis.

Nevertheless, a very alarming, apparently recent development has been the association of community-acquired methicillin resistance (CMRSA) with the production of PVL, giving rise to a frighteningly fulminant necrotizing pneumonia (Gillet *et al.*, 2002). This syndrome, which has usually followed an apparently benign upper respiratory infection, is characterized by the very rapid onset of high fever, respiratory distress, and death within 24-48 hr. Lung tissue at necropsy shows extensive areas of necrosis and leukocyte infiltration. This PVL pneumonia was responsible for the deaths of four children, described as the first examples in the United States of CMRSA (Anonymous, 1999), and with the inclusion of the 12 other recently described cases, has had an overall case-fatality rate of 80%. PVL and *mecA* are both mobile, the former being carried by certain phages, as noted (Zou *et al.*, 2000; Narita *et al.*, 2001), and the latter by SCC*mecs*, and though PVL-expressing MRSA are not rare, they have not previously been associated with necrotizing pneumonia or other clinical conditions different from those associated with PVL in MSSA. However, the syndrome described by Gillet, *et al.* (2002), is associated exclusively with *agr* specificity group III, suggesting that the *agr* group provides a basic biotypical substrate upon which the phenotypes conferred by these two added genetic elements are superimposed, generating a frightening genotypic constellation. The roles of *mecA* and PVL in this hypervirulent pathotype are clear enough; the nature and the role(s) of specific features of the *agr* group III genotype remain to be determined.

Virulence Modulation by Antibiotics

A large body of scientific evidence suggests that *S. aureus* modulates expression of its virulence genes in response to subinhibitory concentrations of antibiotics (Shibl, 1982; Proctor *et al.*, 1983a, 1983b; Shibl, 1984; Sonstein and Burnham, 1993; Ohlsen *et al.*, 1998; Kernodle *et al.*, 1995; Bisognano *et al.*, 2000; Dal Sasso *et al.*, 2002; Herbert *et al.*, 2001; Worlitzsch *et al.*, 2001). Ribosomal and gyrase inhibitors suppress exoprotein expression and causing dramatic attenuation of the disease process in various experimental models. Conversely, cell-envelope targeting antibiotics, such as β-lactams and glycopeptides stimulate exoprotein synthesis and downregulate or have no effect on surface and secreted protein expression. For example, subinhibitory concentrations of β-lactams enhance Hla expression in both MRSA and MSSA strains (O'Reilly, 1986). This may be relevant in the clinical setting as individuals infected with MRSA are often treated with oxacillin, or other β-lactam antibiotics - which expose the organisms to concentrations of antibiotics that are by definition subinhibitory. Thus, It is very possible that these concentrations of β-lactam antibiotics enhance the virulence of MRSA and MSSA *in vivo* (Doss, 1993; Ohlsen *et al.*, 1998) and raises important questions regarding the efficacy of cell-wall targeting antibiotics on the outcome of *S. aureus* infections.

The exact mechanism of the action of subinhibitory concentrations of certain antibiotics is not clear but may involve the stress-activated alternative sigma factor, σ^B (Chan *et al.*, 1998), but apparently not the *agr* system (Kernodle *et al.*, 1995). This phenomenon may represent an evolutionary remnant of existence in the soil, which presumably involves competition with microorganisms that produce antibiotics, where the actual concentration of any such compound in the soil would be very low. In addition, biofilm formation, capsule production and abscess formation, which make intrinsically antibiotic sensitive *S. aureus* strains resistant to antibiotic therapy, perhaps owing to the decreased diffusion of these substances, may be situations in which the bacteria may specifically respond to the low concentrations that eventually make their way to the site of infection. An important area for further investigation would be to determine whether there may be a link between the altered regulation of virulence factor genes induced by subinhibitory antibiotics, severity of pathogenesis or persistence during treatment, and, perhaps, enhanced acquisition of antibiotic resistance determinants.

Sticking Around: *S. aureus* as an Opportunist

An interesting yet hazardous property of staphylococci is their ability to colonize and lead a commensal existence with the host. The frequency of colonization in humans is generally about 30% and nearly 90% of the population is transiently colonized with *S. aureus* at some time. This is particularly relevant with respect to MRSA, since these are very commonly introduced into the institutional environment (*e. g.*, hospitals and nursing homes) by human carriers - indeed, some institutions have established procedures for monitoring incoming patients and treating them to eradicate MRSA (Perl and Golub, 1998; Uehara *et al.*, 2000). Frequency of colonization is affected by risk factors, such as IV drug use, diabetes, etc., and the carrier state is often a predecessor of disease (Harbarth *et al.*, 2000; Peacock *et al.*, 2001;Weems and Beck, 2002): several studies have shown that strains of staphylococci causing disease are most often the strains with which the affected individual was colonized (Tremaine *et al.*, 1993; Fierobe *et al.*, 1999; Bert *et al.*, 2000; Kalmeijer *et al.*, 2000; von Eiff *et al.*, 2001). Very little is known about how *S. aureus* exists in the carrier state and how it avoids the innate immune defenses of the host and mucosal immunity. What molecular signals trigger the progression from colonization to more invasive pathogenesis is also not very clear though disruption of skin integrity, which weakens host defenses, is a prerequisite. Metabolites may be more abundant deeper in host tissues, but the bacterium appears to make a questionable decision to challenge the host immune system head on. Indeed, the normal host usually wins the contest, despite the well-developed staphylococcal virulon.

Virulence of MRSA vs. MSSA

In attempting to address this issue, one must consider two rather different questions: the first is whether the SCC*mec* element has been acquired by strains that are intrinsically more (or less) virulent than strains that have not acquired the element; the second is whether the SCC*mec* element itself confers increased (or decreased) virulence on its host strain. The first question can be addressed by statistical analyses of virulence tests of clinical MRSA vs MSSA strains without any attempt to control for genomic variability. Not surprisingly, therefore, some studies of this type seem to suggest that MRSA are more virulent, others that MSSA are more virulent (Mizobuchi *et al.*, 1994; Papakyriacou *et al.*, 2000; Shopsin *et al.*, 2000; Huesca *et al.*, 2002), and it is doubtful that a definitive answer to this question will ever be obtained. The second question can be answered only by studies with (isogenic) MRSA and MSSA strains differing only in the presence or absence of the SCC*mec* element. No such study has yet been reported.

Nevertheless, we have noted and discussed three areas that may be highly relevant in the context of MRSA and virulence. The first of these is the SCC*mec* element, a large inserted chromosomal segment that is carried by all MRSA. This element is a repository for additional inserted antibiotic resistance units and in this manner contributes to the overall clinical danger of MRSA. For reasons that are not at all understood, however, the SCC*mec* element, like other resistance carriers in bacteria, does not contain virulence genes (with the possible exception of *pls*, which interferes with adhesins and is therefore not obviously a virulence gene), so that its effect on pathogenicity is generally indirect. The second means by which MRSA may have enhanced pathogenicity is related to the CMRSA strains expressing PVL and responsible for fulminant necrotizing pneumonia. Although the SCC*mec* element is always present in these strains, there is presently no reason to believe that *mecA* or other SCC*mec* genes contribute directly to their enhanced virulence. This view, however, could change following more detailed analysis. The third area in which MRSA could be associated with enhanced pathogenicity is related to the observation that expression of certain virulence-associated genes is enhanced by subinhibitory concentrations of β-lactam antibiotics. Again, we are unaware of any direct study of this effect on virulence, but consider it to be an area of great concern and one that would amply repay an experimental investigation.

References

Albus, A., Arbeit, R.D., and Lee, J.C. 1991. Virulence of *Staphylococcus aureus* mutants altered in type 5 capsule production. Infect. Immun. 59:1008-1014.

Anfinsen, C.B., Cuatrecasas, P., and Taniuchi, H. 1971. Staphylococcal nuclease, chemical properties and catalysis. In: The Enzymes, vol. 4. P. Boyer, ed. Academic Press, New York. p. 177-204.

Anonymous. 1999. From the Centers for Disease Control and Prevention. Four pediatric deaths from community-acquired methicillin-resistant *Staphylococcus aureus*—Minnesota and North Dakota, 1997-1999. JAMA 282: 1123-1125.

Arbeit, R.D., Karakawa, W.W., Vann, W.F., and Robbins, J.B. 1984. Predominance of two newly described capsular polysaccharide types among clinical isolates of *Staphylococcus aureus*. Diag. Microbiol. Infect. Dis. 2: 85-91.

Arciola, C.R., Baldassarri, L., and Montanaro, L. 2001. Presence of *icaA* and *icaD* genes and slime production in a collection of staphylococcal strains from catheter-associated infections. J. Clin. Microbiol. 39: 2151-2156.

Arvidson, S. 2000. Extracellular enzymes from *Staphylococcus aureus*. In: Gram-Positive Pathogens. Fischetti, V.A., Novick, R.P., Feretti, J.J., Portnoy, D.A., and Rood, J.I., eds. ASM Press, Washington D.C. p. 745-808.

Arvidson, S., and Tegmark, K. 2001. Regulation of virulence determinants in *Staphylococcus aureus*. Intl. J. Med. Microbiol. 291: 159-170.

Baba, T., Takeuchi, F., Kuroda, M., Yuzawa, H., Aoki, K., Oguchi, A., Nagai, Y., Iwama, N., Asano, K., Naimi, T., Kuroda, H., Cui, L., Yamamoto, K., and Hiramatsu, K. 2002. Genome and virulence determinants of high virulence community-acquired MRSA. Lancet 359: 1819-1827.

Baddour, L.M., Lowrance, C., Albus, A.,. Lowrance, J.H., Anderson, S.K., and Lee, J.C. 1992. *Staphylococcus aureus* microcapsule expression attenuates bacterial virulence in a rat model of experimental endocarditis. J. Infect. Dis. 165: 749-753.

Baddour, L.M., Tayidi, M.M., Walker, E., McDevitt, D., and Foster, T.J. 1994. Virulence of coagulase-deficient mutants of *Staphylococcus aureus* in experimental endocarditis. J. Med. Microbiol. 41: 259-263.

Banbula, A., Potempa, J., Travis, J., Fernandez-Catalan, C., Mann, K., Huber, R., Bode, W., and Medrano, F. 1998. Amino-acid sequence and three-dimensional structure of the *Staphylococcus aureus* metalloproteinase at 1.72 A resolution. Structure 6: 1185-1193.

Bashford, C.L., Alder, G.M., Fulford, L.G., Korchev, Y.E., Kovacs, E., MacKinnon, A., Pederzolli, C., and Pasternak, C.A. 1996. Pore formation by *S. aureus* alpha-toxin in liposomes and planar lipid bilayers: effects of nonelectrolytes. J. Membrane Biol. 150: 37-45.

Bayer, A.S., Ramos, M.D., Menzies, B.E., Yeaman, M.R., Shen, A.J., and Cheung, A.L. 1997. Hyperproduction of alpha-toxin by *Staphylococcus aureus* results in paradoxically reduced virulence in experimental endocarditis: a host defense role for platelet microbicidal proteins. Infect. Immun. 65: 4652-4660.

Bayles, K.W., and Iandolo, J.J. 1989. Genetic and molecular analyses of the gene encoding staphylococcal enterotoxin D. J. Bacteriol. 171: 4799-4806.

Bayles, K.W., Wesson, C.A., Liou, L.E., Fox, L.K., Bohach, G.A., and Trumble, W.R. 1998. Intracellular *Staphylococcus aureus* escapes the endosome and induces apoptosis in epithelial cells. Infect. Immun. 66: 336-342.

Berger-Bächi, B. 1999. Genetic basis of methicillin resistance in *Staphylococcus aureus*. Cell. Mol. Life Sci. 56: 764-770.

Bert, F., Galdbart, J.O., Zarrouk, V., Le Mee, J., Durand, F., Mentre, F., Belghiti, J., Lambert-Zechovsky, N., and Fantin, B. 2000. Association between nasal carriage of *Staphylococcus aureus* and infection in liver transplant recipients. Clin. Infect. Dis. 31: 1295-1299.

Bhakdi, S. 1991. Alpha-toxin of *Staphylococcus aureus*. Microbiol. Rev. 55: 733-751.

Bisognano, C., Vaudaux, P. , Rohner, P., Lew, D.P., and Hooper, D.C. 2000. Induction of fibronectin-binding proteins and increased adhesion of quinolone-resistant *Staphylococcus aureus* by subinhibitory levels of ciprofloxacin. Antimicrob. Agents Chemother. 44: 1428-1437.

Blevins, J.S., Gillaspy, A.F., Rechtin, T.M., Hurlburt, B.K., and Smeltzer, M.S. 1999. The Staphylococcal accessory regulator (*sar*) represses transcription of the *Staphylococcus aureus* collagen adhesin gene (*cna*) in an *agr*-independent manner. Mol. Microbiol. 33: 317-326.

Bronner, S., Stoessel, P., Gravet, A., Monteil, H., and Prevost, G. 2000. Variable expressions of *Staphylococcus aureus* bicomponent leucotoxins semiquantified by competitive reverse transcription-PCR. Appl. Environ. Microbiol. 66: 3931-3938.

Carpenter, D., and Chesbro, W. 1974. Separable forms of nuclease secreted by *Staphylococcus aureus* as a function of growth stage. Can. J. Microbiol. 20: 337-345.

Chamberlain, N.R., and Imanoel, B. 1996. Genetic regulation of fatty acid modifying enzyme from *Staphylococcus aureus*. J. Med. Microbiol. 44: 125-129.

Chan, P.F., and Foster, S.J. 1998. Role of SarA in virulence determinant production and environmental signal transduction in *Staphylococcus aureus* . J. Bacteriol. 180: 6232-6241.

Chan, P.F., Foster, S.J., Ingham, E., and Clements, M.O. 1998. The *Staphylococcus aureus* alternative sigma factor sigma B controls the environmental stress response but not starvation survival or pathogenicity in a mouse abscess model. J. Bacteriol. 180: 6082-6089.

Chavakis, T., Hussain, M., Kanse, S.M., Peters, G., Bretzel, R.G., Flock, J.I., Herrmann, M., and Preissner, K.T. 2002. *Staphylococcus aureus* extracellular adherence protein serves as anti-inflammatory factor by inhibiting the recruitment of host leukocytes. Nat. Med. 8: 687-693.

Cheung, A.L., Heinrichs, J.H., and Bayer, M.G. 1996. Characterization of the *sar* locus and its interaction with *agr* in *Staphylococcus aureus*. J. Bacteriol. 178: 418-423.

Cheung, A.L., Projan, S.J., and Gresham, H. 2002a. The genomic aspect of virulence, sepsis, and resistance to killing mechanisms in *Staphylococcus aureus*. Curr. Infect. Dis. Rep. 4: 400-410.

Cheung, A.L., and Zhang, G. 2002b. Global regulation of virulence determinants in *Staphylococcus aureus* by the SarA protein family. Front. Biosci. 7: d1825-d1842.

Chien, Y., Manna, A.C., Projan, S.J., and Cheung, A.L. 1999. SarA, a global regulator of virulence determinants in *Staphylococcus aureus*, binds to a conserved motif essential for sar-dependent gene regulation. J. Biol. Chem. 274: 37169-37176.

Coleman, D.C., Sullivan, D.J., Russell, R.J., Arbuthnott, J.P., Carey, B.F., and Pomeroy, H.M. 1989. *Staphylococcus aureus* bacteriophages mediating the simultaneous lysogenic conversion of beta-lysin, staphylokinase and enterotoxin A: molecular mechanism of triple conversion. J. Gen. Microbiol. 135: 1679-1697.

Collen, D. 1998. Staphylokinase: a potent, uniquely fibrin-selective thrombolytic agent. Nat. Med. 4: 279-284.

Collen, D., Schlott, B., Engelborghs, Y., Van Hoef, B., Hartmann, M., Lijnen, H.R., and Behnke, D. 1993. On the mechanism of the activation of human plasminogen by recombinant staphylokinase. J. Biol. Chem. 268: 8284-8289.

Cooney, J., Kienle, Z., Foster, T.J., and O'Toole, P.W. 1993. The gamma-hemolysin locus of *Staphylococcus aureus* comprises three linked genes, two of which are identical to the genes for the F and S components of leukocidin. Infect. Immun. 61: 768-771.

Cramton, S.E., Gerke, C., Schnell, N.F., Nichols, W.W., and Götz, F. 1999. The intercellular adhesion (*ica*) locus is present in *Staphylococcus aureus* and is required for biofilm formation. Infect. Immun. 67: 5427-5433.

Cucarella, C., Solano, C., Valle, J., Amorena, B., Lasa, I., and Penades, J.R.. 2001. Bap, a *Staphylococcus aureus* surface protein involved in biofilm formation. J. Bacteriol. 183: 2888-2896.

Cunnion, K.M.,Lee, J.C., and Frank, M.M.. 2001. Capsule production and growth phase influence binding of complement to *Staphylococcus aureus*. Infect. Immun. 69: 6796-6803.

Czech, A., Yamaguchi, T.,Bader, L., Linder, S., Kaminski, K., Sugai, M., and Aepfelbacher, M. 2001. Prevalence of Rho-inactivating epidermal cell differentiation inhibitor toxins in clinical *Staphylococcus aureus* isolates. J. Infect. Dis. 184: 785-788.

Dal Sasso, M., Bovio, C., Culici, M., and Braga, P.C. 2002. Interference of sub-inhibitory concentrations of gatifloxacin on various determinants of bacterial virulence. J. Chemother. 14: 473-482.

Dinges, M.M., Orwin, P.M., and Schlievert, P.M. 2000. Exotoxins of *Staphylococcus aureus*. Clin. Microbiol. Rev. 13: 16-34.

Donvito, B., Etienne, J., Greenland, T., Mouren, C., Delorme, V., and Vandenesch, F. 1997. Distribution of the synergistic haemolysin genes *hld* and *slush* with respect to *agr* in human staphylococci. FEMS Microbiol Lett. 151: 139-144.

Doss, S.A. 1993. Effect of sub-inhibitory concentrations of antibiotics on the virulence of *Staphylococcus aureus*. J. App. Bacteriol. 75: 123-128.

Downer, R., Roche, F., Park, P.W., Mecham, R.P., and Foster, T.J. 2002. The elastin-binding protein of *Staphylococcus aureus* (EbpS) is expressed at the cell surface as an integral membrane protein and not as a cell wall-associated protein. J. Biol. Chem. 277: 243-250.

Dziewanowska, K., Carson, A.R., Patti, J.M., Deobald, C.F., Bayles, K.W., and Bohach, G.A. 2000. Staphylococcal fibronectin binding protein interacts with heat shock protein 60 and integrins: role in internalization by epithelial cells. Infect. Immun. 68: 6321-6328.

Dziewanowska, K., Patti, J.M., Deobald, C.F., Bayles, K.W., Trumble, W.R., and Bohach, G.A. 1999. Fibronectin binding protein and host cell tyrosine kinase are required for internalization of *Staphylococcus aureus* by epithelial cells. Infect. Immun. 67: 4673-4678.

Elasri, M.O., Thomas, J.R., Skinner, R.A. Blevins, J.S., Beenken, K.E., Nelson, C.L., and Smelter, M.S. 2002. *Staphylococcus aureus* collagen adhesin contributes to the pathogenesis of osteomyelitis. Bone 30: 275-280.

Farrell, A.M., Taylor, D., and Holland. K.T. 1995. Cloning, nucleotide sequence determination and expression of the *Staphylococcus aureus* hyaluronate lyase gene. FEMS Microbiol. Lett. 130: 81-85.

Fast, D.J., P. M. Schlievert, and R. D. Nelson. 1988. Nonpurulent response to toxic shock syndrome toxin 1-producing *Staphylococcus aureus*. Relationship to toxin-stimulated production of tumor necrosis factor. J. Immunol. 140: 949-953.

Ferreras, M., Hoper, F., Dalla Serra, M., Colin, D.A., Prevost, G., and Menestrina, G. 1998. The interaction of *Staphylococcus aureus* bi-component gamma-hemolysins and leucocidins with cells and lipid membranes. Biochim. Biophys. Acta. 1414: 108-126.

Fierobe, L., Decre, D., Muller, C., Lucet, J.C., Marmuse, J.P., Mantz, J., and Desmonts, J.M.. 1999. Methicillin-resistant *Staphylococcus aureus* as a causative agent of postoperative intra-abdominal infection: relation to nasal colonization. Clin. Infect. Dis. 29: 1231-1238.

Fink, D. 1989. *Staphylococcus aureus* alpha-toxin activates phospholipases and induces a Ca^{2+} influx in PC12 cells. Cell. Signal. 1: 387-393.

Fitton, J.E., Dell, A., and Shaw, W.V. 1980. The amino acid sequence of the delta haemolysin of *Staphylococcus aureus* . FEBS Lett. 115: 209-212.

Foster, T.J., and Hook, M. 1998. Surface protein adhesins of *Staphylococcus aureus*. Trends Microbiol. 6: 484-488.

Foster, T.J., O'Reilly, M., Phonimdaeng, P., Cooney, J., Patel, A.H., and Bramley, A.J. 1990. Genetic studies of virulence factors of *Staphylococcus aureus*. Properties of coagulase and gamma-toxin, alpha-toxin, beta-toxin and protein A in the pathogenesis of *S. aureus* infections. In: Molecular Biology of the Staphylococci. Novick, R.P., ed. VCH Publishers, New York. p. 403-420.

Fournier, B., Klier, A., and Rapoport, G. 2001. The two-component system ArlS-ArlR is a regulator of virulence gene expression in *Staphylococcus aureus*. Mol. Microbiol. 41: 247-261.

Fowler, V.G., Jr., Fey, P.D., Reller, L.B., Chamis, A.L., Corey, G.R., and Rupp, M.E. 2001. The intercellular adhesin locus ica is present in clinical isolates of *Staphylococcus aureus* from bacteremic patients with infected and uninfected prosthetic joints. Med. Microbiol. Immunol. 189: 127-131.

Francois, P., Vaudaux, P., Foster, T.J., and Lew, D.P. 1996. Host-bacteria interactions in foreign body infections. Infect. Control Hosp. Epidemiol. 17: 514-520.

Garvis, S., Mei, J.M., Ruiz-Albert, J., and Holden, D.W. 2002. *Staphylococcus aureus svrA*: a gene required for virulence and expression of the *agr* locus. Microbiol. 148: 3235-3243.

Gaskill, M.E., and Khan, S.A. 1988. Regulation of the enterotoxin B gene in *Staphylococcus aureus* . J. Biol. Chem. 263: 6276-6280.

Gillaspy, A.F., Lee, C.Y., Sau, S., Cheung, A.L., and Smeltzer. M.S. 1998. Factors affecting the collagen binding capacity of *Staphylococcus aureus*. Infect. Immun. 66: 3170-3178.

Gillaspy, A.F., Patti, J.M., Pratt, Jr., F.L., Iandolo, J.J., and Smeltzer, M.S. 1997. The *Staphylococcus aureus* collagen adhesin-encoding gene (*cna*) is within a discrete genetic element. Gene 196: 239-248.

Gillet, Y., Issartel, B., Vanhems, P., Fournet, J.C., Lina, G., Bes, M., Vandenesch, F., Piemont, Y., Brousse, N., Floret, D., and Etienne, J. 2002. Association between *Staphylococcus aureus* strains carrying gene for Panton-Valentine leukocidin and highly lethal necrotising pneumonia in young immunocompetent patients. Lancet 359: 753-759.

Giraudo, A.T., Calzolari, A., Cataldi, A.A., Bogni, C., and Nagel, R. 1999. The *sae* locus of *Staphylococcus aureus* encodes a two-component regulatory system. FEMS Microbiol. Lett. 177: 15-22.

Giraudo, A.T., Cheung, A.L., and Nagel, R. 1997. The *sae* locus of *Staphylococcus aureus* controls exoprotein synthesis at the transcriptional level. Arch. Microbiol. 168: 53-58.

Graille, M., Stura, E.A., Corper, A.L., Sutton, B.J., Taussig, M.J., Charbonnier, J.B., and Silverman, G.J. 2000. Crystal structure of a *Staphylococcus aureus* protein A domain complexed with the Fab fragment of a human IgM antibody: structural basis for recognition of B-cell receptors and superantigen activity. Proc. Natl. Acad. Sci. USA. 97: 5399-5404.

Gresham, H.D., Lowrance, J.H., Caver, T.E., Wilson, B.S., Cheung, A.L., and Lindberg, F.P. 2000. Survival of *Staphylococcus aureus* inside neutrophils contributes to infection. J. Immunol. 164: 3713-3722.

Hahn, L.H., and Yamada, K.M. 1979. Identification and isolation of a collagen-binding fragment of the adhesive glycoprotein fibronectin. Proc. Natl. Acad. Sci. USA. 76: 1160-1163.

Harbarth, S., Liassine, N., Dharan, S., Herrault, P., Auckenthaler, R., and Pittet, D. 2000. Risk factors for persistent carriage of methicillin-resistant *Staphylococcus aureus*. Clin. Infect. Dis. 31: 1380-1385.

Hartleib, J., Kohler, N., Dickinson, R.B., Chhatwal, G.S., Sixma, J.J., Hartford, O.M., Foster, T.J., Peters, G., Kehrel, B.E., and Herrmann, M. 2000. Protein A is the von Willebrand factor binding protein on *Staphylococcus aureus*. Blood 96: 2149-2156.

Hartman, B.J., and Tomasz, A. 1984. Low-affinity penicillin-binding protein associated with beta-lactam resistance in *Staphylococcus aureus*. J. Bacteriol. 158: 513-516.

Hebert, G.A., and Hancock, G.A. 1985. Synergistic hemolysis exhibited by species of staphylococci. J. Clin. Microbiol.. 22: 409-415.

Herbert, S., Barry, P., and Novick, R.P. 2001. Subinhibitory clindamycin differentially inhibits transcription of exoprotein genes in *Staphylococcus aureus*. Infect. Immun. 69: 2996-3003.

Hook, M., and Foster, T.J. 2000. Staphylococcal surface proteins. In: Gram-Postive Pathogens. Fischetti, V.A., Novick, R.P., Feretti, J.J., Portnoy, D.A., and Rood, J.I., ed. ASM Press, Washington D.C. p. 386-391.

Huesca, M., Peralta, R., Sauder, D.N., Simor, A.E., and McGavin, M.J. 2002. Adhesion and virulence properties of epidemic Canadian methicillin-resistant *Staphylococcus aureus* strain 1: identification of novel adhesion functions associated with plasmin-sensitive surface protein. J. Infect. Dis. 185: 1285-1296.

Jarraud, S. 2001. *egc*, a highly prevalent operon of enterotoxin gene, forms a putative nursery of superantigens in *Staphylococcus aureus*. J. Immunol. 166: 669-677.

Jett, B. D., and Gilmore, M.S. 2002. Internalization of *Staphylococcus aureus* by human corneal epithelial cells: role of bacterial fibronectin-binding protein and host cell factors. Infect. Immun. 70: 4697-4700.

Ji, G., Beavis, R.C., and Novick, R.P. 1995. Cell density control of staphylococcal virulence mediated by an octapeptide pheromone. Proc. Natl. Acad. Sci. USA 92: 12055-12059.

Ji, Y., Marra, A., Rosenberg, M., and Woodnutt, G. 1999. Regulated antisense RNA eliminates alpha-toxin virulence in *Staphylococcus aureus* infection. J. Bacteriol. 181: 6585-6590.

Joh, D., Wann, E.R., Kreikemeyer, B., Speziale, P., and Hook, M. 1999. Role of fibronectin-binding MSCRAMMs in bacterial adherence and entry into mammalian cells. Matrix Biol. 18: 211-223.

Jonsson, K., McDevitt, D., McGavin, M.H., Patti, J.M. and Hook, M. 1995. *Staphylococcus aureus* expresses a major histocompatibility complex class II analog. J. Biol. Chem. 270: 21457-21460.

Josefsson, E., O. Hartford, L. O'Brien, J. M. Patti, and T. Foster. 2001. Protection against experimental *Staphylococcus aureus* arthritis by vaccination with clumping factor A, a novel virulence determinant. J. Infect. Dis. 184: 1572-1580.

Josefsson, E., McCrea, K.W., Ni Eidhin, D., O'Connell, D., Cox, J., Hook, M., and Foster, T.J. 1998. Three new members of the serine-aspartate repeat protein multigene family of *Staphylococcus aureus*. Microbiol. 144: 3387-3395.

Kahl, B.C., Goulian, M., van Wamel, W., Herrmann, M., Simon, S.M., Kaplan, G., Peters, G., and Cheung, A.L. 2000. *Staphylococcus aureus* RN6390 replicates and induces apoptosis in a pulmonary epithelial cell line. Infect. Immun. 68: 5385-5392.

Kalmeijer, M.D., van Nieuwland-Bollen, E., Bogaers-Hofman, D., and de Baere, G.A. 2000. Nasal carriage of *Staphylococcus aureus* is a major risk factor for surgical-site infections in orthopedic surgery. Infect. Control Hosp. Epidemiol. 21: 319-323.

Karlsson, A., Saravia-Otten, P., Tegmark, K., Morfeldt, E., and Arvidson, S. 2001. Decreased amounts of cell wall-associated protein A and fibronectin-binding proteins in *Staphylococcus aureus sarA* mutants due to up-regulation of extracellular proteases. Infect. Immun. 69: 4742-4748.

Katayama, Y., Ito, T., and Hiramatsu, K. 2000. A new class of genetic element, staphylococcus cassette chromosome *mec*, encodes methicillin resistance in *Staphylococcus aureus*. Antimicrob. Agents Chemother. 44: 1549-1555.

Kawabata, S., Morita, T., Iwanaga, S, and Igarashi, H. 1985. Enzymatic properties of staphylothrombin, an active molecular complex formed between staphylocoagulase and human prothrombin. J. Biochem. (Tokyo). 98: 1603-1614.

Kernodle, D.S., McGraw, P.A., Barg, N.L., Menzies, B.E., Voladri, R.K., and Harshman, S. 1995. Growth of *Staphylococcus aureus* with nafcillin *in vitro* induces alpha-toxin production and increases the lethal activity of sterile broth filtrates in a murine model. J. Infect. Dis. 172: 410-419.

Kielian, T., Cheung, A., and Hickey, W.F. 2001. Diminished virulence of an alpha-toxin mutant of *Staphylococcus aureus* in experimental brain abscesses. Infect. Immun. 69: 6902-6911.

Kim, D.M., Lee, S.J., Yoon, S.K., and Byun, S.M. 2002. Specificity role of the streptokinase C-terminal domain in plasminogen activation. Biochem. Biophys. Res. Comm. 290: 585-588.

Konig, B., Prevost, G., and Konig, W. 1997. Composition of staphylococcal bi-component toxins determines pathophysiological reactions. J. Med. Microbiol. 46: 479-485.

Krakauer, T. 1999. Immune response to staphylococcal superantigens. Immunol. Res. 20: 163-173.

Kullik, I., Giachino, P., and Fuchs, T. 1998. Deletion of the alternative sigma factor σ^B in *Staphylococcus aureus* reveals its function as a global regulator of virulence genes. J. Bacteriol. 180: 4814-4820.

Kuroda, M., Ohta, T., Uchiyama, I., Baba, T., Yuzawa, H., Kobayashi, I., Cui, L., Oguchi, A., Aoki, K., Nagai, Y., Lian, J., Ito, T., Kanamori, M., Matsumaru, H., Maruyama, A., Murakami, H., Hosoyama, A., Mizutani-Ui, y., Takahashi, N.K., Sawano, T., Inoue, R., Kaito, C., Sekimizu, K., Hirakawa, H., Kuhara, S., Goto, S., Yabuzaki, J., Kanehisa, M., Yamashita, A., Oshima, K., Furuya, K., Yoshino, C., Shiba, T., Hattori, M., Ogasawara, N., Hayashi, H., and Hiramatsu, K. 2001. Whole genome sequencing of methicillin-resistant *Staphylococcus aureus*. Lancet 357: 1225-1240.

Lebeau, C., Vandenesch, F., Greeland, T., Novick, R.P., and Etienne, J. 1994. Coagulase expression in *Staphylococcus aureus* is positively and negatively modulated by an *agr*-dependent mechanism. J. Bacteriol. 176: 5534-5536.

Lee, C.Y., and Iandolo, J.J. 1986. Integration of staphylococcal phage L54a occurs by site-specific recombination: structural analysis of the attachment sites. Proc. Natl. Acad. Sci. USA. 83: 5474-5478.

Lee, J.C., Betley, M.J., Hopkins, C.A., Perez, N.E., and Pier, G.B. 1987. Virulence studies, in mice, of transposon-induced mutants of *Staphylococcus aureus* differing in capsule size. J. Infect. Dis. 156: 741-750.

Lindsay, J.A., Ruzin, A., Ross, H.F., Kurepina, N., and Novick, R.P. 1998. The gene for toxic shock toxin is carried by a family of mobile pathogenicity islands in *Staphylococcus aureus*. Mol. Microbiol. 29: 527-543.

Lundqvist-Gustafsson, H., Norrman, S., Nilsson, J., and Wilsson, A. 2001. Involvement of p38-mitogen-activated protein kinase in *Staphylococcus aureus*-induced neutrophil apoptosis. J. Leukocyte Biol. 70: 642-648.

Luong, T.T., and Lee, C.Y. 2002. Overproduction of type 8 capsular polysaccharide augments *Staphylococcus aureus* virulence. Infect. Immun. 70: 3389-3395.

Manna, A., and Cheung, A.L. 2001. Characterization of *sarR*, a modulator of *sar* expression in *Staphylococcus aureus*. Infect. Immun. 69: 885-896.

Marshall, M.J., Bohach, G.A., and Boehm, D.F. 2000. Characterization of *Staphylococcus aureus* beta-toxin induced leukotoxicity. J. Nat. Toxins 9: 125-138.

Massimi, I., Park, E., Rice, K., Muller-Esterl, W., Sauder, D.N., and McGavin, M.J. 2002. Identification of a novel maturation mechanism and restricted substrate specificity for the SspB cysteine protease of *Staphylococcus aureus*. J. Biol. Chem. 41770-41777.

Mazmanian, S.K., Liu, G., Jensen, E.R., Lenoy, E., and Schneewind, O. 2000. *Staphylococcus aureus* sortase mutants defective in the display of surface proteins and in the pathogenesis of animal infections. Proc. Natl. Acad. Sci. USA. 97: 5510-5515.

Mazmanian, S.K., Liu, G., Ton-That, H., and Schneewind, O. 1999. *Staphylococcus aureus* sortase, an enzyme that anchors surface proteins to the cell wall. Science 285: 760-763.

Mazmanian, S.K., Ton-That, H., Su, K., and Schneewind, O. 2002. An iron-regulated sortase anchors a class of surface protein during *Staphylococcus aureus* pathogenesis. Proc. Natl. Acad. Sci. USA 99:2293-2298.

McAleese, F.M., Walsh, E.J., Sieprawska, M., Potempa, J., and Foster, T.J. 2001. Loss of clumping factor B fibrinogen binding activity by *Staphylococcus aureus* involves cessation of transcription, shedding and cleavage by metalloprotease. J. Biol. Chem. 276: 29969-29978.

McCormick, J.K., Yarwood, J.M., and Schlievert, P.M. 2001. Toxic shock syndrome and bacterial superantigens: an update. Ann. Rev. Microbiol. 55: 77-104.

McDevitt, D., Francois, P., Vaudaux, P., and Foster, T.J. 1995. Identification of the ligand-binding domain of the surface-located fibrinogen receptor (clumping factor) of *Staphylococcus aureus*. Mol. Microbiol. 16: 895-907.

McDevitt, D., Francois, P., Vaudaux, P., and Foster, T.J. 1994. Molecular characterization of the clumping factor (fibrinogen receptor) of *Staphylococcus aureus*. Mol. Microbiol. 11: 237-248.

McDevitt, D., Vaudaux, P., and Foster, T.J. 1992. Genetic evidence that bound coagulase of *Staphylococcus aureus* is not clumping factor. Infect. Immun. 60: 1514-1523.

McGavin, M.H., Krajewska-Pietrasik, D., Ryden, C., and Hook, M. 1993. Identification of a *Staphylococcus aureus* extracellular matrix-binding protein with broad specificity. Infect. Immun. 61: 2479-2485.

McGavin, M.J., Zahradka, C., Rice, K., and Scott, J.E. 1997. Modification of the *Staphylococcus aureus* fibronectin binding phenotype by V8 protease. Infect. Immun. 65: 2621-2628.

McKenney, D., Pouliot, K.L., Wang, Y., Murthy, V., Ulrich, M., Doring, G., Lee, J.C., Goldmann, D.A., and Pier, G.B. 1999. Broadly protective vaccine for *Staphylococcus aureus* based on an *in vivo*-expressed antigen. Science 284: 1523-1527.

McNamara, P.J., and Iandolo, J.J. 1998. Genetic instability of the global regulator *agr* explains the phenotype of the *xpr* mutation in *Staphylococcus aureus* KSI9051. J. Bacteriol. 180: 2609-2615.

McNamara, P.J., Milligan-Monroe, K.C., Bates, D., and Proctor, R.A. 2001. Presented at the American Society for Microbiology General Meeting, Orlando, FL.

McNamara, P.J., Milligan-Monroe, K.C., Khalili, S., and Proctor, R.A. 2000. Identification, cloning, and initial characterization of *rot*, a locus encoding a regulator of virulence factor expression in *Staphylococcus aureus*. J. Bacteriol. 182: 3197-3203.

Menzies, B.E., and Kourteva, I. 2000. *Staphylococcus aureus* alpha-toxin induces apoptosis in endothelial cells. FEMS Immunol. Med. Microbiol. 29: 39-45.

Mizobuchi, S., Minami, J., Jin, F., Matsushita, O., and Okabe, A. 1994. Comparison of the virulence of methicillin-resistant and methicillin-sensitive *Staphylococcus aureus*. Microbiol. Immunol. 38: 599-605.

Molkanen, T., Tyynela, J., Helin, J., Kalkkinen, N., and Kuusela, P. 2002. Enhanced activation of bound plasminogen on *Staphylococcus aureus* by staphylokinase. FEBS Lett. 517: 72-78.

Morfeldt, E., Taylor, D., von Gabain, A., and Arvidson, S. 1995. Activation of alpha-toxin translation in *Staphylococcus aureus* by the trans-encoded antisense RNA, RNAIII. EMBO J. 14: 4569-4577.

Mortensen, J.E., Shryock, T.R., and Kapral, F.A. 1992. Modification of bactericidal fatty acids by an enzyme of *Staphylococcus aureus*. J. Med. Microbiol. 36: 293-298.

Narita, S., Kaneko, J., Chiba, J., Piemont, Y., Jarraud, S., Etienne, J., and Kamio, Y. 2001. Phage conversion of Panton-Valentine leukocidin in *Staphylococcus aureus*: molecular analysis of a PVL-converting phage, phiSLT. Gene 268: 195-206.

Navarre, W.W., and Schneewind, O. 1999. Surface proteins of gram-positive bacteria and mechanisms of their targeting to the cell wall envelope. Microbiol. Mol. Bio. Rev. (Washington, DC). 63: 174-229.

Nguyen, T., Ghebrehiwet, B., and Peerschke, E.I. 2000. *Staphylococcus aureus* protein A recognizes platelet gC1qR/p33: a novel mechanism for staphylococcal interactions with platelets. Infect. Immun. 68: 2061-2068.

Ni Eidhin, D., Perkins, S., Francois, P., Vaudaux, P., Hook, M., and Foster, T.J. 1998. Clumping factor B (ClfB), a new surface-located fibrinogen-binding adhesin of *Staphylococcus aureus*. Mol. Microbiol. 30: 245-257.

Nilsson, I.M., Lee, J.C., Bremell, T., Ryden, C., and Tarkowski, A. 1997. The role of staphylococcal polysaccharide microcapsule expression in septicemia and septic arthritis. Infect. Immun. 65: 4216-4221.

Novick, R.P. 2003. Autoinduction and signal transduction in the regulation of staphylococcal virulence. Mol. Microbiol. In press.

Novick, R.P. 2000. Pathogenicity factors and their regulation. In: Gram-Postive Pathogens. Fischetti, V.A., Novick, R.P., Feretti, J.J, Portnoy, D.A, and Rood, J.I., ed. ASM Press, Washington D.C. p. 392-407.

Novick, R.P., Ross, H.F., Projan, S.J., Kornblum, J., Kreiswirth, B., and Moghazeh, S. 1993. Synthesis of staphylococcal virulence factors is controlled by a regulatory RNA molecule. EMBO J. 12: 3967-3975.

Novick, R.P., Schlievert, P., and Ruzin, A. 2001. Pathogenicity and resistance islands of staphylococci. Microb. Infect. 3: 585-594.

Novick, R., and Skurray, R. (ed.). 1990. Molecular Biology of the Staphylococci. VCH, New York.

O'Connell, D.P., Nanavaty, T., McDevitt, D., Gurusiddappa, S., Hook, M., and Foster, T.J. 1998. The fibrinogen-binding MSCRAMM (clumping factor) of *Staphylococcus aureus* has a Ca^{2+}-dependent inhibitory site. J. Biol. Chem. 273: 6821-6829.

O'Reilly, M. 1986. Inactivation of the alpha-haemolysin gene of *Staphylococcus aureus* 8325-4 by site-directed mutagenesis and studies on the expression of its haemolysins. Microb. Path. 1:125-38.

Ohlsen, K., Ziebuhr, W., Koller, K.P., Hell, W., Wichelhaus, T.A., and Hacker, J. 1998. Effects of subinhibitory concentrations of antibiotics on alpha-toxin (*hla*) gene expression of methicillin-sensitive and methicillin-resistant *Staphylococcus aureus* isolates. Antimicrob. Agents Chemother. 42: 2817-2823.

Okada, K., Nonaka, T., Matsumoto, H., Fukao, H., Ueshima, S., and Matsuo, O. 1994. Effects of alpha 2-plasmin inhibitor on plasminogen activation by staphylokinase/plasminogen complex. Thrombosis Res. 76: 211-220.

Onogawa, T. 2002. Staphylococcal alpha-toxin synergistically enhances inflammation caused by bacterial components. FEMS Immun. Med. Microbiol. 33: 15-21.

Papakyriacou, H., Vaz, D., Simor, A., Louie, M., and McGavin, M.J. 2000. Molecular analysis of the accessory gene regulator (*agr*) locus and balance of virulence factor expression in epidemic methicillin-resistant *Staphylococcus aureus*. J. Infect. Dis. 181: 990-1000.

Park, P.W., Broekelmann, T.J., Mecham, B.R., and Mecham, R.P. 1999. Characterization of the elastin binding domain in the cell-surface 25-kDa elastin-binding protein of *Staphylococcus aureus* (EbpS). J. Biol. Chem.. 274: 2845-2850.

Patel, A.H., Kornblum, J., Kreiswirth, B., Novick, R., and Foster, T.J. 1992. Regulation of the protein A-encoding gene in *Staphylococcus aureus*. Gene 114: 25-34.

Patel, A.H., Nowlan, P., Weavers, E.D., and Foster, T. 1987. Virulence of protein A-deficient and alpha-toxin-deficient mutants of *Staphylococcus aureus* isolated by allele replacement. Infect. Immun. 55: 3103-3110.

Patti, J.M., Bremell, T., Krajewska-Pietrasik, D., Abdelnour, A., Tarkowski, A., Ryden, C., and Hook, M. 1994. The *Staphylococcus aureus* collagen adhesin is a virulence determinant in experimental septic arthritis. Infect. Immun. 62: 152-161.

Peacock, S.J., de Silva, I., and Lowy, F.D. 2001. What determines nasal carriage of *Staphylococcus aureus*? Trends Microbiol. 9: 605-610.

Perl, T.M., and Golub, J.E. 1998. New approaches to reduce *Staphylococcus aureus* nosocomial infection rates: treating S. aureus nasal carriage. Ann. Pharmacother. 32: S7-S16.

Phonimdaeng, P., O'Reilly, M., Nowlan, P., Bramley, A.J., and Foster, T.J. 1990. The coagulase of *Staphylococcus aureus* 8325-4. Sequence analysis and virulence of site-specific coagulase-deficient mutants. Mol. Microbiol. 4: 393-404.

Plano, L.R.W., Gutman, D.M., Woischnik, M., and Collins, C.M. 2000. Recombinant *Staphylococcus aureus* exfoliative toxins are not bacterial superantigens. Infect. Immun. 68: 3048-3754.

Pohlmann-Dietze, P., Ulrich, M., Kiser, K.B., Doring, G., Lee, J.C., Fournier, J.M., Botzenhart, K., and Wolz, C. 2000. Adherence of *Staphylococcus aureus* to endothelial cells: influence of capsular polysaccharide, global regulator *agr*, and bacterial growth phase. Infect. Immun. 68: 4865-4871.

Pokorny, A., Birkbeck, T.H., and Almeida, P.F. 2002. Mechanism and kinetics of delta-lysin interaction with phospholipid vesicles. Biochem. 41: 11044-11056.

Potempa, J., Dubin, A., Korzus, G., and Travis, J. 1988. Degradation of elastin by a cysteine proteinase from *Staphylococcus aureus*. J. Biol. Chem. 263: 2664-2667.

Potempa, J., Watorek, W., and Travis, J. 1986. The inactivation of human plasma alpha 1-proteinase inhibitor by proteinases from *Staphylococcus aureus*. J. Biol. Chem.. 261: 14330-14334.

Prevost, G., Couppie, P., Prevost, P., Gayet, S., Petiau, P., Cribier, B., Monteil, H., and Piemont, Y. 1995. Epidemiological data on *Staphylococcus aureus* strains producing synergohymenotropic toxins. J. Med. Microbiol. 42: 237-245.

Proctor, R.A., Hamill, R.J., Mosher, D.F., Textor, J.A., and Olbrantz, P.J. 1983a. Effects of subinhibitory concentrations of antibiotics on *Staphylococcus aureus* interactions with fibronectin. J. Antimicrob. Chemother. 12 Suppl. C: 85-95.

Proctor, R.A., Olbrantz, P.J., and Mosher, D.F. 1983b. Subinhibitory concentrations of antibiotics alter fibronectin binding to *Staphylococcus aureus*. Antimicrob. Agents Chemother. 24: 823-826.

Proctor, R.A., van Langevelde, P., Kristjansson, M., Maslow, J.N., and Arbeit, R.D. 1995. Persistent and relapsing infections associated with small-colony variants of *Staphylococcus aureus*. Clin. Infect. Dis. 20: 95-102.

Que, Y.A., Francois, P., Haefliger, J.A., Entenza, J.M., Vaudaux, P., and Moreillon, P. 2001. Reassessing the role of *Staphylococcus aureus* clumping factor and fibronectin-binding protein by expression in *Lactococcus lactis*. Infect. Immun. 69: 6296-6302.

Recsei, P., Kreiswirth, B., O'Reilly, M., Schlievert, P., Gruss, A., and Novick, R.P. 1986. Regulation of exoprotein gene expression in *Staphylococcus aureus* by *agr*. Mol. Gen. Genet. 202: 58-61.

Regassa, L.B., Couch, J.L., and Betley, M.J. 1991. Steady-state staphylococcal enterotoxin type C mRNA is affected by a product of the accessory gene regulator *agr* and by glucose. Infect. Immun. 59: 955-962.

Rhem, M.N., Lech, E.M., Patti, J.M., McDevitt, D., Hook, M., Jones, D.B., and Wilhelmus, K.R. 2000. The collagen-binding adhesin is a virulence factor in *Staphylococcus aureus* keratitis. Infect. Immun. 68: 3776-3754.

Rice, K., Peralta, R., Bast, D., de Azavedo, J., and McGavin, M.J. 2001. Description of staphylococcus serine protease (*ssp*) operon in *Staphylococcus aureus* and nonpolar inactivation of sspA-encoded serine protease. Infect. Immun. 69: 159-169.

Rogolsky, M., Warren, R., Wiley, B., Nakamura, H.T., and Glasgow, L.A. 1974. Nature of the genetic determinant controlling exfoliative toxin production in *Staphylococcus aureus*. J. Bacteriol. 117: 157-165.

Rousseaux, J., Rousseaux-Prevost, R., Bazin, H., and Biserte, G. 1983. Proteolysis of rat IgG subclasses by *Staphylococcus aureus* V8 proteinase. Biochimica et Biophysica Acta. 748: 205-212.

Saravia-Otten, P., H. P. Muller, and S. Arvidson. 1997. Transcription of *Staphylococcus aureus* fibronectin binding protein genes is negatively regulated by *agr* and an *agr*-independent mechanism. J. Bacteriol. 179: 5259-5263.

Sato, H., Watanabe, T., Murata, Y., Ohtake, A., Nakamura, M., Aizawa, C., Saito, H., and Maehara, N. 1999. New exfoliative toxin produced by a plasmid-carrying strain of *Staphylococcus hyicus*. Infect. Immun. 67: 4014-4018.

Sawai, T., Tomono, K., Yanagihara, K., Yamamoto, Y., Kaku, M., Hirakata, Y., Koga, H., Tashiro, T., and Kohno, S. 1997. Role of coagulase in a murine model of hematogenous pulmonary infection induced by intravenous injection of *Staphylococcus aureus* enmeshed in agar beads. Infect. Immun. 65: 466-471.

Scheifele, D.W., Bjornson, G.L., Dyer, R.A., and Dimmick, J.E. 1987. Delta-like toxin produced by coagulase-negative staphylococci is associated with neonatal necrotizing enterocolitis. Infect. Immun. 55: 2268-2273.

Schlievert, P., Osterholm, M., Kelly, J., and Nishimura, R. 1982. Toxin and enzyme characterization of *Staphylococcus aureus* isolates from patients with and without toxic-shock syndrome. Ann. Intern. Med. 96: 937-940.

Schmidt, K.A., Manna, A.C., Gill, S., and Cheung, A.L. 2001. SarT, a repressor of alpha-hemolysin in *Staphylococcus aureus*. Infect. Immun. 69: 4749-4758.

Shapiro, R.L., Duquette, J.G., Nunes, I., Roses, D.F., Harris, M.N., Wilson, E.L., and Rifkin, D.B. 1997. Urokinase-type plasminogen activator-deficient mice are predisposed to staphylococcal botryomycosis, pleuritis, and effacement of lymphoid follicles. Am. J. Pathol. 150: 359-369.

Sheehan, B.J., Foster, T.J., Dorman, C.J., Park, S., and Stewart, G.S. 1992. Osmotic and growth-phase dependent regulation of the *eta* gene of *Staphylococcus aureus*: a role for DNA supercoiling. Mol. Gen. Genet. 232: 49-57.

Shibl, A.M. 1982. Subcutaneous staphylococcal infections in mice: the influence of antibiotics on staphylococcal extracellular products. Chemother. 28: 46-53.

Shibl, A.M. 1984. Selective inhibition of enzyme synthesis by lincomycin in *Staphylococcus aureus*. J. Antimicrob. Chemother. 13: 625-7.

Shinefield, H., Black, S., Fattom, A., Horwith, G., Rasgon, S., Ordonez, J., Yeoh, H., Law, D., Robbins, J.B., Schneerson, R., Muenz, L., Fuller, S., Johnson, J., Fireman, B., Alcorn, H., and Naso, R. 2002. Use of a *Staphylococcus aureus* conjugate vaccine in patients receiving hemodialysis. N. Engl. J. Med. 346: 491-496.

Shopsin, B., Mathema, B., Zhao, X., Martinez, J., Kornblum, J., and Kreiswirth, B.N. 2000. Resistance rather than virulence selects for the clonal spread of methicillin-resistant *Staphylococcus aureus*: implications for MRSA transmission. Microbial Drug Resistance-Mechanisms Epidemiol. Dis. 6: 239-244.

Siboo, I.R., Cheung, A.L., Bayer, A.S., and Sullam, P.M. 2001. Clumping factor A mediates binding of *Staphylococcus aureus* to human platelets. Infect. Immun. 69: 3120-3127.

Silence, K., Collen, D., and Lijnen, H.R. 1993. Interaction between staphylokinase, plasmin(ogen), and alpha 2-antiplasmin. Recycling of staphylokinase after neutralization of the plasmin-staphylokinase complex by alpha 2-antiplasmin. J. Biol. Chem. 268: 9811-9816.

Simons, J.W., Adams, H., Cox, R.C., Dekker, N., Götz, F., Slotboom, A.J., and Verheij, H.M. 1996. The lipase from *Staphylococcus aureus*. Expression in *Escherichia coli*, large-scale purification and comparison of substrate specificity to *Staphylococcus hyicus* lipase. Eur. J. Biochem. 242: 760-769.

Sinha, B., Francois, P.P., Nusse, O., Foti, M., Hartford, O.M., Vaudaux, P., Foster, T.J., Lew, D.P., Herrmann, M., and Krause, K.H. 1999. Fibronectin-binding protein acts as *Staphylococcus aureus* invasin via fibronectin bridging to integrin alpha5beta1. Cell. Microbiol. 1: 101-117.

Sinha, P., Ghosh, A.K., Das, T., Sa, G., and Ray, P.K. 1999. Protein A of *Staphylococcus aureus* evokes Th1 type response in mice. Immunol. Lett. 67: 157-165.

Smeltzer, M.S., Hart, M.E., and Iandolo, J.J. 1993. Phenotypic characterization of *xpr*, a global regulator of extracellular virulence factors in *Staphylococcus aureus*. Infect. Immun. 61: 919-925.

Smeltzer, M.S., Hart, M.E., and Iandolo, J.J. 1994. The effect of lysogeny on the genomic organization of *Staphylococcus aureus*. Gene 138: 51-57.

Smyth, C.J., Mollby, R., and Wadstrom, T. 1975. Phenomenon of hot-cold hemolysis: chelator-induced lysis of sphingomyelinase-treated erythrocytes. Infect. Immun. 12: 1104-1111.

Song, L., Hobaugh, M.R., Shustak, C., Cheley, S., Bayley, H., and Gouaux, J.E. 1996. Structure of staphylococcal alpha-hemolysin, a heptameric transmembrane pore. Science 274: 1859-1866.

Sonstein, S.A., and Burnham, J.C. 1993. Effect of low concentrations of quinolone antibiotics on bacterial virulence mechanisms. Diagn. Microbiol. Infect. Dis. 16: 277-289.

Starovasnik, M.A., Skelton, N.J., O'Connell, M.P., Kelley, R.F., Reilly, D., and Fairbrother, W.J. 1996. Solution structure of the E-domain of staphylococcal protein A. Biochem. 35:15558-69.

Stutzmann Meier, P., Entenza, J.M., Vaudaux, P., Francioli, P., Glauser, M.P., and Moreillon, P. 2001. Study of *Staphylococcus aureus* pathogenic genes by transfer and expression in the less virulent organism *Streptococcus gordonii*. Infect. Immun. 69: 657-664.

Switalski, L.M., Patti, J.M., Butcher, W., Gristina, A.G., Speziale, P., and Hook, M. 1993. A collagen receptor on *Staphylococcus aureus* strains isolated from patients with septic arthritis mediates adhesion to cartilage. Mol. Microbiol. 7: 99-107.

Szmigielski, S., Sobiczewska, E., Prevost, G., Monteil, H., Colin, D.A., and Jeljaszewicz, J. 1998. Effect of purified staphylococcal leukocidal toxins on isolated blood polymorphonuclear leukocytes and peritoneal macrophages *in vitro*. Intl. J. Med. Microbiol. 288: 383-394.

Tappin, M.J. 1988. High-resolution 1H NMR study of the solution structure of delta-hemolysin. Biochem. 27: 1643-1647.

Tegmark, K., Karlsson, A., and Arvidson, S. 2000. Identification and characterization of SarH1, a new global regulator of virulence gene expression in *Staphylococcus aureus*. Mol. Microbiol. 37: 398-409.

Thakker, M., Park, J.S., Carey, V., and Lee, J.C. 1998. *Staphylococcus aureus* serotype 5 capsular polysaccharide is antiphagocytic and enhances bacterial virulence in a murine bacteremia model. Infect. Immun. 66: 5183-5189.

Throup, J.P., Zappacosta, F., Lunsford, R.D., Annan, R.S., Carr, S.A., Lonsdale, J.T., Bryant, A.P., McDevitt, D., Rosenberg, M., and Burnham. M.K. 2001.

The *srhSR* gene pair from *Staphylococcus aureus*: genomic and proteomic approaches to the identification and characterization of gene function. Biochem. 40: 10392-10401.

Toshkova, K., Annemuller, C., Akineden, O., and Lammler, C. 2001. The significance of nasal carriage of *Staphylococcus aureus* as risk factor for human skin infections. FEMS Microbiol. Lett. 202: 17-24.

Tremaine, M.T., Brockman, D.K., and Betley, M.J. 1993. Staphylococcal enterotoxin A gene (*sea*) expression is not affected by the accessory gene regulator (*agr*). Infect. Immun. 61: 356-359.

Tzianabos, A. O., J. Y. Wang, and J. C. Lee. 2001. Structural rationale for the modulation of abscess formation by *Staphylococcus aureus* capsular polysaccharides. Proc. Natl. Acad. Sci. USA 98: 9365-9370.

Uehara, Y., Nakama, H., Agematsu, K., Uchida, M., Kawakami, Y., Abdul Fattah, A.S., and Maruchi, N. 2000. Bacterial interference among nasal inhabitants: eradication of *Staphylococcus aureus* from nasal cavities by artificial implantation of *Corynebacterium* sp. J. Hosp. Infect. 44: 127-133.

van der Vijver, J.C., van Es-Boon, M.M., and Michel, M.F. 1975. A study of virulence factors with induced mutants of *Staphylococcus aureus*. J. Med. Microbiol. 8: 279-287.

Vaudaux, P.E., Monzillo, V., Francois, P., Lew, D.P., Foster, T.J., and Berger-Bächi, B. 1998. Introduction of the *mec* element (methicillin resistance) into *Staphylococcus aureus* alters *in vitro* functional activities of fibrinogen and fibronectin adhesins. Antimicrob. Agents Chemother. 42: 564-570.

Veldkamp, K.E., Heezius, H.C., Verhoef, J., van Strijp, J.A., and van Kessel, K.P. 2000. Modulation of neutrophil chemokine receptors by *Staphylococcus aureus* supernate. Infect. Immun. 68: 5908-5913.

Vojtov, N., Ross, H.F., and Novick, R.P. 2002. Global repression of exotoxin synthesis by staphylococcal superantigens. Proc. Natl. Acad. Sci. USA 99: 10102-10107.

von Eiff, C., Becker, K., Machka, K., Stammer, H., and Peters, G. 2001. Nasal carriage as a source of *Staphylococcus aureus* bacteremia. Study Group. N. Engl. J. Med. 344: 11-16.

Wagner, P.L., and Waldor, M.K. 2002. Bacteriophage control of bacterial virulence. Infect. Immun. 70: 3985-3993.

Walev, I., Weller, U., Strauch, S., Foster, T., and Bhakdi, S. 1996. Selective killing of human monocytes and cytokine release provoked by sphingomyelinase (beta-toxin) of *Staphylococcus aureus*. Infect. Immun. 64: 2974-2979.

Wann, E.R., Gurusiddappa, S., and Hook, M. 2000. The fibronectin-binding MSCRAMM FnbpA of *Staphylococcus aureus* is a bifunctional protein that also binds to fibrinogen. J. Biol. Chem. 275: 13863-13871.

Weems, J.J., and Beck, L.B. 2002. Nasal Carriage of *Staphylococcus aureus* As a Risk Factor for Skin and Soft Tissue Infections. Curr. Infect. Dis. Rep. 4: 420-425.

Wesson, C.A., Liou, L.E., Todd, K.M., Bohach, G.A., Trumble, W.R., and Bayles, K.W. 1998. *Staphylococcus aureus* Agr and Sar global regulators influence internalization and induction of apoptosis. Infect. Immun. 66: 5238-5243.

Wilde, C., Chhatwal, G.S., Schmalzing, G., Aktories, K., and Just, I. 2001. A novel C3-like ADP-ribosyltransferase from *Staphylococcus aureus* modifying RhoE and Rnd3. J. Biol. Chem. 276: 9537-9542.

Wolz, C., McDevitt, D., Foster, T.J., and Cheung, A.L. 1996. Influence of *agr* on fibrinogen binding in *Staphylococcus aureus* Newman. Infect. Immun. 64: 3142-3147.

Worlitzsch, D., Kaygin, H., Steinhuber, A., Dalhoff, A., Botzenhart, K., and Doring, G. 2001. Effects of amoxicillin, gentamicin, and moxifloxacin on the hemolytic activity of *Staphylococcus aureus in vitro* and *in vivo*. Antimicrob. Agents Chemother. 45: 196-202.

Yamaguchi, T., Hayashi, T., Takami, H., Ohnishi, M., Murata, T., Nakayama, K., Asakawa, K., Ohara, M., Komatsuzawa, H., and Sugai, M. 2001. Complete nucleotide sequence of a *Staphylococcus aureus* exfoliative toxin B plasmid and identification of a novel ADP-ribosyltransferase, EDIN-C. Infect. Immun. 69: 7760-777.

Yamaguchi, T., Nishifuji, K., Sasaki, M., Fudaba, Y., Aepfelbacher, M., Takata, T., Ohara, M., Komatsuzawa, H., Amagai, M., and Sugai, M. 2002. Identification of the *Staphylococcus aureus* etd pathogenicity island which encodes a novel exfoliative toxin, ETD, and EDIN-B. Infect. Immun. 70: 5835-5845.

Yarwood, J.M., McCormick, J.K., and Schlievert, P.M. 2001. Identification of a novel two-component regulatory system that acts in global regulation of virulence factors of *Staphylococcus aureus*. J. Bacteriol. 183: 1113-1123.

Zhang, S., and Stewart, G.C. 2000. Characterization of the promoter elements for the staphylococcal enterotoxin D gene. J. Bacteriol. 182: 2321-2325.

Zou, D., Kaneko, J., Narita, S., and Kamio, Y. 2000. Prophage, phiPV83-pro, carrying panton-valentine leukocidin genes, on the *Staphylococcus aureus* P83 chromosome: comparative analysis of the genome structures of phiPV83-pro, phiPVL, phi11, and other phages. Biosci. Biotechnol Biochem. 64: 2631-2643.

From: *MRSA: Current Perspectives*
Edited by: A.C. Fluit and F.-J. Schmitz

Chapter 10

Small Colony Variants: Another Mechanism By Which *Staphylococcus aureus* Can Evade the Immune Response and Antimicrobial Therapy

Christof von Eiff and Karsten Becker

Abstract

In the past, most studies have addressed the problem of antibiotic-resistant infectious diseases from the standpoint of classic forms of antibiotic resistance mainly founded on the possession of resistance genes. However, bacteria such as *Staphylococcus aureus* may have additional mechanisms for resisting therapy that extend beyond these classic mechanisms. In patients whose acute infection initially responded to antimicrobial treatment and which recured after long disease-free intervals or with infections that persisted despite appropriate antibiotic treatment, small colony variants (SCVs) of *S. aureus* were recovered. A wide variety of bacterial species are known to form SCVs, but in particular in staphyloccoal infections a renewed interest due to SCVs emerged in the last decade following the first description as dwarf-colony or G variants, since an association of the occuurence of *S. aureus* SCVs and persistent and relapsing infection was described. SCVs are a naturally occuring subpopulation which may be identified in the microbiological laboratory as nonpigmented, nonhemolytic, slow-growing pinpoint colonies. In addition, the often relatively unstable SCVs demonstrate a number of other characteristics that are atypical for *S. aureus* making the correct identification difficult, including: reduced α-toxin production, delayed coagulase activity, failure to use mannitol and an increased resistance to aminoglycosides and cell-wall active antibiotics. Most characteristics of the SCVs can be tied together by a common thread, which is alterations in electron transport. *S. aureus* SCVs from clinical material are commonly auxotrophic for menadione and hemin, which are key co-factors for the formation of menaquinone and cytochromes, respectively, and are thus important components of the electron transport chain. While studies with clinical isolates of SCVs suggested a link between persistent infections and electron transport defective strains or thymidine-auxotrophic SCVs, a defined *hemB* mutant mimicking the SCV phenotype provided additional evidence for these connections. In a model of endovascular infection to determine the intracellular persistence, it was demonstrated that > 200-fold more *hemB*-mutant cells persisted intracellularly after 24 or 48 h incubation relative to the parent strain. The intracellular location of the SCV phenotype may shield this subpopulation from host defenses and antibiotics, thus providing one explanation for the difficulty in clearing *S. aureus* SCVs from host tissues.

Introduction

Staphylococci have been recognized as serious pathogens for over a century, yet despite efforts to halt their spread, they remain one of the most common causes of both, endemic nosocomial infections and epidemics of hospital-acquired infections (Lowy, 1998). In US-hospitals that participate in the National Nosocomial Infections Surveillance (NNIS) system, *Staphylococcus aureus* accounted for 12.6 percent of isolates recovered from bloodstream infections in patients in intensive care units during 1992 through 1999, with the percentage increasing in recent years (National Nosocomial Infections Surveillance, 1999). In addition, community-acquired *S. aureus* infections are common, causing a wide range of infections. The disease spectrum includes bacteremia, endocarditis, central nervous system infections, osteomyelitis, abscesses, pneumonia, urinary tract infections, and a host of syndromes caused by exotoxins, including food poisoning, scalded skin syndrome, and toxic shock syndrome (Lowy, 1998).

A renewed interest in staphylococcal infections has emerged in recent years, since staphylococci resistant to multiple antibiotics have been reported with increasing frequency worldwide (Jones, 2001). Depending on local epidemiological conditions, a significant number of isolates are resistant to methicillin, lincosamides, macrolides, aminoglycosides and/or fluoroquinolones (Moellering, Jr., 1998). New agents give a renewed opportunity for control, however, the emergence of *S. aureus* strains with intermediate resistance to glycopeptides has heightened the fears of pan-antibiotic-resistant strains (Hiramatsu *et al.*, 1997).

In the past, most studies have addressed the problem of antibiotic resistance from the standpoint of classic forms of resistance mainly founded on the possession of genes which are responsible for a wide range of resistance (Daum and Seal, 2001; Hiramatsu *et al.*, 2001). However, bacteria such as *Staphylococcus aureus* may have additional mechanisms for resisting therapy that extend beyond these classic mechanisms. Such mechanisms include an ability to evade the effect of a given substance despite having tested susceptible to it: evasion could possibly be mediated by the production of diffusion barriers, e.g. biofilm production, or by withdrawal into the intracellular milieu (Proctor *et al.*, 1997; Proctor and Peters, 1998; von Eiff *et al.*, 1999). In patients whose acute infection initially seems to respond to antimicrobial treatment and which recurs after long disease-free intervals or with infections that persist despite appropriate antibiotic treatment, small colony variants (SCVs) of the usually extracellularly existing *S. aureus* have been isolated (Goudie and Goudie, 1955; Proctor *et al.*, 1995; von Eiff *et al.*, 1997a; Kahl *et al.*, 1998; Seifert *et al.*, 1999; Looney, 2000; McNamara and Proctor, 2000; Abele-Horn *et al.*, 2000; von Eiff *et al.*, 2001). These variants, which are adapted to the intracellular habitat, have been isolated from patients who have been receiving antibiotic treatment, however, SCVs have also been recovered from humans with staphylococcal infections in the absence of antibiotic pressure (Proctor *et al.*, 1997; McNamara and Proctor, 2000).

SCVs: A Subpopulation as Electron Transport Variants

SCVs represent a naturally occurring subpopulation of *S. aureus* and are defined by nonpigmented and nonhemolytic colonies about 10 times smaller than the parent strain, hence, their name. The tiny size, of clinical and experimental-derived SCVs on solid agar, is often due to auxotrophy for thymidine, menadione, or hemin (Proctor *et al.*, 1995, 1997; von Eiff *et al.*, 1997a; Kahl *et al.*, 1998). When the growth media are supplemented with these compounds, most SCVs grow as rapidly as the parent strains. Thiamine, menadione, and hemin are required for the biosynthesis of electron transport chain components. However the mechanism for thymidine auxotroph formation is not likely to

Table 1. Typical features of *S. aureus* SCV phenotype compared to *S. aureus* isolates of normal morphotype

Characteristics	Typical features	
	Normal phenotype	SCV phenotype
Colony morphology		
Size	Up to 6-8 mm in diameter	Approximately 1 mm in diameter
Pigmentation	Yellow-orange	White
Hemolysis	β-hemolysis	No or reduced β-hemolysis
Growth rate on solid media	Normal (12-18 h)	Delayed (pinpoint colonies following 24 h incubation)
Auxotrophism	No	Hemin, menadione, thymidine
Coagulase activity	Normal	3-4-fold delayed compared to the normal phenotype
Biochemical reactions	Normal	Delayed and/or changed (reduced sugar utilization)
Resistance to antibiotics	Varying	Reduced susceptibility to aminoglycosides and TMP-SMZ
Typical clinical association	Acute infections	Chronic, persistent infections

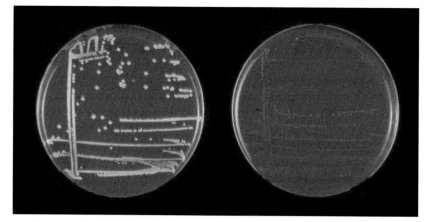

Figure 1. Blood agar plates showing the normal (left) and small colony variant (right) phenotype of *S. aureus*.

be related to that for the electron transport type SCVs. Menadione is isoprenylated to form menaquinone, the acceptor of electrons from nicotinamide adenine dinucleotide (NADH)/flavin adenine dinucleotide (FADH$_2$) in the electron transport chain. Hemin is required for the biosynthesis of cytochromes, which accept electrons from menaquinone and transport them to the ATP synthesis complex in the cell membrane. Thiamine is required for menadione biosynthesis; hence, thiamine auxotrophs are also menadione auxotrophs. Many previous reports also noted decreased respiratory activity in staphylococcal SCVs, which is also consistent with reduced electron transport activity (Acar *et al.,* 1978; Proctor *et al.,* 1997; von Eiff *et al.,* 1997b; Proctor and Peters, 1998).

The characteristics of SCVs, that are atypical for normal *S. aureus* morphotypes, can be tied together by a common thread, which is alterations in electron transport. Specifically, the following findings are very likely linked to an interruption of electron transport: (i) slow growth because cell wall synthesis requires large quantities of ATP, (ii) decreased pigment formation because carotenoid biosynthesis requires electron transport, (iii) resistance to aminoglycosides as their uptake requires the large membrane potential generated by electron transport, and (iv) mannitol fermentation negative because utilization of mannitol (sugar alcohol) is decreased when electron transport is not used (see also Table 1 and Figure 1) (Lewis *et al.,* 1990; Proctor *et al.,* 1997; von Eiff *et al.,* 1997b; Proctor and Peters, 1998).

Identification of SCVs in the Microbiological Laboratory

Identification of SCVs in the microbiological laboratory may be difficult because of their fastidious growth characteristics. A prerequisite for the recovery and isolation of these variants is the application of extended conventional culture and identification techniques. SCVs are rapidly overgrown and are easily missed when the normal *S. aureus* is present since SCVs grow about nine times more slowly than *S. aureus* with normal phenotype. Because SCVs and *S. aureus* with normal phenotype have the same appearance on gram-staining, there is no reason to suspect a mixed culture (Proctor *et al.,* 1997; Proctor and Peters, 1998).

The variants may be identified mostly as nonpigmented, nonhemolytic pinpoint colonies, slow-growing after 24 - 72 hours incubation on rabbit blood agar. In addition to the atypical colonial morphology, unusual biochemical reactions, and reduced coagulase activity (SCVs are often only coagulase-positive in the tube test after incubation for > 18 hours) are typical features of the variants compared to the *S. aureus* isolates with normal phenotype. Furthermore, some SCVs grow more rapidly in the presence of CO_2 and on

Table 2. Isolation of staphylococcal SCVs directly from human clinical specimens (Note: unless stated otherwise all SCVs are SCVs of *S. aureus*)*

No. of patients	Type or site of infection	Comments	Reference
1	Abscess	Detection also from anterior nares	Hale, 1947
2	Cutaneous abscess		Sherris, 1952
8	Urine, throat, blood, pleural fluid	In one case, SCVs were found by prospectively examining urine cultures performed during treatment with erythromycin	Wise and Spink, 1954
1	Hand abscess	Detection over an one-year-period from a total of seven separate purulent fluid cultures as well as from several nares specimens	Goudie and Goudie, 1955
2	Cutaneous abscess	Detection also from nose of index patient; patient 2 was a household contact	Thomas and Cowlard, 1955
2	Infected burn, chronic bronchitis		Wise, 1956
1	Prosthetic valve endocarditis	Microcolonies (G-variants)	Quie, 1969
2	Cutaneous abscess		Slifkin et al., 1971
3	Osteomyelitis		Borderon and Horodniceanu, 1976
8	Osteomyelitis, blood, cutaneous abscess, CSF	Seven additional isolates possibly were SCVs but were unstable	Acar et al., 1978
1	Pneumonia	Infection following chest trauma	Spagna et al., 1978

1	Bacteremic cellulitis	Patient with erythema multiforme developing septicemia	Sompolinsky et al., 1985
1	Prosthetic valve endocarditis	S. epidermidis SCVs	Baddour and Christensen, 1987
1	Pacemaker electrode infection	S. epidermidis SCV	Baddour et al., 1990
5	Osteomyelitis, septic arthritis, muscle abscess, sinusitis	Persisting and relapsing infections over a period of several months to years (maximum of 54 years)	Proctor et al., 1995
1	Sternoclavicular joint septic arthritis	First report of invasive disease due to SCV in a child (11-year-old boy)	Spearman et al., 1996
4	Osteomyelitis	Followed gentamicin bead placement	von Eiff et al., 1997
26	Bronchial secretions, throat swabs	Cystic fibrosis patients	Kahl et al., 1998
1	Deep hip abscess	Methicillin-resistant SCV in a patient with AIDS with long-term trimethoprim/sulfamethoxazole prophylaxis	Seifert et al., 1999
2	Pacemaker electrode infection	S. epidermidis and S. capitis SCVs	von Eiff et al., 1999
1	Abscess and fistula	Persistent wound infection 13 months after herniotomy; initially misidentification of the S. aureus SCV as coagulase-negative Staphylococcus	Abele-Horn et al., 2000
1	Skin lesions	Isolation of several SCV genotypes including MRSA-SCV from a patient with Darier's disease (keratosis follicularis) over a period of 28 months	von Eiff et al., 2001

* modified and updated according to (Proctor et al., 1995)

rich media, such as Schaedler's agar, which contains hemin. Hence, these SCVs can be construed to be anaerobic organisms as these conditions are often used for anaerobic blood cultures. As oxygen does not enhance growth of electron transport deficient SCVs, then their growth in an anaerobic chamber on Schaedler's agar can make the laboratory personnel believe that anaerobes are present (von Eiff *et al.*, 1999; Kahl *et al.*, 1998). Isolates suspected to be *S. aureus* SCVs should be confirmed as *S. aureus* molecularly by testing for the species-specific *nuc* and *coa* genes (Brakstad *et al.*, 1992). Auxotrophy for hemin may be tested for by using standard disks, and auxotrophy for thymidine or menadione by impregnating disks with 15 µL of thymidine at 100 µg/mL or menadione at 10 µg/mL. Test isolates should be inoculated on chemically defined medium as described previously (Kahl *et al.*, 1998).

When it comes to susceptibility testing, SCVs also present another challenge. A small percentage of normally growing organisms will rapidly replace the SCVs in liquid medium in an overnight culture because the doubling time of normal *S. aureus* is about 20 minutes, whereas SCVs double in about 180 minutes; hence, the SCVs may be overgrown to such an extent that they may not be included in the inoculum used for susceptibility testing (Chuard *et al.*, 1997).

A further pitfall in susceptibility testing of SCVs, irrespectively of their auxotrophisms, may occur if these variants are resistant to methicillin and oxacillin, respectively. Phenotypical tests for susceptibility testing using disc diffusion test, Etest, microdilution test, determination of MICs by VITEK© 2 automated susceptibility testing system as well as MRSA slide latex agglutination tests may fail to detect SCVs as methicillin-resistant. Because of the slow growth of the SCVs, results of testing by use of disk diffusion or by automated overnight methods are invalidated, since the colonies may be too small to be seen on agar or to be detected by optical density measurements in automated systems (Proctor *et al.*, 1997; Chuard *et al.*, 1997; Proctor and Peters, 1998). As a consequence, isolates with SCV phenotype might be tested misleadingly as methicillin-susceptible. The detection of the *mecA* gene (Murakami *et al.*, 1991) combined with the detection of genes specific for *S. aureus* such as the *nuc* gene (Brakstad *et al.*, 1992) by molecular methods should be used for reliable diagnosis or validation of *S. aureus* SCVs.

Association of SCVs with Alterations in Electron Transport and Persistent Infections

The ability to interrupt electron transport and to form a variant subpopulation affords *S. aureus* a number of survival advantages that extends beyond simply increased resistance to antibiotics. Although SCVs have been recognized for many decades following the first description as dwarf-colony or G variants

(Hale, 1947; Goudie and Goudie, 1955), connecting this phenotype to persistent and recurrent infections has only recently been appreciated (see also Table 2) (Proctor *et al.,* 1994, 1995; Proctor and Peters, 1998; Looney, 2000; McNamara and Proctor, 2000). R. A. Procter *et al.* (1995) first reported a model that showed a relationship between the multiple changes in phenotypic characteristics of *S. aureus* SCVs, to alterations in electron transport, and the clinical pattern of persistent, and relapsing infection. They described five patients with unusually persistent and/or antibiotic-resistant infections due to SCVs of *S. aureus*. All SCV strains were nonhemolytic and nonpigmented and grew very slowly on routine culture media in an ambient atmosphere. All four strains available for further studies were shown to be auxotrophs that reverted to normal colony forms in the presence of menadione, hemin, and/or CO_2 supplement. Similarly, these isolates were resistant to gentamicin, but susceptible in the presence of metabolic supplements. To evaluate clonality of isolates with different colony morphotypes, which were derived from the same patients, genotyping by restriction endonuclease analysis was performed. The resulting patterns showed clonal identity of the isolates, indicating phenotypic variants within individual clones (Proctor *et al.,* 1995).

In the past, the frequency of *S. aureus* SCVs among clinical isolates has not been established by prospective studies, except for the one blood culture study where stable SCVs were recovered from 1% of patients (Acar *et al.,* 1978). Since then, two groups of patients who have had extended exposure to antibiotics were evaluated in particular for the presence of *S. aureus* SCVs, i.e. patients with cystic fibrosis (CF) and patients with chronic osteomyelitis (von Eiff *et al.,* 1997a; Kahl *et al.,* 1998).

Recovery of *S. aureus* SCVs in Respiratory Secretions of Cystic Fibrosis Patients

The prevalence of *S. aureus* SCVs in respiratory secretions of cystic fibrosis patients, who frequently have staphylococcal infections and receive large quantities of antibiotics, was prospectively studied (Kahl *et al.,* 1998). In a 34-month period, encompassing 78 patients, 53 patients (67. 9%) harbored *S. aureus* in their respiratory specimens. Twenty-seven patients (50. 9%) had *S. aureus* with normal phenotype and 26 (49. 1%) had normal plus SCVs, SCVs alone, or pure cultures with a normal phenotype alternating with pure cultures of SCVs. In consecutive specimens from 19 of these 26 patients, the variants were isolated over a period of 2 - 31 months indicating persistence of these bacteria. Using pulsed-field gel electrophoresis (PFGE), clonal identity of the isolates exhibiting both morphotypes was shown from 16 of 19 patients with persistent SCV colonization. In a total of 78 isolates with SCV phenotype, auxotrophism for thymidine was demonstrated in 41, for hemin in 10, and for menadione in 2 SCVs. Double auxotrophy for thymidine plus hemin was found in 25 isolates. While the normal, methicillin-susceptible *S. aureus* isolates were trimethoprim-sulfamethoxazole-susceptible (MIC < 0. 125 µg/mL), all SCV isolates were resistant (MIC > 32 µg/mL) to trimethoprim-sulfamethoxazole. In the 34-month period, only one MRSA strain was isolated. This strain was also resistant to erythromycin, lincomycin, gentamicin, netilmicin, and trimethoprim-sulfamethoxazole. Of 12 SCV-normal strain pairs, 11 of 12 SCVs had higher gentamicin MICs than did their corresponding normal *S. aureus* strain. Of interest, all 26 patients with SCVs had received prophylaxis with trimethoprim-sulfamethoxazole, whereas only 10 of 27 patients with normal *S. aureus* received this treatment (P < 0. 001). In addition, patients with SCVs were treated for longer with trimethoprim-sulfamethoxazole (median, 23. 5 months) than patients without SCVs but with normal *S. aureus* (median 18 months), but this difference was not significant. The phenotypic variants were isolated from patients even after extended trimethoprim-sulfamethoxazole-free intervals (3-31 months) and remained *in vitro* as stable SCV phenotype after primary culture and multiple passages in antibiotic-free medium. Notably, 11 of 19 patients with persistent SCV colonization received interventional aminoglycoside therapy. Thus, while antibiotic exposure of CF patients may contribute to SCV selection or induction, the *S. aureus* subpopulation then persists, even in the absence of selective antibiotic pressure. After transformation of normal *S. aureus* phenotype into an SCV, these bacteria acquire phenotypic resistance against antimicrobial agents such as aminoglycosides by decreased antibiotic uptake or against antifolates by acquiring the ability to use exogenous nucleotide sources. This decreased susceptibility may then contribute to clinical persistence despite continued use of various antibiotics (Kahl *et al.,* 1998).

Recovery of SCVs in Bone Specimens or Deep Tissue Aspirates from Patients with Osteomyelitis

SCVs are more resistant to aminoglycosides and can be regularly recovered *in vitro* from cultures of normal *S. aureus* strains exposed to aminoglycosides. Selection for SCVs readily occur within 72 hours in clinical isolates of normal *S. aureus* that are exposed to gentamicin at 1 µg/mL Mueller Hinton broth (Musher *et al.,* 1977; Lewis *et al.,* 1990; Balwit *et al.,* 1994).

Beads containing gentamicin are used as an adjunct to debridement and antibiotic therapy for the treatment of osteomyelitis. The beads slowly release the antibiotic over a period of weeks to months providing a sustained local level of drug (Walenkamp *et al.,* 1986; Evans and Nelson, 1993). While the patient milieu is clearly more complex, gentamicin bead placement might be an efficient way to select for SCVs. Thus, in order to test whether the slow release into the local environment may be an efficient way to select for SCVs *in vivo*, a case-control study was performed over an eighteen month period. Bone specimens or deep tissue aspirates from patients with suspected osteomyelitis who had received gentamicin beads were carefully screened for *S. aureus* SCVs (von Eiff *et al.,* 1997a). Patients were divided on the basis of previous placement of gentamicin beads in their bone, and their charts were reviewed. Only those patients with cultures that contained *S. aureus* were included. In the 18-months study period, fourteen patients fulfilled these criteria (clinical signs of osteomyelitis, and a *S. aureus*-positive culture). *S. aureus* SCVs were recovered from four patients, i. e. only from those who had previously been treated with gentamicin beads. Three patients with SCVs had large and small colony types isolated from simultaneous or from sequential cultures. Restriction fragments of total bacterial DNA of isolates with different colonial morphologies resolved with PFGE showed clonal identity of the different phenotypes, respectively. All SCV strains were shown to be auxotrophs that reverted to normal colony forms in the presence of menadione or hemin. Regarding the susceptibility to antimicrobial agents, MICs for gentamicin were up to 32-fold higher for the SCVs (1 µg/mL) as compared to the parent strain (<0. 031 µg/mL) whereas no differences were found in susceptibilities against other antimicrobial agents. Methicillin-resistant strains were not isolated.

In the patients, in which SCVs were isolated from bone specimens or deep tissue aspirates, therapy failed despite using antibiotics with *in vitro* activity. In contrast, the other ten patients with normal *S. aureus* had no relapses of osteomyelitis occurring more than one year after primary diagnosis once active antibiotics were given for at least four weeks intravenously. Clinical evaluation showed no other major differences between the two groups. These results provide evidence that *S. aureus* SCVs may be selected *in vivo* by antimicrobials, in particular following gentamicin bead placement in patients with osteomyelitis (von Eiff *et al.,* 1997a).

Recovery of SCVs in a Patient with Persistent and Antibiotic Resistant Skin Infection

Recently, the first case of persistent and antibiotic resistant skin infection with different phenotypes and genotypes of *S. aureus*, including clonally different *S. aureus* SCVs was described (von Eiff *et al.,* 2001). The 39-year-old patient with Darier's disease (Dyskeratosis follicularis) was hospitalized during a period of several years because of continuous worsening of his skin condition. At the age of 22, the first typical skin lesions appeared and the diagnosis of Darier's Disease was confirmed by histology. Although the patient received different combinations of antibiotics, the skin condition did not improve significantly. Since MRSA was isolated from skin and anterior nares, antimicrobial agents as vancomycin, rifampicin and clindamycin were given intravenously for several weeks. Furthermore, a topical mupirocin ointment for the nasal mucosa was given. Additional therapeutic approaches included several topical treatments with steroids and antiseptics as polyvidon-iodine, chlorhexidine, and chlorquinaldol.

In this patient, SCVs of *S. aureus* and isolates with normal colony size isolated from simultaneous or sequential cultures were recovered over a period of 28 months. All together, 119 isolates were derived from 53 different clinical specimens, which were predominantly taken from different areas of the affected skin (swabs as well as needle biopsies) and from the anterior nares. Phenotypic characterization of the isolates showed that different *S. aureus* strains with normal colony size as well as hemin auxotrophic SCVs were associated with the infection of the skin. One MRSA clone was found in isolates with normal morphotype as well as with SCV phenotype. In comparison to clonally identical strains with normal phenotype, SCVs revealed up to 32-fold higher MIC values of gentamicin.

With use of PFGE and arbitrarily-primed PCR, molecular typing revealed seven genotypes involved over this 28-months period, including four genotypes of SCVs. PFGE-revealed bands of *S. aureus* and SCVs strains with the same profile were considered to be clonal. One clone, growing alternately in both phenotypes persisted over a period of 18 months, the other clones were isolated over periods of one week, and 5, 7, 13, 16 months, respectively. One clone was detected only one time in a single specimen. One of the two SCVs selected for the intracellular persistence assay belonged to the clone which was isolated over a period of 18 months (von Eiff *et al.,* 2001).

This report suggests that in patients suffering from chronic exacerbating skin diseases like Darier's Disease with negative culture results for *S. aureus* obtained under routine culture conditions these pathogens must be actively sought by use of appropriate selective media and growth conditions. In addition, the decreased susceptibility of *S. aureus* SCVs against antimicrobials typically used for *S. aureus* treatment requires the identification of these variants even in the presence of normal *S. aureus* in the specimens (Proctor and Peters, 1998).

Intracellular Persistence and Ultrastructural Examination

Intracellular persistence assays were performed with various strains of *S. aureus* SCVs to determine whether these variants could persist intracellularly within different kinds of cell lines more efficiently than the corresponding *S. aureus* parent strains with a normal phenotype (Balwit *et al.,* 1994; Vesga *et al.,* 1996; Kahl *et al.,* 1998; von Eiff *et al.,* 2001). For example, the intracellular survival

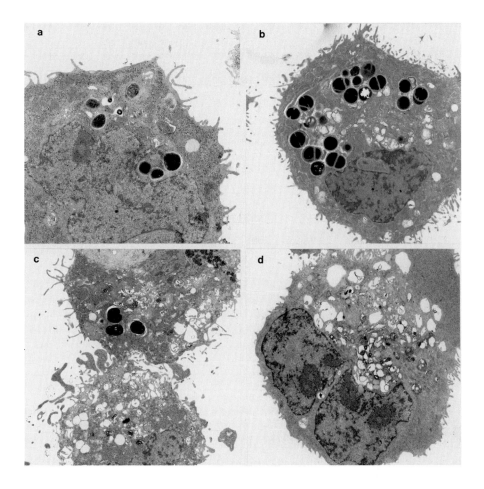

Figure 2. Electron micrographs of the keratinocyte cell line HaCaT infected with a clinical isogenic normal and *S. aureus* SCV isolate pair (SCV1, NP1). After incubation of the infected HaCaT cells in the presence of lysostaphin for 30 min (a, c) and 48 hours (b, d), respectively (analogous to the intracellular persistence assay), the cells were washed, dehydrated, and embedded in epon. Ultrathin sections were counterstained and examined in a Philips EM 10 electron microscope.**(a, b.)** Intracellular persistence of SCVs (SCV1) within viable HaCaT cells, (a) 30 min, (b) 48 hours incubation time. The epithelial cells appear viably and show no signs of degeneration (original magnification, X3400). **(c, d.)** *S. aureus* with normal phenotype (NP1) is incorporated after 30 min by intact HaCaT cells (c). However, after 48 hours incubation time most of the epithelia show severe lytic degeneration and release of bacteria (d), (original magnification, X3400).

of SCVs within a spontaneously transformed human keratinocyte cell line (HaCaT cell line) was studied using two clonally different SCVs derived from the patient described above with the chronic skin infection (von Eiff *et al.*, 2001). After an initial 3. 5-h coincubation to allow uptake of *S. aureus* SCVs by HaCaT cells, lysostaphin treatment revealed that a greater than 100-fold more SCV cells persisted intracellularly after 24 or 48 h of incubation relative to the normal phenotypes. The difference in intracellular persistence was not due to differential susceptibilities of the strains to lysostaphin, which was maintained in the culture medium during the assay to eliminate extracellular organisms.

Ultrastructural examination of the HaCaT cells confirmed persistence of *S. aureus* SCVs within the epithelial cells. After 30 min and 48 hours, SCVs appeared to be within the cytoplasm (Figure 2a,b). The epithelial cells appeared viable and showed no signs of degeneration. Likewise, the corresponding parent strain with normal phenotype was incorporated after 30 min by intact HaCaT cells (Figure 2c). However, after 48 hrs incubation time, most of the epithelial cells harboring the parent strain underwent severe lytic degeneration and release of bacteria (Figure 2d). The very few remaining cells were vacuolated and showed no *S. aureus* when the parent strain was used (Figure 2d), whereas the SCVs showed many intracellular bacteria at 48 hrs (Figure 2b) (von Eiff *et al.*, 2001).

Thus, as shown *in vitro* with the human keratinocyte cell line HaCaT, *S. aureus* SCVs may persist in chronic skin infection, probably due to persistence within keratinocytes. In addition, the methicillin resistance of one of the clones might have contributed to the recurrence of the infection in the patient described above, as vancomycin is not very efficient in treating skin infections. When an infection is particularly resistant to therapy, persists for a long period, or fails to respond to apparently adequate therapy, one should consider special efforts to search for SCVs.

Antibiotic Resistance and Therapy for Infections Due to *S. aureus* SCVs

To test whether SCVs acquire chromosomally encoded resistance phenotypes differently from parent strains with a normal phenotype, the mutation rates and the accumulation of mutations in the target genes of isolates exposed to ciprofloxacin, rifampicin and mupirocin were investigated. The *in vitro* activities of these three compounds were measured in SCVs and their corresponding parent strains before and after 10 serial passages in antibiotic-containing medium, followed by sequencing of the target genes (Schmitz *et al.*, 2001). All isolates tested became resistant to ciprofloxacin, rifampicin and mupirocin. Rates of appearance of colonies with higher MICs were in the range 10^{-5} to 10^{-6} for ciprofloxacin, 10^{-6} to 10^{-7} for rifampicin and 10^{-7} for mupirocin. Differences in mutation rates or MICs were not detected between SCVs exhibiting different auxotrophisms and their clonally identical parent strains with normal phenotype, indicating that this phenotype does not affect the development of ciprofloxacin, rifampicin or (low-level) mupirocin resistance in *S. aureus* (Schmitz *et al.*, 2001).

The interruption in electron transport - as described above - reduces the electrochemical gradient across the bacterial membrane, resulting in decreased uptake of antimicrobial agents that require a charge differential to be active such as aminoglycosides. Therefore, these substances should not be used, although single strains with SCV phenotype might be susceptible to gentamicin or other aminoglycosides. In addition, the slow growth of SCVs reduces the effectiveness of cell-wall-active antibiotics such as β-lactams (Proctor *et al.*, 1997).

However, the development of the SCV phenotype may not always be a clinically disadvantageous situation. Because of their low exotoxin production, SCVs cause less tissue damage than do rapidly growing staphylococci. The use of drugs that interfered with electron transport might be particularly valuable as a short-term measure if such drugs could rapidly turn off toxin production. There are currently investigations studying the effect of an electron transport inhibitor on *S. aureus*. It was found that α-toxin production and damage to cultured endothelial cells was reduced by this compound, which would represent a new class of anti-virulence factor drug (R. A. Proctor, 1997).

Optimal therapy for infections due to *S. aureus* SCVs has not been defined. Reversal of the auxotrophy *in vitro* is encouraging, because reversion to the normal colony form makes these organisms more susceptible to antibiotics. In the case of menadione auxotrophs, this reversal can be easily accomplished by administering vitamin K to patients. However, whether this will prove to be of benefit remains to be determined by clinical trials. It was found, that trimethoprim-sulfamethoxazole combined with rifampin was the most active therapeutic regimen in a tissue-culture system where the SCVs were inside endothelial cells, but still more research is necessary to define the optimal therapy (Proctor *et al.*, 1997; Proctor and Peters, 1998). Nowadays, the oxazolidinones, which represent a new chemical class of synthetic antimicrobial agents with an unique mechanism of inhibiting bacterial protein synthesis and which are known for their excellent tissue penetration, provide an alternative strategy for therapy of chronic persistent *S. aureus* infections. Linezolid, the first oxazolidinone to be approved for clinical use, displays *in vitro* activity against many gram-positive pathogens, including MRSA (von Eiff and Peters, 1999; Clemett and Markham, 2000).

A Site-Directed *Staphylococcus aureus hemB* Mutant Mimics the SCV Phenotype

The clinical strains with a SCV phenotype exhibited under *in vitro* culture conditions a high rate of reversion to the large colony form. Although the strains examined so far were not genetically characterized, the features of the variants suggested a link between electron transport defective strains and persistent infections. Genetically undefined strains might carry mutations in more than one virulence factor, especially since the clinical SCVs show multiple phenotypic changes as compared to the parent strains. In order to address questions concerning possible roles of respiratory-defective *S. aureus* in the pathogenesis of staphylococcal infection, a stable mutant in electron transport was generated by interrupting *hemB* in *S. aureus* (von Eiff *et al.,* 1997b). Heme is the prosthetic group of cytochromes, which plays an essential role in electron transport and the *hemB* gene is a member of the family of genes encoding enzymes of the porphyrin biosynthetic pathway. This mutant allowed us to characterize the phenotype of a genetically defined SCV of *S. aureus* and to test the hypothesis that defects in electron transport promote the development of intracellular persistence (von Eiff *et al.,* 1997b).

The *S. aureus hemB* mutant showed the typical characteristics of clinical SCVs: (i) pinpoint colonies on solid agar, that are >10 fold smaller than the parent strain and slow growth of the *hemB* mutant was also observed in liquid medium (TSB or CDM). The *hemB* mutant reached stationary phase at a 10-fold lower level of total growth in comparison to the wild type. (ii) Decreased pigment formation: whitish colonies versus golden yellow colored colonies of the parent strain. (iii) Reduced hemolytic activity: the *hemB* mutant showed > 90-fold reduction in percentage of lysis of rRBC compared with the parent strain 8325-4 (0. 25% versus 89%). (iv) Decreased coagulase activity: in the tube coagulase test, the mutant showed a delayed coagulase reaction, being positive after 22 h incubation at 37°C whereas the parent strain was positive after 2 h. (v) Resistance to aminoglycosides: MIC for gentamicin was 16-fold higher for the mutant (MIC = 0. 5 µg/mL) compared to the wild-type strain (MIC = < 0. 031 µg/mL) and MIC for kanamycin was 8-fold higher for the mutant (MIC = 2. 0 µg/mL) compared to the wild-type strain (MIC = 0. 25 µg/ mL). (vi) In contrast to the parent strain and the plasmid complemented mutant, the *hemB* mutant showed biochemical characteristics that were atypical for *S. aureus,* such as reduced lactose-, turanose- and mannitol-fermentation, no nitrate reduction, and reduced N-acetyl-glucosamine utilization as analyzed in the API-systems and with conventional biochemicals. The SCV phenotype of the *hemB* mutant was essentially reversed by growing with hemin at a concentration of 1 µg/mL or by complementation with intact *hemB*.

Western blot analysis showed that α-toxin is produced in the parent strain, in the mutant supplemented with hemin, and in the complemented mutant, however, α-toxin was not detectable in the *hemB* mutant. Northern blot analysis performed to determine whether reduced protein levels correlated with reduced transcription, showed that transcription of *hla* was high in the parent strain as well as in the plasmid complemented mutant but was not detectable in the non-complemented mutant.

In a model of endovascular infection to determine the intracellular persistence, higher numbers of the mutant were seen following an initial 3. 5 h coincubation and a 20 min. incubation in the presence of lysostaphin compared to the parent strain and to the plasmid complemented mutant. Further coincubation in the continuous presence of lysostaphin revealed that > 200-fold more *hemB*-mutant cells persisted intracellularly after 24 or 48 h incubation relative to the other strains tested.

Thus, while the studies with clinical isolates of SCVs suggested a link between electron transport defective strains and persistent infections, the defined *hemB* mutant with the SCV phenotype provided strong additional definitive evidence for these connections. The *hemB* mutant was phagocytized by cultured endothelial cells, but did not lyse these cells, because the mutant produced very little α-toxin (as shown on protein and transcription level). The intracellular location may shield the SCVs from host defenses and antibiotics, thus providing one explanation for the difficulty in clearing *S. aureus* SCVs from host tissues.

Recently, *S. aureus* SCVs were found to cause a persistent and antibiotic resistant mastitis in mice (Martinez *et al.,* 1999). The investigators made a *hemB* mutant for these investigations essentially identical to that used by our group (von Eiff *et al.,* 1997b). This mutant invaded the epithelial cells of the mouse as well as the parent strain, but it produced a more persistent and antibiotic refractory infection. Albeit these were defined in a murine experimental model, this clinical picture is very typical of bovine mastitis caused by *S. aureus*, which is a major problem for the dairy industry.

References

Abele-Horn, M., Schupfner, B., Emmerling, P., Waldner, H., and Göring, H. 2000. Persistent wound infection after herniotomy associated with small-colony variants of *Staphylococcus aureus*. Infection 28: 53-54.

Acar, J. F., Goldstein, F. W., and Lagrange, P. 1978. Human infections caused by thiamine- or menadione-requiring *Staphylococcus aureus*. J. Clin. Microbiol. 8: 142-147.

Baddour, L. M., Barker, L. P., Christensen, G. D., Parisi, J. T., and Simpson, W. A. 1990. Phenotypic variation of *Staphylococcus epidermidis* in infection of transvenous endocardial pacemaker electrodes. J. Clin. Microbiol. 28: 676-679.

Baddour, L. M. and Christensen, G. D. 1987. Prosthetic valve endocarditis due to small-colony staphylococcal variants. Rev. Infect. Dis. 9: 1168-1174.

Balwit, J. M., van Langevelde, P., Vann, J. M., and Proctor, R. A. 1994. Gentamicin-resistant menadione and hemin auxotrophic *Staphylococcus aureus* persist within cultured endothelial cells. J. Infect. Dis. 170: 1033-1037.

Borderon, E. and Horodniceanu, T. 1976. Mutants deficients a colonies naines de *Staphylococcus*: etude de trois souches isolees chez des malades porteurs d'osteosyntheses. Annales de L'Institut Pasteur Microbiologie (Paris) 127A: 503-514.

Brakstad, O. G., Aasbakk, K., and Maeland, J. A. 1992. Detection of *Staphylococcus aureus* by polymerase chain reaction amplification of the *nuc* gene. J. Clin. Microbiol. 30: 1654-1660.

Chuard, C., Vaudaux, P., Proctor, R. A., and Lew, D. P. 1997. Decreased susceptibility to antibiotic killing of a stable small colony variant of *Staphylococcus aureus* in fluid phase and on fibronectin-coated surfaces. J. Antimicrob. Chemother. 39: 603-608.

Clemett, D. and Markham, A. 2000. Linezolid. Drugs 59: 815-827.

Daum, R. S. and Seal, J. B. 2001. Evolving antimicrobial chemotherapy for *Staphylococcus aureus* infections: Our backs to the wall. Crit. Care Med. 29: N92-N96.

Evans, R. P. and Nelson, C. L. 1993. Gentamicin-impregnated polymethylmethacrylate beads compared with systemic antibiotic therapy in the treatment of chronic osteomyelitis. Clin. Orthop. 295: 37-42.

Goudie, J. G. and Goudie, R. B. 1955. Recurrent infections by a stable dwarf-colony variant of *Staphylococcus aureus*. J. Clin. Pathol. 8: 284-287.

Hale, J. H. 1947. Studies on *Staphylococcus* mutation: characteristics of the "G" (gonidial) variant and factors concerned in its production. Br. J. Exp. Pathol. 28: 202-210.

Hiramatsu, K., Aritaka, N., Hanaki, H., Kawasaki, S., Hosoda, Y., Hori, S., Fukuchi, Y., and Kobayashi, I. 1997. Dissemination in Japanese hospitals of strains of *Staphylococcus aureus* heterogeneously resistant to vancomycin. Lancet 350: 1670-1673.

Hiramatsu, K., Cui, L., Kuroda, M., and Ito, T. 2001. The emergence and evolution of methicillin-resistant *Staphylococcus aureus*. Trends Microbiol. 9: 486-493.

Jones, R. N. 2001. Resistance patterns among nosocomial pathogens: Trends over the past few years. Chest 119: 397S-404S.

Kahl, B., Herrmann, M., Schulze-Everding, A., Koch, H. G., Becker, K., Harms, E., Proctor, R. A., and Peters, G. 1998. Persistent infection with small colony variant strains of *Staphylococcus aureus* in patients with cystic fibrosis. J. Infect. Dis. 177: 1023-1029.

Lewis, L. A., Li, K., Bharosay, M., Cannella, M., Jorgenson, V., Thomas, R., Pena, D., Velez, M., Pereira, B., and Sassine, A. 1990. Characterization of gentamicin-resistant respiratory-deficient (res-) variant strains of *Staphylococcus aureus*. Microbiol. Immunol. 34: 587-605.

Looney, W. J. 2000. Small-colony variants of *Staphylococcus aureus*. Br. J. Biomed. Sci. 57: 317-322.

Lowy, F. D. 1998. *Staphylococcus aureus* infections. New Engl. J. Med. 339: 520-532.

Martinez, A., Boyll, B. J., and Allen, N. E. 1999 The role of small-colony variants of *Staphylococcus aureus* in chronic bovine mastitis. Abstracts of the 99th Meeting of the American Society for Microbiology, Abstract D/B100.

McNamara, P. J. and Proctor, R. A. 2000. *Staphylococcus aureus* small colony variants, electron transport and persistent infections. Int. J. Antimicrob. Agents 14: 117-122.

Moellering, R. C., Jr. 1998. Problems with antimicrobial resistance in gram-positive cocci. Clin. Infect. Dis. 26: 1177-1178.

Murakami, K., Minamide, W., Wada, K., Nakamura, E., Teraoka, H., and Watanabe, S. 1991. Identification of methicillin-resistant strains of staphylococci by polymerase chain reaction. J. Clin. Microbiol. 29: 2240-2244.

Musher, D. M., Baughn, R. E., Templeton, G. B., and Minuth, J. N. 1977. Emergence of variant forms of *Staphylococcus aureus* after exposure to gentamicin and infectivity of the variants in experimental animals. J. Infect. Dis. 136: 360-369.

National Nosocomial Infections Surveillance. 1999. National nosocomial infections surveillance (NNIS) system report: Data summary from January 1990-May 1999. issued June 1999. Am. J. Infect. Control. 27: 520-532.

Proctor, R. A., Balwit, J. M., and Vesga, O. 1994. Variant subpopulations of *Staphylococcus aureus* as cause of persistent and recurrent infections. Infect. Agents Dis. 3: 302-312.

Proctor, R. A., Kahl, B., von Eiff, C., Vaudaux, P. E., Lew, D. P., and Peters, G. 1997. Staphylococcal small colony variants have novel mechanisms for antibiotic resistance. Clin. Infect. Dis. 27 (Suppl. 1): S68-S74.

Proctor, R. A. and Peters, G. 1998. Small colony variants in staphylococcal infections: diagnostic and therapeutic implications. Clin. Infect. Dis. 27: 419-423.

Proctor, R. A., van Langevelde, P., Kristjansson, M., Maslow, J. N., and Arbeit, R. D. 1995. Persistent and relapsing infections associated with small-colony variants of *Staphylococcus aureus*. Clin. Infect. Dis. 20: 95-102.

Quie, P. G. 1969. Microcolonies (G variants) of *Staphylococcus aureus*. Yale J. Biol. Med. 41: 394-403.

Schmitz, F. J., Fluit, A. C., Beeck, A., Perdikouli, M., and von Eiff, C. 2001. Development of chromosomally encoded resistance mutations in small-colony variants of *Staphylococcus aureus*. J. Antimicrob. Chemother. 47: 113-115.

Seifert, H., von Eiff, C., and Fätkenheuer, G. 1999. Fatal case due to methicillin-resistant *Staphylococcus aureus* small colony variants in an AIDS patient. Emerg. Infect. Dis. 5: 450-453.

Sherris, J. C. 1952. Two small colony variants of *Staphylococcus aureus* isolated in pure culture from closed infected lesions and their carbon dioxide requirements. J. Clin. Pathol. 5: 354-355.

Slifkin, M., Merkow, L. P., Kreuzberger, S. A., Engwall, C., and Pardo, M. 1971. Characterization of CO_2 dependent microcolony variants of *Staphylococcus aureus*. Am. J. Clin. Pathol. 56: 584-592.

Sompolinsky, D., Schwartz, D., Samra, Z., Steinmetz, J., and Siegman-Igra, Y. 1985. Septicemia with two distinct strains of *Staphylococcus aureus* and dwarf variants of both. Isr. J. Med. Sci. 21: 434-440.

Spagna, V. A., Fass, R. F., Prior, R. B., and Slama, T. G. 1978. Report of a case of bacterial sepsis caused by a naturally occuring variant form of *Staphylococcus aureus*. J. Infect. Dis. 138: 277-288.

Spearman, P., Lakey, D., Jotte, S., Chernowitz, A., Claycomb, S., and Stratton, C. 1996. Sternoclavicular joint septic arthritis with small-colony variant *Staphylococcus aureus*. Diagn. Microbiol. Infect. Dis. 26: 13-15.

Thomas, M. E. M. and Cowlard, J. H. 1955. Studies on a CO_2-dependent *Staphylococcus*. J. Clin. Pathol. 8: 288-291.

Vesga, O., Groeschel, M. C., Otten, M. F., Brar, D. W., Vann, J. M., and Proctor, R. A. 1996. *Staphylococcus aureus* Small Colony Variants are induced by the endothelial cell intracellular milieu. J. Infect. Dis. 173: 739-742.

von Eiff, C., Becker, K., Metze, D., Lubritz, G., Hockmann, J., Schwarz, T., and Peters, G. 2001. Intracellular persistence of *Staphylococcus aureus* small-colony variants within keratinocytes: A cause for antibiotic treatment failure in a patient with Darier's disease. Clin. Infect. Dis. 32: 1643-1647.

von Eiff, C., Bettin, D., Proctor, R. A., Rolauffs, B., Lindner, N., Winkelmann, W., and Peters, G. 1997a. Recovery of small colony variants of *Staphylococcus aureus* following gentamicin bead placement for osteomyelitis. Clin. Infect. Dis. 25: 1250-1251.

von Eiff, C., Heilmann, C., and Peters, G. 1999. New aspects in the molecular basis of polymer-associated infections due to staphylococci. Eur. J. Clin. Microbiol. Infect. Dis. 18: 843-846.

von Eiff, C., Heilmann, C., Proctor, R. A., Wolz, C., Peters, G., and Götz, F. 1997b. A site-directed *Staphylococcus aureus hemB* mutant is a small colony variant which persists intracellularly. J. Bacteriol. 179: 4706-4712.

von Eiff, C. and Peters, G. 1999. Comparative *in vitro* activities of moxifloxacin, trovafloxacin, quinupristin/dalfopristin and linezolid against staphylococci. J. Antimicrob. Chemother. 43: 569-573.

von Eiff, C., Vaudaux, P., Kahl, B., Lew, D. P., Schmidt, A., Peters, G., and Proctor, R. A. 1999. Blood stream infections caused by small colony variants of coagulase-negative staphylococci following pacemaker implantation. Clin. Infect Dis. 29: 932-934.

Walenkamp, G. H., Vree, T. B., and van Rens, T. J. 1986. Gentamicin-PMMA beads. Pharmacokinetic and nephrotoxicological study. Clin. Orthop. : 171-183.

Wise, R. I. 1956. Small colonies (G variants) of staphylococci: isolation from cultures and infections. Ann. N. Y. Acad. Sci. 65: 169-174.

Wise, R. I. and Spink, W. W. 1954. The influence of antibiotics on the origin of small colonies (G variants) of *Micrococcus pyogenes* var *aureus*. J. Clin. Invest. 33: 1611-1622.

From: *MRSA: Current Perspectives*
Edited by: A.C. Fluit and F.-J. Schmitz

Chapter 11

Treatment of MRSA Infections

Debby Ben-David and Ethan Rubinstein

Abstract

Staphylococcus aureus is a major cause of nosocomial infections. In recent years, the prevalence of methicillin-resistant *S. aureus* (MRSA) has increased considerably. Glycopeptides, the agents of choice for treatment of MRSA, have a relatively slow bactericidal effect and have been associated with clinical and bacteriological failures. Fusidic acid, rifampin, fosfomycin, quinolones and trimethoprim-sulfamethoxazole all have potential activity against MRSA. However, the increasing cross-resistance to these agents has restricted their treatment efficacy. The emergence of clinical isolates of *S. aureus* with reduced susceptibility to vancomycin (VISA) is of great concern. Several new antimicrobial agents, including streptogramins, oxazolidinones, daptomycin, glycylcyclines, oritavancin, and peptides, have proved useful against MRSA and VISA strains and are under rapid development. Currently, there are no recommended therapy guidelines for VISA infections. Treatment options include new *in vitro* active antimicrobial agents against VISA strains, and different existing antibiotic combinations.

Introduction

Methicillin-resistant *Staphylococcus aureus* (MRSA) was isolated soon after the drug was introduced in 1960s and has subsequently become a persistent and worldwide problem. In recent years, the prevalence of MRSA has increased in an alarming rate in many countries (Panlilio *et al.*, 1992; Voss *et al.*, 1994; Diekema *et al.*, 2001).

Methicillin resistance mediates resistance to all currently available β-lactam antibiotics. MRSA isolates have also been multiple resistant to other unrelated antimicrobial agent that do not target the cell wall, including aminoglycosides, clindamycin, fluoroquinolones, and macrolides (Diekema *et al.*, 2001). The management of infections caused by MRSA poses a challenge, due to the fact that there are presently no potent antibiotics that have an unequivocal rapid bacterial-killing potency. Therefore, the development of resistance of *S. aureus* to multiple antibiotics is a real threat in surviving mutants.

Glycopeptides have traditionally been considered as the agents of choice for MRSA infections. However, the emergence of clinical isolates of *S. aureus* with reduced susceptibility to vancomycin has been reported since 1996 (Hiramatsu *et al.*, 1997; Centers for Disease Control and Prevention, 1997; Ploy *et al.*, 1998; Smith *et al.*, 1999) and has caused growing concern, as therapeutic options have become increasingly limited. In view of the high toxicity of current antimicrobials used to treat MRSA, and the emergence of glycopeptide resistance, decisions regarding therapy of gram-positive infections, must include consideration not only of efficacy but also toxicity, selection of resistant organisms and the cost involved to the patient, to the society and to the ecology.

The widespread emergence of antibiotic resistance among Gram-positive bacteria has stimulated the search for effective new alternative antimicrobial agents in recent years. Several new agents, including streptogramins, oxazolidinones, daptomycin, oritavancin, some new β-lactams, and peptides, have proved useful against MRSA strains and are under rapid preclinical and clinical development.

This chapter will discuss the efficacy of antibiotic options currently available for MRSA infections and recommendations for treatment of common clinical syndromes.

Multidrug-Resistant MRSA Strains

MRSA strains are more likely to be resistant to other antimicrobial agents than MSSA isolates. High rates of multidrug resistance have been reported from many regions of the world. The SENTRY Antimicrobial Surveillance Program, a global network of sentinel hospitals in the USA, Canada, Europe and the Western Pacific region, has reported high rates of co-resistance of MRSA isolates between 1997 and 1999 (Diekema *et al.*, 2001). Rates of co-resistance vary significantly between different regions. Latin American MRSA isolates were resistant to a median of 6 antimicrobial classes, whereas USA and Canadian strains demonstrated resistance to a median of 3 antimicrobial agents. High levels of resistance to erythromycin, clindamycin were found among MRSA in all regions (up to 94.7%; 88%; 89.6%, respectively). Antimicrobial agents with significant regional differences included rifampin (4.9% in Canada vs. 44.4% in Europe), gentamicin (25.9% in Canada vs. 91.2% in Latin America) and tetracycline (14.8% in Canada vs. 82% in Western Pacific). Vancomycin remained highly active against MSRA isolates. Only one strain had vancomycin MIC of 8 µg/mL. The *in vitro* activity of quinupristin-dalfopristin and linezolid demonstrated excellent activity against *S. aureus*.

In conclusion, the spread of multidrug resistant MRSA strains limits the use of most antimicrobial agents, except glycopeptides. However, the relatively slow bactericidal effect of the glycopeptides, the high rate of adverse events and the possible emergence of glycopeptide-resistant strains in the future, highlights the need of for alternative antimicrobial agents with antistaphylococcal activity.

Antibiotics for Infections Caused by MRSA

Glycopeptides

Glycopeptides are currently still the antibiotics of choice for infections caused by MRSA. Glycopeptides are large, rigid molecules that inhibit a late step in bacterial cell wall peptidoglycan synthesis and cross-linking. Until recently, *S. aureus* was highly susceptible to glycopeptides, including vancomycin and teicoplanin. However, strains isolated from patients that were intermediate resistant to glycopeptides were detected in Japan in 1996, in the United States in 1997, and soon after in other countries (Ploy *et al.*, 1998; McManus, 1999; Mi-Na, *et al.*, 2000) and raised serious concerns about the continued effectiveness of vancomycin in the treatment of MRSA infections.

Mechanism of Action

Glycopeptides exhibit bactericidal activity through inhibition of cell wall synthesis. Glycopeptides bind to D-alanyl-D-alanine terminus of the N-acetyl-muramyl-pentapeptide subunit of the nascent cell wall in gram-positive bacteria. The glycopeptide thus prevents both the transglycosylation and transpeptidation reactions, which determine the formation of the mature cell wall in *S. aureus*.

There are two classes of binding targets in the staphylococcal cell wall. The D-alanyl-D-alanine residues in the completed peptidoglycan layers; and the murein monomer located in the cytoplasmic membrane that serves as the substrate for the glycosyltransferase. The binding of glycopeptides to the former targets does not inhibit peptidoglycan synthesis, although it may interfere with cross-bridge formation mediated by penicillin binding proteins (PBP). When glycopeptides bind to murein monomers, peptidoglycan synthesis is completely blocked, and the cells undergone osmotic rupture and death. However, for the glycopeptide molecules to bind to such targets, they have to pass through about 20 peptidoglycan layers without been trapped by the first targets. Since there are many D-alanyl-D-alanine targets in the peptidoglycan layers, many glycopeptide molecules are trapped, compromising their effectiveness.

Vancomycin

Vancomycin is the prototype of glycopeptide antibiotic that was isolated in 1956 from the actinomycete *Streptomyces orientalis* from soil samples from Borneo. Vancomycin was introduced into clinical use in 1958, but its use was limited for 2 decades, until the emergence of MRSA in the late 1970s. With the spread of MRSA in the United States, vancomycin underwent a marked increase in frequency of use and became the drug of choice for treatment of serious MRSA infections.

Antibacterial Activity

Vancomycin shows good *in-vitro* activity against Gram-positive aerobic and anaerobic bacteria, including staphylococci, streptococci, enterococci, *Clostridium* spp. and *Corynebacterium* spp. The National Committee for Clinical Laboratory Standards (NCCLS, 2000) defines staphylococci requiring concentration of vancomycin 4 µg/mL for growth inhibition as susceptible, those requiring 8 µg/mL to 16 µg/mL as intermediate, and those requiring concentrations 32 µg/mL as resistant (NCCLS, 2000). Vancomycin is only slowly bactericidal against both *S. aureus* and *Staphylococcus epidermidis* (Gopal *et al.*, 1976; Levine *et al.*, 1991).

In vitro synergy of vancomycin with aminoglycosides has been demonstrated against some *S. aureus* strains (Watanakunakorn *et al.*, 1982). *In vitro* and *in vivo* studies, concerning the efficacy of the combination vancomycin- rifampin against *S. aureus* have yielded conflicting results. Indifference or antagonism has been reported with the checkerboard technique, whereas synergistic activity has been demonstrated by time-kill curves or in *in vivo* models (Watanakunakorn *et al.*, 1981; *Bayer et al.*, 1984; Bayer *et al.*, 1985).

Pharmacokinetics and Pharmacodynamics

Vancomycin should be given by intravenous route. Pain on injection, and erratic absorption from tissues prevent the intramuscular use of vancomycin. Vancomycin is minimally absorbed from the gastrointestinal tract, and is used orally only for treatment of severe clostridial entrocolitis. Vancomycin does not undergo metabolism and is cleared by the kidney through glomerular filtration of the active drug. The half-life of vancomycin in serum is 6 to 8 hours in patients with normal renal function. Vancomycin has a large volume of distribution, and achieves therapeutic levels in blood, pleural, pericardial and synovial and ascitic fluids. Vancomycin does not achieve regularly therapeutically effective levels in cerebrospinal fluid (CSF) in the absence of inflamed meninges. In meningitis, CSF concentrations are significantly higher (Albanese *et al.*, 2000). In the vitreous and acquous humor of the eye vancomycin levels are also subtherapeutic following systemic administration. Vancomycin antibacterial activity, similarly to penicillin, is time-dependent. The length of time that concentrations are maintained above the pathogen's MIC (T>MIC) is critical for bacterial eradication. Vancomycin continues to exert its antibacterial activity after concentrations have fallen below its inhibition levels (MIC), with a post-antibiotic effect of about two hours.

Dosage

The usual daily dose of vancomycin in adults with normal renal function is 15 mg/kg every 12 hours or 8 mg/kg every 6-8 hours. The dose should be adjusted for renal insufficiency and other special situations such as burns, fluid overload and high fat mass. Recent data suggests that single daily dose might also be appropriate in selected patients (Cohen *et al.*, 2002). Since vancomycins antimicrobial activity is time-dependent, continuous infusion has potential befits over intermittent infusions. However, currently there is no clinical data demonstrating higher potency for the continuous infusion administration.

Peak and through serum concentrations should be monitored to ensure effective dosing and to avoid ototoxicity and nephrotoxicity. Recommended peak serum levels (drawn one hour after completion of the infusion) and trough levels are 20-40 µg/ml and 5-10 µg/ml, respectively. However, the current practice of achieving a minimal trough concentration and maximal peak concentration is not based on solid clinical evidence but rather on information inferred from *in vitro* data.

Toxicity and Adverse Reactions

Earlier formulations of vancomycin contained a number of impurities that contributed to it's toxicity. Current formulations that have been purified by high-pressure liquid chromatography have reduced considerably vancomycin-related toxicity but not the allergic adverse events. The most common side effects are fever, chills and phlebitis at the site of infusion. These occur less often if the drug is infused slowly in a large volume of fluid. A reaction known as 'red man' syndrome is a well-recognized complication of vancomycin infusion. It is believed to result from vancomycin-induced histamine release.

Nephrotoxicity and vestibular toxicity were relatively common with early, impure preparations of vancomycin. With new preparations now available, nephrotoxicity has become uncommon. Some studies have demonstrated that the rate of vancomycin-related nephrotoxicity is in the range 1.5%-15% (Mellor *et al.*, 1985; Sorrell *et al.*, 1985; Downs *et al.*, 1989). The risk of nephrotoxicity appears to be enhanced when nephrotoxic drugs such as aminoglycosides are given concomitantly (Rybak *et al.*, 1990; Goetz *et al.*, 1993). Vestibular toxicity is rare. Manifestations include tinitus, high tone hearing loss and permanent deafness.

Therapeutic Use in MRSA infections

Vancomycin is the drug of choice for treating severe MRSA infections, including bacteremia, endocarditis, pneumonia, and complicated surgical site infections.

Teicoplanin

Teicoplanin is a glycopeptide produced by *Actinoplanes teichomyceticus*. It is active against gram-positive organisms, including MRSA. Its efficacy is similar to that of vancomycin in the treatment of gram-positive infections. Compared with vancomycin, teicoplanin is associated with less nephrotoxicity, fewer anaphylactoid reactions, requires less monitoring, and is more convenient to administer. Although not approved by the Food and Drug Administration for use in the USA, teicoplanin had been extensively used in Europe.

Antibacterial Activity

Teicoplanin has an antibacterial spectrum similar to vancomycin. In *in vitro* studies, teicoplanin has excellent inhibitory activity against *S. aureus*, including methicillin-resistant isolates. Inhibitory concentrations range from 0.025 to 3.1 mg/L .*In vitro* synergy of teicoplanin and aminoglycosides has been demonstrated against some strains of *S. aureus* (Neu *et al.*, 1983).

Pharmacokinetics and Pharmacodynamics

Teicoplanin is poorly absorbed from the gastrointestinal tract and is administrated by either the intravenous or the intramuscular routes. The later may be painful. The drug is highly protein-bound, which may account for its slow renal clearance and the frequently reported failure of therapy of endocarditis when used in low doses (Wilson *et al.*, 1994). Teicoplanin's half-life is longer than that of vancomycin, allowing for once daily dosing. Teicoplanin does not undergo extensive metabolism and is excreted almost entirely unchanged by the kidneys. Renal clearance declines with decreasing renal function. The typical dose range is 6-15 mg/kg/day. Wilson et al explored the relationship between dosage and clinical outcome and found that most infections responded to a teicoplanin dose of 6 mg/kg per day (Wilson *et al.*, 1994). For MRSA endocarditis a dose of 12 mg/kg per day seems to be necessary to provide a cure rate similar to vancomycin.

Adverse Events

Teicoplanin is generally well tolerated. In comparative trials, teicoplanin appeared to have a lower potential to cause nephrotoxicity than vancomycin (Kureishi *et al.*, 1991; Van der Auwera *et al.*, 1991). The 'red man' syndrome, which typically occurs with vancomycin infusions, has rarely been reported among teicoplanin recipients.

Therapeutic Use in MRSA Infections

Teicoplanin is an effective alternative to vancomycin, with the advantage of less frequent dosing and less adverse events. Non-comparative and comparative trials have demonstrated the efficacy of teicoplanin in the treatment of gram-positive infections, including bacteremia, endocarditis, pneumonia, osteomyelitis and soft tissue infections. Clinical trials have demonstrated similar efficacy for both antibiotics (Smith *et al.*, 1989; Coni-Makhoul *et al.*, 1990; Van der Auwera *et al.*, 1991). Teicoplanin could be used for patients who have had allergic reactions to vancomycin as no cross-allerginicity between these two agents exists.

Limitations of Glycopeptides in Treatment of MRSA Infections

Although MRSA is susceptible *in vitro* to vancomycin and teicoplanin, these agents are handicapped by a relatively slow bactericidal effect, that may lower their effectiveness *in vivo,* as demonstrated repeatedly in delayed defervesence, prolonged period to obtain sterile blood cultures in staphylococcal infections and high rate of relapse (Chambers *et al.*, 1990). Vancomycin, in comparison with anti-staphylococcus penicillins, exhibited *in vitro* slower killing rate against staphylococci (Gopal *et al.*, 1976). Its activity against deep-seated staphylococcal infections in patients is not always optimal and often requires administration of additional antimicrobial agents (Massanari *et al.*, 1978). Accumulating evidence suggests that in treatment of *S. aureus* endocarditis, glycopeptides are slowly bactericidal and clinical response is slow and is associated with a high clinical and bacteriological failure (Small *et al.*, 1990; Fortun *et al.*, 1995, 2001). The mean duration of *Staphylococcus aureus* bacteremia treated with vancomycin is significantly longer (9 days) when compared to therapy with β-lactam antibiotics (3 days) (Korzeniowski *et al.*, 1976).

Trimethoprim-Sulphamethoxazole

Trimethoprim was first used for the treatment of infections in humans in 1962, and was registered for clinical use, in combination with sulfonamides, in 1968. Both drugs affect folic acid synthesis. The combination has a synergistic effect (Bushby *et al.*, 1968). Trimethoprim-sulphamethoxazole (TMP/SMX) is active *in vitro* against many strains of *S. aureus*, with MICs ranging from 0.05-1.0 to 0.4-8.0 mg/L (Yeladandi *et al.*, 1988; Lawrence *et al.*, 1993). Time killing studies, show this combination to have a rapid bactericidal effect at concentrations four times the MIC. During the last few years reports have appeared demonstrating increased resistance to TMP/SMX among MRSA strains (Then *et al.*, 1992; Alonso *et al.*, 1997; Diekema *et al.*, 2001)

A controlled comparative clinical trial of TMP/SMX versus vancomycin for severe *S. aureus* infections in intravenous drug users demonstrated that patients with methicillin-susceptible *S. aureus* (MSSA) had a significantly lower survival rate with TMP/SMX (73%) compared with vancomycin (97%). However, in MRSA infected cases, both drugs achieved similar rates of cure (100%) (Markowitz *et al.*, 1992). In contrast to this clinical report, in an experimental rabbit MRSA endocarditis model, treatment with TMP/SMX appeared to be significantly less effective than vancomycin (Gorgolas *et al.*, 1995).

Therefore, most experts do not recommend the use of TMP/SMX in the treatment of severe MRSA infections (e.g; endocarditis, nosocomial pneumonia). Furthermore, concomitant resistance may limit the use of TMP/

SMX. There are a few clinical trials demonstrating some benefit of TMP/ SMX in orthopedic implant infections (Stein *et al.*, 1998) but prospective clinical trials are needed to demonstrate the efficacy of TMP/SMX in bone and joint infections.

Rifampin

Rifampin is a semisynthetic derivative of rifamycin B, a macrocyclic antibiotic compound produced by the mold *Streptomyces mediterranei*. First isolated from fermentation culture of soil in 1957. Rifampin, a derivative of rifamycin, is more active and soluble than its parent compound. Rifampin has a bactericidal activity against a wide range of organisms.

Mechanism of Action

The rifamycins exert a bactericidal effect by inhibition of DNA dependent RNA polymerase at the β-subunit, preventing chain initiation.

Antibacterial Activity

Rifampin is extremely active against *S. aureus*. MICs range from 0.003 to 0.3 mg/L against most isolates of MSSA and MRSA. Rifampin is bactericidal against both actively dividing and stationary-phase bacteria. However, staphylococci rapidly develop resistance to rifampin by a single point mutation in the β-subunit of the DNA-dependent RNA polymerase. Intrinsic resistance to rifampin occurs naturally among staphylococci with a frequency of 1 in 10^7 colony-forming units.

Pharmacokinetics

Rifampin can be administered orally or intravenously. When given orally, rifampin is absorbed rapidly and almost completely. Plasma clearance is through hepatic uptake, deacetylation to an active drug, and biliary exertion. Rifampin should be avoided in patients with hepatic failure. Dosage adjustment is however unnecessary in renal failure. Rifampin penetrates well into almost all body tissues, having an excellent penetration into soft tissues, bone, abscess cavities, heart valve vegetations and into PMN, allowing killing of phagocytised intra-cellular bacteria.

Therapeutic Use in MRSA Infections

In view of rapid development of resistance during monotherapy, rifampin should never be used alone for treating staphylococcal infections. The companion drug should exhibit pharmacokinetics characteristics similar to rifampin without an antagonistic interaction. Drugs that penetrate poorly into tissues may thus not be optimal, as demonstrated by the failure of vancomycin to prevent the emergence of rifampin resistance in patients with staphylococcal infections treated with combination therapy (Simon *et al.*, 1983).

Some investigators have demonstrated improved cure rates of endocarditis and orthopedic device infections with rifampin containing regimens (Acar *et al.*, 1983; Swanberg *et al.*, 1984; Zimmerli *et al.*, 1998). On the other hand, no advantage was demonstrated for the vancomycin-rifampin combination when compared to monotherapy with vancomycin for the treatment of MRSA endocarditis (Levine *et al.*, 1991). A retrospective study (Burnie *et al.*, 2000) demonstrated a lower mortality among patients with rifampin-susceptible MRSA blood stream infections treated with vancomycin-rifampin combination compared to patients with rifampin-resistant MRSA strains or in whom rifampin was contraindicated. However, most patients in the latter group had liver failure that could influence their prognosis.

In conclusion, there are no well-designed prospective studies demonstrating a benefit from addition of rifampin to vancomycin therapy for severe MRSA infections (e.g endocarditis or catheter related blood stream infection). However, if there is inadequate response to vancomycin alone, then the addition of gentamicin or rifampin or both should be considered. For orthopedic implant infections, rifampin combined with another antimicrobial agent is recommended for early post-surgical MRSA infections (Widmer, 2001).

Fusidic Acid

Fusidic acid is derived from the fungus *Fusidium coccineum* originally isolated from monkey faeces. Fusidic acid has a narrow spectrum of activity principally against gram-positive bacteria, including MRSA. Fusidic acid is chemically related to cephalosporin P1 and helvolic acid. The fusidic acid nucleus has properties common to other tetracyclic structures such as the adrenocorticoids and bile salts. Despite its steroid- like structure, fusidic acid does not have steroid activity.

Mechanism of Action

Fusidic acid exerts its antibacterial activity by inhibiting protein synthesis by interference with elongation factor G (EF-G). EF-G is an essential bacterial protein that promotes translocation on the ribosome after peptide bond formation. Fusidic acid binds to the EF-G- ribosome complex in combination with either GTP or GDP, preventing further elongation by inhibiting the GFPase function of EF-G. Fusidic acid is usually bacteriostatic, but at higher concentration it may be bacteriocidal.

Antibacterial Activity

Fusidic acid has a good *in vitro* activity against *Staphylococcus* species, including methicillin-resistant strains. There appears to be little difference between the activity of fusidic acid against MRSA and MSSA. MICs for *S. aureus* range from 0.03 to 0.25 mg/L. Experimental *in vivo* an *in vitro* investigations have shown indifference between vancomycin and fusidic acid used in combination against MRSA (Simon *et al.*, 1990; Fantin *et al.*, 1993; Drugeon *et al.*, 1994).

Pharmacokinetics and Pharmacodynamics

Fusidic acid may be used parenterally as an intravenous formulation, or orally. Fusidic acid is rapidly absorbed after oral administration, with a mean bioavailability of approximately 90% (Taburet *et al.*, 1990). Fusidic acid is highly protein bound, but has a good penetration into a number of tissues, including synovial fluid, bone, bronchial secretions and burns. It is metabolized by the liver and is eliminated primarily by biliary exertion. Dosage modification is not necessary for patients with renal failure.

Adverse Reactions

Oral use is generally well tolerated with the main side effects being gastrointestinal tract discomfort, diarrhea and headache (Coombs *et al.*, 1987; Carr *et al.*, 1994). With the intravenous formulation, thrombophlebitis and reversible jaundice have been reported (Humble *et al.*, 1980; Portier *et al.*, 1990).

Clinical Use

Fusidic acid has been used in systemic (intravenous, oral) and topical formulations primarily for the treatment of soft tissue infections, chronic osteomyelitis, septic arthritis and endocarditis. Clinical efficacy has been demonstrated in a series of small studies and case reports. Unfortunately, only a few randomized, controlled studies using fusidic acid in the treatment of patients with clinically significant infections have been published. A number of case reports and small case series have suggested that fusidic acid is clinically efficacious in combinations with other antimicrobial agents for a variety if MRSA infections (Sorrel *et al.*, 1982; Portier *et al.*, Cox 1990; Cox *et al.*, 1995). However, clinical failures with the use of fusidic acid either in combination or alone have been reported (Besneir *et al.*, 1991). No prospective, controlled trials have been designed to evaluate the efficacy of fusidic acid in endocarditis or blood stream infections. There have been a few published studies on the efficacy of fusidic acid in orthopedic implant infections (Johnson *et al.*, 1986; Coombs *et al.*, 1987). However, there is only one prospective randomized study which compared the efficacy of fusidic acid -rifampin combination to ofloxacin-rifampin (Drancourt *et al.*, 1997). The overall success rate was similar in both groups (55% and 50%, respectively). The authors did not state whether any of the staphylococci were methicillin resistant. Further clinical studies are needed to clarify the role of fusidic acid in orthopedic implant infection. In conclusion, the use of fusidic acid in monotherapy or in combination with antibacterial agents for therapy of MRSA infections is not well established.

Quinolones

The antistaphylococcal activity of the "old" fluoroquinolones is marginal. The rapid emergence of resistance limits the use of fluoroquinolones in staphylococcal infections. During therapy, blood levels fall below inhibitory concentrations, leading to emergence of resistance. A high rate of resistance to fluoroquinolones among MRSA strains has been reported in many countries (Smith *et al.*, 1990; Diekema *et al.*, 2001). The use of fluoroquinolones in combination with other antistaphylococcal agents could increase therapeutic activity and prevent selection of quinolone resistant strains (Zimmerli *et al.*, 1998).

New fluoroquinolones with enhanced activity against staphylococci, including MRSA, are currently being evaluated, These include moxifloxacin, sparfloxacin, levofloxacin, trovofloxacin, gatifloxacin, clinafloxacin, grepafloxacin, garifloxacin and BMS 284756 (Maple *et al.*, 1991; Piddock *et al.*, 1994; Palmer *et al.*, 1996; Esposito *et al.*, 2002). Unfortunately, the development of some of these agents was discontinued. *In vitro* and experimental animal models have shown high efficacy of several new fluoroquinolones in MRSA infections (Aldrige *et al.*, 1992; Cagni *et al.*, 1995; Frosco *et al.*, 1996). However, their clinical efficacy may be hampered by cross-resistance with ciprofloxacin-resistant strains.

Thus, monotherapy with 'old' quinolones for MRSA infections should not be recommended, due to the high risk for selecting resistant strains. Combinations of quinolone with rifampin were demonstrated to be highly effective in orthopedic implant infections (Zimmerli *et al.*, 1998) and right-sided endocarditis among IV drug users (Dworkin *et al.*, 1989; Heldman *et al.*, 1996). This combination has prevented the selection of quinolone resistant strains (Zimmerli *et al.*, 1998). The potential high potency of the 'new' quinolones in a combination with a second antimicrobial agent is still unknown.

Aminoglycosides

Aminoglycosides are potent bactericidal antibacterial agents with activity against many strains of staphylococci. Their mode of action is through inhibition of protein synthesis by binding to the 30S ribosomal subunit. Time-kill studies have demonstrated synergism between aminoglycosides and glycopeptides against some strains of staphylococci (Watanakunakorn *et al.*, 1982). However, aminoglycosides should not be used alone for staphylococci infections, since monotherapy with aminoglycosides has been associated with high mortality rates (Cafferkey *et al.*, 1985). Gentamicin and tobramycin are often used in combination with vancomycin or teicoplanin in the treatment of MRSA endocarditis. However, there is only limited clinical data, demonstrating a benefit of this combination. In a prospective randonized study among drug abusers with right sided endocarditis, a short course with vancomycin or teicoplanin, even when combined therapy with an aminoglycoside, was inferior to treatment with cloxacillin (Fortun *et al.*, 2001).

Resistance to aminoglycosides is common among isolates of MRSA, limiting their use for treatment of MRSA infections. The main mechanism of aminoglycoside resistance in staphylococci is drug inactivation by cellular aminoglycoside-modifying enzymes. During the last decade a significant increase in the prevalence of resistance rates to aminoglycosides among *S. aureus* has been demonstrated (Dornbusch *et al.* 1990, Schmitz *et al.*, 1999). Resistance rates range between 25% and 91% in different regions (Diekema *et al.*, 2001). Resistance to aminoglycosides is associated with methicillin resistance, with prevalence rates 15-18 higher than that of MSSA (Schmitz *et al.* 1999). However, arbekacin, a derivate of kanamycin B, is active against many MRSA isolates (Inoue *et al.*, 1994). Currently, it is frequently used for treatment of MRSA infections in Japan.

In conclusion, the following factors limit the routine use of aminoglycosides for MRSA infections. There is sparse clinical data of the efficacy of aminoglycosides in MRSA infections; the risk of increased rates of adverse events (as the combination of vancomycin and aminoglycosides is more toxic than either drug alone); and spread of resistant strains.

Fosfomycin

Fosfomycin is a fermentation product of *Streptomyces* spp. that is now produced synthetically. Chemically it is a (1R, 2S)-1,2-epoxypropylphosphonic acid. It contains an epoxide ring and a phosphorous-carbon bond.

Mechanism of Action

The antibiotic interferes with bacterial peptidoglycan synthesis by disrupting cell-wall synthesis. In more detail: fosfomycin enters the bacterial cell via an active transport mechanism for the uptake of L-α-glycerophosphate or hexose phosphate, this patway can be induced by the presence of glucose-6-phosphate. After binding to the target site, the oxygen bonds of fosfomycin opens to form a reactive intermediate that alkylates the active site of the bacterial enzyme phosphoenolpyruvate transferase. This stops the transformation of N-acetyl glucosamine to N-acetyl muramic acid so that the synthesis of acetyl-muramic pentapeptide, the building block for the bacterial peptidoglycan polymer, is prevented (Patel, 1997).

Antibacterial Activity

It has a broad spectrum of activity including both gram-negative and gram-positive bacteria. It is more active against gram-negative bacteria. Fosfomycin is however active against most strains of *S. aureus* including MRSA strains, and against enterococcal strains including VRE. With MIC's (in agar dilution) ranging from 32->200 mg/L. There seems however to be regional difference in susceptibility patterns of various pathogens to fosfomycin (Shrestha, 2001). Fosfomycin exhibits a synergistic activity with β-lactams and aminoglycosides against MRSA and other gram-positive bacteria. The breakpoints of fosfomycin (200 µg disc) are: < 64 µg/mL for susceptible, 128 µg/mL for intermediate strains and > 256 µg/mL for resistant strains.

Pharmacokinetics

Following a single 3 g oral dose of fosfomycin peak serum levels range between 22-32 mg/L with a mean of 26.1 mg/L and are reached 2 h after the dose. The agent is not bound to serum proteins and is excreted unmetabolized in the urine through glomerular filtration. The antibiotic is well distributed in the body and crosses the CSF barrier and the placenta.. Renal impairment decreases the excretion of fosfomycin and an increase in the T1/2 from 11 h in normal kidney function to 50 h is observed when the creatinine clearance varies between 7-54 mL/min.

Therapeutic Use

For severe staphylococcal infection only the parenteral form should be used and in combination with β-lactam or aminoglycosides. There are reports of cure of severe staphylococcal pneumonia, of staphylococcal meningitis and of endocarditis caused by MRSA and infections caused by VISA as well. How ever most of these descriptions are of case reports or very small series with no comparison.

Linezolid

Linezolid is the first of a new class of antibacterial drugs, the oxazolidinones. It has inhibitory activity against a broad range of gram-positive bacteria, including MRSA, GISA, vancomycin-resistant enterococci and penicillin-resistant *Streptococcus pneumoniae*.

Mechanism of Action

The oxazolidinones are inhibitors of protein synthesis and are bacteriostatic against a variety of bacteria. The mechanism of activity is unique and involves the inhibition of the initiation phase of bacterial protein synthesis, by inhibiting formation of the initiation complex formed with 30S ribosomal subunit, initiation factors IF2, IF3 and fMet-tRNA (Swaney *et al.*, 1998). Linezolid inhibits the expression of virulence factors derived from *S. aureus*. The production of α-haemolysin was markedly reduced in the presence of linezolid at concentration 12.5 to 50% of the MIC (Gemell *et al.*, 1999). The *in vivo* correlates of this activity have not yet been documented.

Antibacterial Activity

Linezolid is active against staphylococci, including strains resistant to methicillin, glycopeptides and other antibacterial agents. The breakpoint for linezolid against *Staphylococcus* spp. is ≤4 µg/mL (susceptible). A recent study that tested the activity of linezolid against 1,707 strains of MRSA isolated between 1997-1999 reported 100% susceptibility (Borek *et al.*, 2000). *In vitro*, staphylococci resistant to linezolid can be selected only with difficulty (Swaney *et al.*, 1998). Rapid development of resistance was not observed after serial passages (Zurenko *et al.*, 1996). The first report of clinical isolate resistant to linezolid was from a patient treated with linezolid for peritoneal-dialysis associated peritonitis (Tsiodras *et al.*, 2001).

Pharmacokinetics and Pharmacodynamics

Linezolid is rapidly absorbed after oral administration, with a mean bioavailability of 100% (Stalker *et al.*, 1997; Welshman *et al.*, 1998). Peak concentration (C_{max}) is achieved 1 to 2 hours after administration. Through concentration (C_{min}) is close to or below the susceptibility break point. The elimination half-life is approximately 5.5 hours. Plasma protein binding has been estimated to be 32%. Linezolid is metabolized by oxidation of the morpholine ring, resulting in the formation of 2 inactive metabolites (Slatter *et al.*, 2001). The pharmacodynamic parameter that correlates with the antibacterial efficacy of linezolid is time above the MIC. The time above the MIC should be 40-50% of the dosing time (Craig, 2001). *In vitro* linezolid has a postantibiotic effect that ranges between 2 and 3 hours (Munckhof *et al.*, 2001).

Dosage and Administration

The recommended dosage of linezolid for severe infections caused by VRE or MRSA (e.g nosocomial pneumonia, complicated soft tissue infections, etc.) is 600 mg every 12 hours. The drug may be given via IV infusion or orally. In patients with uncomplicated skin and soft tissue infections, the recommended dosage is 400 mg every 12 hours. Dosage adjustment is not necessary in mild to moderate impaired renal and hepatic failures.

Toxicity and Adverse Events

Volunteer studies and phase II and III studies have shown that linezolid is a relatively safe drug. The most frequently reported adverse events are diarrhea, headache, nausea and vomiting (Kalamazoo, 2000). Thrombocytopenia occurred at a rate of approximately 1-2 %. The thrombocytopenia appears to be related to duration of treatment (>2 weeks of administration). In most patients, platelet counts returned to normal after drug discontinuation. A recent report however suggested that the incidence of thrombocytopenia may be up to a third of the patients if treatment period exceeds two weeks (Attassi 2002). It is advisable to monitor platlet counts in patients at risk for bleeding complications, patients with previous thrombocytopenia and patients treated for longer than 2 weeks with linezolid.

Linezolid is a weak reversible inhibitor of the human monoamine oxidase. Clinical effects, however, have not been observed during clinical studies (that excluded the use of monoamine oxidase inhibitors). The manufacturer recommends that linezolid should not be used in patients taking with monoamine oxidase blocking agents.

Therapeutic use in MSSA and MRSA Infections

The efficacy of linezolid in treatment of infections caused by MRSA has been assessed in several phase II and phase III clinical trials, that included nosocomial pneumonia, community acquired pneumonia (Cammarata *et al.*, 2000; Rubinstein *et al.*, 2001) and complicated and uncomplicated skin and skin structure infections (Duvall *et al.*, 2000; Stevens *et al.*, 2000) and VRE infections (Hartaman *et al.* 2000). Hospitalized patients presenting with known or suspected methicillin resistant staphylococcal infections, including bacteremia, skin and soft tissue infections, urinary tract infections or pneumonia, were enrolled in a multicenter, randomized, non-blinded comparison with vancomycin (Leach *et al.*, 2000). Linezolid treatment achieved similar clinical and microbiological success rates compared to vancomycin. In an open-label, randomized controlled trial, linezolid had similar success rate to teicoplanin in suspected or proven gram-positive infections (Wilcox *et al.*, 2001).

In a randomized comparative study of linezolid (600 mg IV q12h) versus oxacillin (2 g IV every 6 h), with a step-down to linezolid 600 mg orally q12h versus dicloxacillin (500 mg orally q6h) in 826 hospitalized adult patients (Stevens *et al.*, 2000). In an *intention to treat* analysis cure rates were 69.8% and 64.9% respectively (p=0.141; 95% confidence interval –1.58 to 11.25). In clinically evaluable patients, who were treated with linezolid, the cure rate was 88.6% compared with 85.8% in the beta-lactam treated patients. In patients, who were also microbiologically evaluable, the cure rates were 88.1% and 86.1%, respectively, for linezolid and beta-lactam treated patients. These data support the use of linezolid also for patients with MSSA infections. In a study that compared linezolid to teicoplanin in 430 hospitalized patients, intention to treat analysis demonstrated clinical success in 95.5% of linezolid treated patients vs. 87.6% in teicoplanin treated patients (95% confidence interval: 0.025-0.132). There were 4.5% failures in the linezolid group and 12.4% in the teicoplanin group. Among patients with pneumonia the success rates were 96.2% and 92.9% for the linezolid and teicoplanin groups, respectively, for skin and soft tissue infections 96.6 vs. 92.8%, respectively, and for bacteremia 88.5 vs. 56.7%, respectively. Long-term followup resulted in similar success rates (95.3 vs. 92.8%, respectively) (Wilcox *et al.*, 2001).

In a study that compared the length of hospital stay (LOS) in patients with MRSA infections that were treated with either linezolid or vancomycin, recruited from phase III clinical trials worldwide, 460 patients with MRSA infection were included. Linezolid recipients median LOS was shorter in patients with complicated skin and soft tissue infections (both in the intent to treat analysis and the clinically evaluable patients, than vancomycin treated patients. This difference was slight, but not significantly different in the overall intent to treat analysis (Li *et al.* 2001). Linezolid treated patients were more often discharged than vancomycin treated patients during the first week of therapy.

Quinupristin/Dalfopristin

Quinupristin/dalfopristin is the first parenteral streptogramin antibacterial agent. The streptogramins are isolated from *Streptomyces pristinaespiralis*. The family is divided into group A and Group B based on molecular structure. Dalfopristin is a derivative of a group A and quinupristin is a group B streptogramins. These two water-soluble agents have been combined into an injectable form at a 30:70 weight-to-weight ratio. Quinupristin/dalfopristin belongs to the macrolide, lincosamide and streptogramin (MLS) family of antibiotics. The combination has a bactericidal effect against a wide range of Gram-positive bacteria, including MRSA, vancomycin-resistant *Enterococcus faecium*, and penicillin resistant *S. pneumoniae*.

Mechanisms of Action

Quinupristin and dalfopristin bind to sequential sites located on the 50S subunit of the bacterial ribosome. Dalfopristin binding causes a conformational change in the ribosome that subsequently increases the binding of quinupristin (Aumercier *et al.*, 1992). The combined actions of the two agents creates a stable drug-ribosome complex that causes inhibition of protein synthesis by several mechanisms, including prevention of peptide-chain formation, blockade of the extrusion of newly formed peptide chains and a subsequent bacterial cell death (Nadler *et al.*, 1999).

Antibacterial activity

Both quinupristin and dalfopristin possess inhibitory antibacterial activity; however, as combination they demonstrate markedly increased or synergistic activity against a wide range of bacteria, compared with either agent alone (Neu *et al.*, 1992; Jamjian *et al.*, 1997). The NCCLS suggested breakpoints of quinupristin/dalfopristin for *Staphylococcus* spp. are: <1 µg/mL, suscptible; 2 µg/mL, intermediate; and >4 µg/mL, resistant. Quinupristin/dalfopristin demonstrates good activity against *S. aureus*, with MIC_{90} values ranging from 0.5 to 1 µg/mL. In a large North American study, 99% of MRSA were susceptible to quinupristin/dalfopristin (Jones *et al.*, 1998). MRSA isolates with reduced susceptibility to vancomycin have been reported to be *in vitro* still susceptible to quinupristin/dalfopristin (Cohen *et al.*, 1999). Most MLS-resistant staphylococci remain susceptible to quinupristin/dalfopristin (Lina *et al.*, 1999). However, quinupristin/dalfopristin is not consistently bactericidal against staphylococcal strains with constitutive MLS_B resistance (Leclercq *et al.*, 1992).

Pharmacokinetics

Quinupristin/dalfopristin undergoes a rapid elimination from the blood and demonstrates a wide tissue distribution (Etienne *et al.*, 1992). Quinupristin/dalfopristin does not penetrate the central nervous system and the CSF barrier. Quinupristin and dalfopristin undergo hepatic metabolism, and biliary exertion is the primary route of elimination for both compounds. Urinary exertion accounts only for approximately 15 to 19% of quinupristin/dalfopristin elimination (Bergeron *et al.*, 1997).

Quinupristin/dalfopristin is taken up by macrophages, and is active intracellularly, where concentrations are 30 to 50 times the extracellular concentration. Quinupristin/dalfopristin has demonstrated an extended postantibiotic effect against *S. aureus*. After a single hour of exposure of *S. aureus* to quinupristin/dalfopristin at a concentration of 2 to 4 times the MIC, the post antibiotic effect against ranged from 2 to 8 hours, longer than that observed with vancomycin, oxacillin or erythromycin (Bouanchaud, 1997). *In vitro* synergy was demonstrated for quinupristin/dalfopristin with rifampin against MRSA (Sambatakou *et al.*, 1998).

Dosage

The recommended dosage is 7.5 mg/kg administered intravenously every 8 hours for life threatening infections and nosocomial pneumonia, and every 8 or 12 hours for complicated skin and skin structure infections. Dose adjustment may be necessary in hepatic insufficiency and the combination should be used in caution in patients with impaired renal function.

Adverse Events

Adverse events associated with quinupristin/dalfopristin are usually mild to moderate in severity in over 80% of cases (Rubinstein *et al.*, 1999) The most common adverse events are infusion-site reactions including inflammation, pain, edema and thrombophlebitis. Other common adverse events are nausea, vomiting, diarrhea, rash and headache.

Clinical Use

The efficacy of intravenous quinupristin/dalfopristin in the treatment of MRSA infections has been documented in a comparative pilot study (Raad *et al.*, 1999) and phase III studies conducted as part of the compassionate-use program (Drew *et al.*, 2000). Quinupristin/dalfopristin has been investigated in the treatment of gram-positive bacterial infections when no other antibacterial treatment was available (i.e. causative pathogen with *in vitro* resistance to alternative antibacterial agents or treatment failure or intolerance to alternative agents in particular vancomycin). A multicenter, open-label, non-comparative, emergency use clinical study enrolled patients with culture-confirmed infections with MRSA (Drew *et al.*, 2000). The most common indications were bone and joint infections and skin and skin structure infections. The overall success rate was 71.1% in all-treated population and 66.7% in patients with who were both clinically and bacteriologically evaluable. Clinical success rates in patients with bacteremia, endocarditis and respiratory infections were lower (overall success rate: 70.5%, 54.4% and 40.0%, respectively). The presence of MLSb did not appear to affect the clinical response.

Quinupristin/dalfopristin showed good efficacy in complicated skin and soft tissue infections in two large randomized trials (Nichols *et al.*, 1999). Quinupristin/dalfopristin achieved clinical success rate of 68% and a rate of 71% with the comparator agents. The combination also demonstrated equivalent efficacy to vancomycin in the treatment of patients with gram-positive pneumonia (Fagon *et al.*, 2000).

At present, the US Food and Drug Administration-approved indications include treatment of serious or life-threatening infections associated with VREF bacteremia and complicated skin and skin-structure infections. The greatest potential benefit of quinupristin/dalfopristin is likely to be in the management of patients with multiple-resistant MRSA or VREF infections.

Daptomycin

The cyclic lipopeptide daptomycin is a fermentation product of *Streptomyces roseosporus*. It was discovered in the early 1980s. Daptomycin has been shown to be effective in a number of *in vivo* animal models including soft tissue, bloodstream, kidneys, lung and bone infections caused by gram-positive strains resistant to standard therapies. On the basis of animal models, phase 1 and phase 2 clinical studies, daptomycin shows promise for the treatment of a wide variety of gram-positive infections, including those resistant to standard therapy.

Mechanism of Action

The precise mechanism of action of daptomycin is not completely understood. Daptomycin has a unique mechanism of action, killing gram-positive bacteria by disrupting multiple aspects of the bacterial plasma membrane function without penetrating into the cytoplasm. Possible mechanisms include inhibition of peptidoglycan synthesis, inhibition of lipoteichoic acid synthesis and alterations of the cytoplasmic membrane potential (Allen *et al.*, 1989; Canepari *et al.*, 1990; Alborn *et al.*, 1991).

Antibacterial Activity

Daptomycin shows excellent *in vitro* activity against a wide range of gram-positive bacteria, including those resistant to various antibiotics. The drug is bactericidal against MRSA, vancomycin resistant entrococci, and penicillin-resistant *S. pneumoniae* (Lamp *et al.*, 1992; Louie *et al.*, 1993; King *et al.*, 2001).

Pharmacokinetics and Pharmacodynamics

Daptomycin is administrated intravenously, with once-daily administration, Daptomycin exhibits linear pharmacokinetics and minimal accumulation with doses up to 6 mg/kg in healthy volunteers (Tally *et al.*, 1999). Excertion of the drug occurs primarily via the kidney. The antibacterial activity of daptomycin is concentration-dependent (Lamp *et al.*, 1992). Against *Staphylococcus aureus* daptomycin produces a PAE lasting from 1 to 6 hours (Bush *et al.*, 1989; Hanberger *et al.*, 1991).

Adverse Events

In phase 2 trials, adverse events and rates of discontinuation of due to adverse events were comparable between patients receiving daptomycin and patients receiving conventional treatments (Tally *et al.*, 1999). However, daptomycin was associated, in early trials with mild, reversible skeletal muscle adverse events, at high multiple-dose level (4 mg/kg every 12 hours) however, at the once daily dose of 6 mg/kg these adverse events did not appear.

Clinical Studies

Two multicenter, randomized phase 2 clinical trials evaluated the efficacy of daptomycin compared with conventional therapy (β-lactam agents or vancomycin) in a total of 285 patients (Tally *et al.*, 1999; Kotra, 2000). Daptomycin demonstrated efficacy against skin and soft tissue infections comparable to that of conventional therapy. Daptomycin was administered to patients with bacteremia at a 6 mg/kg loading dose followed by 3 mg/kg every 12 hours. It produced a favorable clinical outcome and bacteriological cure. This antibiotic drug is being compared currently with standard therapy for the treatment of complicated skin and soft tissues infection, complicated urinary tract infections and bloodstream infections, in randomized, prospective, phase 2 and phase 3 clinical trials.

Glycylcyclines

Those compounds are derivatives of tetracycline with an ability to circumvent the resistance mechanisms that confer resistance to the tetracyclines. Tigecycline (GAR-936) is a derivative of minocycline (Hooper, 2002).

Antibacterial Activity

Tigecycline has activity against resistant gram-positive cocci including MRSA and VISA strains as well as resistant enterococci and pneumococci. In the rat endocarditis model, tigecycline reduced the number of staphylococci in the vegetation. In the neutropenic mouse model with a thigh infection time over MIC and AUC predicted efficacy.

Pharmacokinetics

In humans the volume of distribution of tigecycline is high (12 L/kg) with a Cmax in serum of 0.5-1.0 mg/L and a long half-life allowing twice daily dosing.

Therapeutic Use

Efficacy has been demonstrated in skin and soft tissue infections in phase II studies in humans. Other indications are awaiting study results.

Drugs in Pre-clinical Studies

There are numerous investigational antibiotics that have demonstrated excellent antibacterial activity *in vitro* and in animal models against gram-positive bacteria, including multiple resistant pathogens.

Oritavancin

Oritavancin (LY333328) is a new semi-synthetic glycopeptide derived from N-alkylation of LY264826, a naturally occurring vancomycin-like drug. Oritavancin is active against gram-positive bacteria, including staphylococci, enterococci and *S. pneumoniae*. The drug is highly effective against strains of MRSA, penicillin-resistant *S. pneumoniae* and vancomycin resistant-enterococci. The *in vitro* activity of the drug was studied in comparison with other glycopeptides against 637 isolates. Oritavancin was effective against all the gram-positive organisms tested, including those that are resistant to vancomycin and teicoplanin (Noviello *et al.*, 2001).

Cationic Peptides

Cationic antimicrobial peptides are a new class of natural antibiotics (Hancock *et al.*, 2001). Cationic peptides are produced by all organisms, as a major part of their immediately effective, nonspecific defense against infections (innate immunity). Cationic peptides have an extraordinary spectrum of antibacterial activity, including gram-positive and gram-negative bacteria, fungi, enveloped viruses and eukaryotic parasites, including antibacterial activity against multiple resistant bacteria, including MRSA. Their unique mechanism of action is by their interaction with the outer membrane and cytoplasmic bacteria cell membrane leading to break down of the membrane (Matsuzaki, 1998; Huang, 2000). Because of their mode of action, cationic peptides are not affected by common antibiotic resistance mechanisms. Some cationic peptides are currently being investigated in phase II and phase III clinical trials as topical treatments in skin and mucosal infections.

Treatment of MRSA Infections

Catheter Related Staphylococcal Infections

Gram-positive bacteria are the predominant pathogens in intravascular catheter related infections (Schaberg *et al.*, 1991). Vancomycin is considered to be the treatment of choice for nosocomial MRSA infections associated with vascular catheters. However, there are several reasons to search for alternative agents, including the relative high toxicity, slow bactericidal effect of vancomycin and the risk for emergence of glycopeptide resistance after prolonged treatment with vancomycin. When treatment of catheter related staphylococcal infections has failed, addition of gentamicin, rifampin or both should be considered.

There is no data to support the routine addition of rifampin to vancomycin therapy in catheter related staphylococcal infections. The addition of rifampin should be considered if there is inadequate response to vancomycin alone. However, the development of rifampin resistance during therapy for MRSA infections with vancomycin plus rifampin has been reported (Simon *et al.*, 1983). In a pilot study of quinupristin/dalfopristin versus vancomycin in the treatment of catheter related staphylococcal bacteremia, quinupristin/dalfopristin achieved similar response rates and safety comparable to vancomycin (Raad *et al.*, 1999).

Endocarditis

Assessing the clinical efficacy of antibacterial agents in endocarditis is distinctively difficult, as controlled studies are rare. Vancomycin is recommended for therapy of *S. aureus* endocarditis when MRSA strains are involved. However, vancomycin is less rapidly bactericidal *in vitro* and *in vivo* than β-lactam antistaphylococcal agents, especially when high inocula of bacteria are present. There have been several reports of therapeutic failure in patients with staphylococcal endocarditis treated with vancomycin (Small *et al.*, 1990; Fortun *et al.*, 1995, 2001). Clinical trials also demonstrated that teicoplanin has been associated with a high rate of failure in patients treated with the standard low dose regimen (6 mg/kg/day) of this antibiotic. A dose of 12 mg/kg/day appears to be necessary to provide a cure rate similar to that of vancomycin (Wilson *et al.*, 1994).

For patients with MRSA endocarditis not responding to vancomycin, several choices are available, including the addition of rifampin or gentamicin (or both). Other regimens including minocycline, TMP-SMX, and ciprofloxacin-rifampin have been used with a different degree of success. When staphylococcal bacteremia persists during appropriate treatment, the possibility of emergence of vancomycin intermediate *S. aureus* strains should be raised.

There is a controversy about the role of rifampin in the treatment of MRSA endocarditis. Owing to the rapid emergence of resistance, rifampin should not be used alone. Results of *in vitro* studies and animal models on rifampin combinations with vancomycin are contradictory. Some studies have suggested that incubation of vancomycin *in vitro* with rifampin suppresses the emergence of rifampin resistance (Watanakunakorn *et al.*, 1981). However, in other studies rifampin resistance was observed in MRSA strains during combination treatment of rifampin and vancomycin (Simon *et al.*, 1983). A randomized controlled study of vancomycin compared with vancomycin plus rifampin for MRSA endocarditis showed slow clearance of bacteremia in both groups, and no benefit for the combination (Levine *et al.*, 1991).

It is generally recommended to add rifampin to vancomycin treatment in cases of MRSA endocarditis with metastatic abscesses (renal, splenic, myocardial or cerebral), or because of failure of monotherapy with vancomycin. The optimal regimen should include at least two other drugs added to vancomycin such as gentamicin, or a fluoroquinolone, to minimize the probability of developing rifampin resistance.

Pneumonia

Over the past 15 years, the incidence of nosocomial pneumonia due to gram-positive pathogens relative to gram-negative bacteria has steadily increased (Rello, 1991). The National Nosocomial Infections Surveillance System reported that *S. aureus* is the most common cause of nosocomial infections and accounts for 20% of reported pathogens in adult patients with nosocomial pneumonia in ICUs (Richards *et al.* 1999).

Vancomycin is considered to be the standard treatment for MRSA respiratory infections based on susceptibility rates. Although MRSA is sensitive *in vitro* to vancomycin, the agent's moderate extravascular diffusion, low epithelial lining fluid concentration and absent pulmonary macrophage concentration combined with a relatively slow bactericidal effect may lower vancomycin effectiveness in pneumonia. Clinical studies in critically ill patients with severe nosocomial, but also community-acquired pneumonia due to MRSA and MSSA have described a high mortality rate among patients treated with vancomycin for pneumonia (Rello *et al.*, 1994). A recent study, documented a high mortality rate among patients treated with vancomycin for pneumonia caused by MRSA and MSSA (Gonzalez *et al.*, 1999). Mortality was significantly higher among MSSA infected patients treated with vancomycin than among those treated with cloxacillin (47% vs. 0%). As vancomycin has bactericidal activity that is time-dependent, some specialists suggests that vancomycin should be used in accordance with its pharmacokinetics/pharmacodynamic properties, to minimize exposure to suboptimal antibiotic concentrations (Rello *et al.*, 2001). The way to optimize the antibiotic efficacy is to maximize the time that the drug levels present at the site of infection exceed the pathogen's MIC, this is best achieved through continuous infusions, which provide longer time >MIC rather than intermittent dosing which is presently practiced. However, there are no controlled trials showing superiority of continuous infusions over intermittent dosing. Another approach for treating MRSA pneumonia is combination treatment, including vancomycin with rifampin. To date as previously mentioned, there is only little data on the penetration of vancomycin into various pulmonary compartments and fluids in infection. There is no information on the ability of vancomycin to exert its bactericidal activity in the presence of sputum, pus, acid pH and anaerobic conditions that are prevalent in infected pulmonary regions.

In light this poor effectiveness of current treatments, new compounds for effective therapy of nosocomial pneumonia with multiresistant gram-positive pathogens are under rapid clinical development. These compounds include the streptogramins, fluoroquinolones, oxazolidinones, daptomycin and ketolides.

A prospective, randomized, open-label, multicenter clinical trial in patients with nosocomial pneumonia compared quinupristin/dalfopristin with vancomycin (aztreonam was used in both groups to treat potential gram-negative pathogens) (Fagon *et al.*, 2000). The study focused on patients with suspected or proven gram-positive nosocomial pneumonia. Quinupristin/dalfopristin was shown to be equivalent to vancomycin. Clinical response, bacteriologic response and adverse events were similar in the two groups. Clinical success rates among bacteriologic evaluable patients with MRSA pneumonia were also comparable. However, response rates in both treatment groups were relatively low (30.9% in the quinupristin/dalfopristin group and 44.4% in the vancomycin group).

A multinational, randomized, double blind, controlled trial compared the efficacy of linezolid with vancomycin (combined with aztreonam) in the treatment of nosocomial pneumonia (Rubinstein *et al.*, 2001). Clinical cure rates and microbiological success rates for evaluable patients were equivalent between the treatment groups, however in this study the number of proven MRSA was small.

Infections of Orthopedic Device Implants

S. aureus and coagulase-negative staphylococci are the most frequently encountered microorganisms isolated from patients with orthopedic device infections, accounting for 50% of the cases (Fitzgerald, 1989). Such infections are extremely difficult to treat. The microenvironment and biofilm created around the implant is responsible for the difficulty in eradicating microorganisms from the site of infection (Costerton *et al.*, 1999). Bacteria embedded in biofilm are less susceptible to antibacterial agents due to failure of the antimicrobial agent to penetrate to biofilm. The biofilm made of bacterial glycocalyx forms a protective environment that allows the bacteria to persist in the stationary phase of growth. β-lactam antimicrobial agents and glycopeptides have reduced activity against bacteria in stationary-phase (Farber *et al.*, 1990). Therefore, successful treatment should contain the removal of the infected implant and long term intravenous antibiotic therapy. Debridement with retention of the prosthesis is an attractive therapeutic option, avoiding multiple surgical interventions and prolonged periods of immobility. However, the incidence of treatment failure is higher among patients treated with debridement and prosthesis retention than among those treated with removal of the infected prosthetic joint. An increased risk of treatment failure has been associated with *S. aureus* (Wilson *et al.*, 1990; Schoifet *et al.*, 1990; Tsukayama *et al.*, 1991; Brandt *et al.*, 1997). A retrospective cohort study determined the probability of treatment failure for patients with *S. aureus* prosthetic joint infections treated with debridement, prosthesis retention and traditional parenteral antibiotics, such as vancomycin or β-lactams. The failure rate was 54% after 1 year and 69% after 2 years (Brandt *et al.*, 1997).

Treatment with antimicrobial agents active against stationary-phase bacteria could possibly eradicate the infection and enable retention of the device. Rifampin has excellent efficacy against stationary-phase *S. aureus*, trough concentrations exceeding the MICs by a factor of 10-100 are easily achieved, and the antimicrobial as previously described is orally well absorbed. Rifampin has been shown to eliminate stationary-phase staphylococci *in vitro* and in clinical trials (Widmer *et al.*, 1992; Drancourt *et al.*, 1993; Zimmerli *et al.*, 1998).

A double blind, randomized controlled study was conducted among patients with stable implants, short duration of infection, and initial debridement (Zimmerli *et al.*, 1998). The investigators randomized 33 patients with proven staphylococcal implant infection to receive either rifampin or placebo; both groups also received vancomycin or cloxacillin for two weeks followed by oral ciprofloxacin for 3 to 6 months. Cure was achieved in 100% (12/12) of patients in the rifampin group and in 58% (7/12) in the placebo group. However, in both study groups, there were no patients with MRSA infections and the groups contained only a very limited number of patients. The rapid emergence of quinolone resistance among strains of MRSA may the limit the usefulness of quinolone-rifampin combinations (Diekema *et al.*, 2001). Treatment with ciprofloxacin and rifampin of a quinolone-resistant staphylococcus would likely lead to failure and secondary emergence of rifampin resistance during treatment. Fusidic acid could be an alternative to the combination of rifampin and quinolone when the isolate is resistant to quinolones. A prospective, randomized study compared the efficacy of rifampin combined with either fusidic acid or ofloxacin for the treatment of staphylococcal orthopedic implant infections

(Drancourt *et al.*, 1997). Overall treatment was successful in 55% of patients treated with rifampin and fusidic acid and 50% of the patients treated with rifampin and ofloxacin. An effective alternative for the treatment of multidrug resistant staphylococci (including rifampin and quinolones) is TMP/SMX. A prospective, non-comparative study was conducted among patients with multidrug-resistant staphylococci organisms (susceptible only to glycopeptides and TMP/SMX) infected orthopedic implants (Stein *et al.*, 1998). All patients were treated with high dose TMP/SMX for 6-9 months (trimethoprim, 20 mg/ kg /day; sulfamethoxazole, 100 mg/kg/day). The overall treatment success rate was 66.7% (26 of 39 patients). The cure rate among patients with *S. aureus* was also 66.7% (16 of 24 patients). Nevertheless, a high percentage of patients had to discontinue their therapy because of severe side effects. The presence of high rate of side effects suggests the TMP/SMX should be prescribed only for treatment of multidrug resistance staphylococci for patients unfit for surgery.

Early limited experience in patients with early postoperative orthopedic device infection showed that they can be successfully treated with retention of prosthesis and long-term treatment with antimicrobial agents. Patients eligible for possible salvage of the prosthesis should have a stable implant, symptoms less than 28 days and a pathogen susceptible to oral antimicrobial agent (Widmer, 2001). The recommended regimen for MRSA infection includes IV vancomycin for 4 weeks followed with oral therapy for a period of 3-6 months. Rifampin should be included in the treatment of MRSA if the strain is susceptible *in vitro*. Rifampin must however be combined with another antimicrobial agent (quinolone or fusidic acid). The newer quinolones, such as moxifloxacin, have lower MICs for staphylococci and might be favored over other quinolones when combined with rifampin. However, no clinical data are yet available.

Patients with chronic implant infection are not likely to respond to antimicrobial therapy alone and require the removal of the implant and 6 weeks of IV vancomycin. Re-implantation of a new device, either immediately after removal of the infected implant (one-step arthroplasty) or at a later stage, following antimicrobial therapy (two-step arthroplasty).

Prolonged suppressive antibiotic treatment is an alternative to surgery in patients who cannot undergo surgical interventions (Stein *et al.*, 1998). Two cases with MRSA prosthetic hip replacement were treated successfully with a long course of linezolid (Bassetti *et al.*, 2001). However, long courses of therapy with linezolid are not approved.

Treatment of Infections Caused by VISA

VISA has emerged recently worldwide as individual cases and also as limited outbreaks. Currently, there are no recommended therapies or guidelines for VISA infections. The emergence of VISA clinical isolates has prompted a search for new drugs as well of assessment of 'old' antibiotics and antibiotic combination therapies.

Clinical Data

Currently, the frequency of infection caused by VISA isolates appears to be very low. Several countries, including the USA (Centers for Disease Control and Prevention, 1997; Smith *et al.*, 1999), and from Europe (Ploy *et al.*, 1998; Hood *et al.*, 2000), Asia (Hiramatsu *et al.*, 1997; McManus, 1999; Mi-Na *et al.*, 2000) and South Africa (Ferraz *et al.*, 2000) have already reported a handful of cases of VISA. The majority of patients infected with VISA strains had received prolonged therapy with vancomycin in the months preceding VISA infections. Most of the patients had severe underlying diseases, mainly chronic renal failure (requiring hemodialysis) and malignancies. It is not clear what is the exact clinical significance of the VISA isolates. Although VISA strains were isolated from patients who apparently failed vancomycin therapy, several of the patients had complicated factors such as undrained foci of infection and foreign bodies, which reduce the effectiveness of vancomycin. These VISA strains showed resistance to many other antimicrobial agents. However, all the isolates remain susceptible to certain antibiotics including rifampin, TMP/SMX and gentamicin. Patients were treated with various antibiotics, including new drugs (linezolid, quinupristin/dalfopristin) or combination of established antimicrobial agents (e.g: rifampin plus TMP/SMX plus gentamicin). The mortality of these patients was very high, but it is not clear that death was directly attributed to VISA infections, as most patients had severe underlying disease or undrained foci of infection.

In Vitro and In Vivo Studies

New agents are being developed against resistant gram-positive bacteria. Some are expected to have considerable activity against VISA. Quinupristin/dalfopristin, daptomycin and linezolid have been demonstrated to possess potent activity *in vitro* and in experimental models of VISA infections (Akins *et al.*, 2000). Nonetheless, linezolid and other oxazolidinones are merely bacteriostatic against *S. aureus*.

The synergistic activity of several combinations of antimicrobial agents against VISA strains has been investigated (Hershberger *et al.*, 1999; Akins *et al.*, 2000). *In vitro* and *in vivo* studies have demonstrated synergy between vancomycin and β-lactams in treating VISA (Sieradzki *et al.*, 1997; Aritaka *et al.*, 2001). Ampicillin/sulbactam has also demonstrable activity due to increased production of penicillin binding proteins of the isolates. However, currently there is no clinical data demonstrating the efficacy of antibiotic combinations. In conclusion, at the present time, treatment of VISA infections should be patient specific, including new antistaphylococcal antimicrobial agents or various drug combinations. The treatment strategy should be based on susceptibility profiles and *in vitro* data.

References

Acar, J.F., Goldstien, E.W., and Duval, J. 1983. Use of rifampin for the treatment of serious staphylococcal and gram-negative bacillary infections. Rev. Infect. Dis. 5 (Suppl.): 502-506.

Aeschlimann, J.R., Hershberger E., and Rybak, M.J. 2000. Activities of trovafloxacin and ampicillin-sulbactam alone or in combination versus three strains of vancomycin-intermediate *Staphyloccus aureus* in an *in vitro* pharmacodynamic infection model. Antimicrob. Agents. Chemother. 44: 1153-1158.

Akins, R.L. and Rybak, M.J. 2000. *In vitro* activities of daptomycin, arbekacin, vancomycin and gentamicin alone and/or in combination against glycopeptide intermediate-resistant *Staphylococcus aureus* in an infection model. Antimicrob. Agents Chemother. 44: 1925-1929.

Albanese, J., Leone, M., Bruguerolle, B., Ayem, M.L. Lacarelle, B., and Martin, C. 2000. Cerebrospinal fluid penetration and pharmacokinetics of vancomycin administered by continuous infusion to mechanically ventilated patients in an intensive care unit. Antimicrob. Agents Chemother. 44: 1356-1358.

Alborn, W.E. Jr, Allen, N.E., and Preston, D.A. 1991. Daptomycin disrupts membrane potential in growing *Staphylococcus aureus*. Antimicrob. Agents Chemother. 35: 2282-2287.

Aldrige, K.E., Jones, R.N., Barry, A.L, and Gelfand, M.S. 1992. *In vitro* activity of various antimicrobial agents against *Staphylococcus aureus* isolates including fluoroquinolones and oxacillin-resistant strains. Diagn. Microbiol. Infect. Dis. 15: 517-521.

Alonso, R., Padilla, B., Sanchez-Carrillo, C., Munoz, P., Rodriguez-Creixems, M., and Bouza, E. 1997. Outbreak among HIV-infected patients of *Staphylococcus aureus* resistant to cotrimoxazole and methicillin. Infect. Control. Hosp. Epidemiol.18: 617-621.

Allen, N.E., Hobbs, J.N., and Alborn, W.E. 1987. Inhibition of peptidoglycan biosynthesis in gram-positive bacteria by LY146032. Antimicrob. Agents Chemother. 31: 1093-1099.

Attassi, K., Hershberger, E., Alam, R., and Zervos, M. 2002.Thrombocytopenia associated with linezolid therapy. Clin. Infect. Dis. 34: 695-698.

Aumercier, M., Bouhallab., S., and Capmau M-L, le Goffic, F. 1992. RP 59500: a proposed mechanism for its bactericidal activity. J. Antimicrob. Chemother. 30: 9-14.

Bassetti, M., Biagio, D.A., Cenderello, G., Del Bono, V., Palermo, A., Cruciani, M., and Bassetti, D. 2001. Linezolid treatment of prosthetic hip infection due to methicillin-resistant *Staphylococcus aureus* (MRSA). J. Infect. Dis. 43: 148-149.

Bayer, A.S. and Lam, K. 1985. Efficacy of vancomycin plus rifampin in experimental aortic-valve endocarditis due to methicillin-resistant *Staphylococcus aureus*: *in vitro-in vivo* correlations. J. Infect. Dis. 151: 157-165.

Bayer, A.S., and Morrison, J.O. 1984. Disparity between timed-kill and checkerboard methods for determination of *in vitro* bactericidal interactions of vancomycin plus rifampin versus methicillin-susceptible and resistant *Staphylococcus aureus*. Antimicrob. Agents Chemother. 26: 220-223.

Bergeron, M., and Montay, G. 1997. The pharmacokinetics of quinupristin/ dalfopristin in laboratory animals and in humans. J. Antimicrob. Chemother. 39 (Suppl. A): 129-138.

Besneir, J.M., Kanoum, F., Martin, C., Cotty, P., Fenneteau, A., and Choutet, P. 1991. Failure of combination of fusidic acid and acid in a patient with staphylococcal infection. J. Antimicrob. Chemother. 27: 560-562.

Borek, A.P., Peterson, L.R., and Noskin, G.A.. 2000. Activity of linezolid (LIN) against medically important Gram-positive bacteria from 1997 to 1999 40th Interscience Conference on Antimicrobial Agents and Chemotherapy Sep. 17-20: Toronto. Abstract 2299, p.185.

Bouanchaud, D.H. 1997. *In vitro* and *in vivo* antibacterial activity of quinuprisitin/dalfopristin. J. Antimicrob. Chemother. 39: 15-21.

Brandt, C.M., Sistrunk, W.W., Duffy, M.C., Hanssen, A.D., Steckelberg, J.M. Ilstrup, D.M. and Osmon, D.R. 1997. *Staphylococcus aureus* prosthetic joint infection treated with debridement and prosthesis retention. Clin. Infect. Dis. 24: 914-919.

Burnie, J., Matthews, R., Jiman-Fatme, A., Gottardello, P., Hodgetts, S., and D'arcy S. 1999. Analysis of 42 cases if septicemia caused by an epidemic strain of methicillin-resistant *Staphylococcus aureus*: evidence of resistance to vancomycin, Clin. Infect. Dis 31: 684-689.

Bush, L.M., Boscia, J.A., Wendeler, M., Pitsakis, P.G., and Kaye, D. 1989. *In vitro* postantibiotic effect of daptomycin (LY 146032) against *Enterococcus faecalis* and methicillin-susceptible and methicillin-resistant *Staphyloccus aureus* strains. Antimicrob. Agents Chemother. 33: 1198-1200.

Bushby, S.R.M., and Hitchings, G.H.. 1968. Trimethoprim, a sulphonamide potentiator. Br. J. Pharmacol. 33: 72-90.

Cagni, A., Chuard, C, Vaudaux, P.E., Schrenzel, J., and Lew, D.P. 1995. Comparison of sparfloxacin, temafloxacin and ciprofloxacin for prophylaxis and treatment of experimental foreign-body infection by methicillin-resistant *Staphyloccus aureus*. Antimicrob. Agents Chemother. 39: 1014-1016.

Cafferkey, M.T., Hone, R., and Keane, C.T. 1985. Antimicrobial chemotherapy of septicemia due to methicllin-resistant *Staphylococcus aureus*. Antimcrob. Agents Chemother. 28: 819-823.

Cammarata, S.K., Schueman, L.K., and Timm, J.A. 2000. Oral linezolid in the treatment of community-acquired pneumonia: a phase III trial [abstract]. Am. J. Res. Crit. Care. Med. 161 (Suppl.): A654.

Canepari, P., Boaretti, M,. del Mar Lleo, M., and Satta, G. 1990. Lipoteichoic acid as a new target for activity of antibiotics: mode of action of daptomycin (LY 146032) Antimicrob. Agents Chemother. 34: 1220-1226.

Carr, W.D., Wall, A.R. , Georgala-Zervogiani S, Stratrigos, J., and Gouriotou, K. 1994. Fusidic acid tablets in patients with skin and soft-tissue infection. A dose-finding study. Eur. J. Clin. Res. 5: 87-95.

Centers for Disease Control and Prevention. 1997. *Staphylococcus aureus* with reduced vancomycin susceptibility. MMWR. 46: 765-766.

Cohen, E., Dadashev, A., Drucker M., Samra Z., Rubinstein E., and Garty M. 2002. Once daily versus twice-daily intravenous administration of vancomycin for infections in hospitalized patients. J. Antimicrob. Chemother. 49: 155-160.

Cohen, M.A., and Huband, M.D. 1999. Activity of clinafloxacin, trovafloxacin, quinupristin/dalfopristin, and other antimicrobial agents versus *Staphylococcus aureus* 33: 43-46.

Cony-Makhoul, P., Brossard, G., Marit, G., Pellegrin, J.L., Texier-Maugein, and Reiffers, J. 1990. A prospective study comparing vancomycin and teicoplanin as second-line empiric therapy for infection in neutropenic patients. Br. J. Haematol. 76: 35-40.

Coombs, R.R.H., Mehtar, S., and Menday, A.P. 1987. Fusidic acid in orthopaedics. Curr. Ther. Res. Clin. Exp. 42: 501-508.

Costerton, J.W., Stewart, P.S., and Greenberg, E.P. 1999. Bacterial biofilms: a common cause of persistent infections. Science. 284; 1318-1322.

Cox, R.A., Mallaghan, C., Conquest, C., and King, J. 1995. Epidemic methicillin-resistant *Staphylococcus aureus*: controlling the spread outside hospital. J. Hosp. Inf. 29: 107-119.

Diekema, D.J., Pfaller, M.A., Schmitz, F., Smayevsky, J., Bell, J., Jones, R.N., Beach, M., and the SENTRY Participants Group. 2001. Survey of infections due to staphylococcus species: Frequency and occurrence and antimicrobial susceptibility of isolates collected in the United States, Canada, Latin America, Europe, and the Western Pacific Region for the SENTRY antimicrobial surveillance program, 1997-1999. Clin. Infect. Dis. 32 (Suppl 2): S114-S132.

Dornbusch, K., Miller, G.H., Hare, R.S., and Shaw, K.J. 1990. Resistance to aminoglycisides antibiotics in gram-negative bacilli and staphylococci isolated from blood. Report from a European collaborative study. The ESGAR study group (European Study Group on Antibiotic Resistance). J. Antimicrob. Chemother. 26: 131-144.

Downs, N.J., Neihart, R.E., Dolezal, J.M., and Hodges, G.R. 1989. Mild nephrotoxicity associated with vancomycin use. Arch. Inter. Med. 149: 1777-1781.

Drancourt, M., Stein, A., and Argenson, J.N., Zannier A, Curvale G, and Raoult D. 1993. Oral rifampin plus ofloxacin for treatment of staphylococcus-infected orthopedic implants. Antimicrob. Agents Chemother. 37: 1214-218.

Drancourt, M., Stein, A., Argenson, J.N., Roiron, R., Groulier, P., and Raoult, D. 1997. Oral treatment of *Staphylococcus* spp. infected orthopaedic implants with fusidic acid or orfloxacin in combination with rifampicin. J. Antimicrob. Chemother. 39: 235-240.

Drugeon, H.B., Caillon, J., and Juvin, M.E. 1994. *In vitro* antibacterial activity of fusidic acid alone and in combination with other antibiotics against methicillin-sensitive and resistant *Staphylococcus aureus*. J. Antimicrob. Chemother. 34: 899-907.

Drew, R.H. Perfect, J.R., Srinath, L., Kurkimilis, E., Dowzicky, M. and Talbot, G.H. for the Synercid Emergency-use Study Group. 2000. Treatment of methicillin-resistant *Staphylococcus aureus* infections with quinupristin-dalfopristin in patients intolerant of or failing prior therapy. J. Antimicrob. Chemother. 46: 775-784.

Duvall, S.E., Seas, C., and Bruss, J.B. 2000. Comparison of oral linezolid to oral clarithromycin in the treatment of uncomplicated skin infection: results from a multinational phase III trial [abstract]. 9th International Congress on Infectious Diseases. Buenos Aires, 184.

Esposito, S., Noviello, S., and Ianniello, F. 2002. Antimicrobial activity of BMS-284756 against staphylococci and respiratory pathogens. Clin. Microbiol. Infect. 8 (Suppl. 1): Abstract P767.

Etienne, S.D., Montay, G., Le Liboux, A., Frydman, A., and Garaud, J.J. 1992. A phase I, double-blind, placebo-controlled study of the tolerance and pharmacokinetic behaviour of RP 59500. J. Antimcrob. Chemother. 30: 123-131.

Fagon, J., Patrick, H,. Haas, D.W., Torres, A., Gibert, C., Cheadle, W.G., Falcone R.E., Anholm, J.D., Paganin, F., Fabian, T.C., and Lilienthal, F. 2000. Treatment of gram-positive nosocomial pneumonia. Prospective randomized comparison of quinupristin/dalfopristin versus vancomycin. Nosocomial Pneumonia Group. Am. J. Respir. Crit. Care Med. 161: 1759-1760.

Fantin, B., Leclerq, R., Duval, J., and Carbon, C. 1993. Fusidic acid alone or in combination with vancomycin for therapy of experimental endocarditis due to methicillin-resistant *Staphylococcus aureus*. Antimicrob Agents Chemother. 37: 2466-2469.

Farber, B.F., Kaplan, H., and Clogston, A.G. 1990. *Staphylococcus epidermidis* extracted slime inhabits the antimicrobial action of glycopeptide antibiotics. J. Infect. Dis. 161: 37-40.

Ferraz, V., Duse, A.G, Kassel, M., and Black, A.D., Ito, T., Hiramatsu, K. 2000. Vancomycin-resistant *Staphylococcus aureus* occurs in South Africa. S. Afr. Med. J. 90: 1113.

Fitzgerald, R.J. 1989. Infections of hip prosthesis and artificial joints. Infect. Dis. Clin. North Am. 3: 329-338.

Fortun, J., Navas, E., Martinez-Beltran, J., Perez-Molina, J., Martin-Davila, P., Guerrero A., and Moreno S. 2001. Short course therapy for right-side endocarditis due to *Staphylococcus aureus* in drug abusers: cloxacillin versus glycopeptides in combination with gentamicin. Clin. Infect. Dis. 33: 120-125.

Fortun, J., Perez-Molina, J.A., Anon, M.T., Martinez-Beltran, J., Loza, E., and Guerrero, A. 1995. Right-sided endocarditis caused by *Staphylococcus aureus* in drug abusers. Antimicrob. Agents Chemother. 39: 525-528.

Frosco, M.B., Melton, J.L., Stewart, F.P., Kulwich B.A., Licata, L., and Barrett, J.F. 1996. *In vivo* efficacies of levofloxacin and ciprofloxacin in acute murine hematogenous pyelonephritis induced by methicillin-susceptible and resistant *Staphylococcus aureus* strains. Antimicrob. Agents. Chemother. 40: 2529-2534.

Gemell, C.G., and Ford, C.W. 1999. Expression of virulence factors by gram-positive cocci exposed to sub-MIC levels of linezolid 39th Interscience Conference on Antimicrobial Agents and Chemotherapy. Sep 26-29; San Francisco, Abstract B118.

Goetz, M.B. and Sayers, J. 1993. Nephrotoxicity of vancomycin and aminoglycoside therapy separately and in combination. J. Antimicrob. Chemother. 32: 325-334.

Gonzalez, C., Rubio, M., Romero-Vivas, J., Gonzalez, M, and Picazo, J. 1999. Bacteremic pneumonia due to *Staphylococcus aureus*: A comparison of disease caused by methicillin-resistant and methicillin-susceptible organisms. Clin. Infect. Dis. 29: 1171-1177.

Gopal, V., Bisno, A.L., and Silverblatt, F.J. 1976. Failure of vancomycin in treatment in *Staphylococcus aureus* endocarditis. *In vivo* and *in vitro* observations. JAMA 236: 1604-1606.

Gorgolas de M., Aviles, P., Verdejo, C., and Fernandez-Guerrero, M.L. 1995. Treatment of experimental endocarditis due to methicillin-susceptible or methicillin-resistant *Staphylococcus aureus* with trimethoprim-sulfamethoxazole and antibiotics that inhibit cell wall synthesis. Antimicrob. Agents Chemother. 39: 953-957.

Hanberger, H., Milsson, L.E., Maller, R. and Isaksson B. 1991. Pharmacodynamics of daptomycin and vancomycin on *Enterococcus faecalis* and *Staphylococcus aureus* demonstrated by studies of initial killing and postantibiotic effect and influence of Ca^{2+} and albumin on these drugs. Antimicrob. Agents Chemother. 35: 1710-1716.

Hartman, C.S., Leach, T.S., and Kaja, R.W. 2000. Linezolid in the treatment of vancomycin-resistant enterococcus: a dose comparative, multicenter phase III trial [abstract]. 40th Interscience Conference on Antimicrobial Agents and Chemotherapy. Toronto, 448.

Hershberger, E., Aeschlimann, J.R., Moldovan, T., and Rybak, M.J. 1999. Evaluation of bactericidal activities of LY333328, vancomycin, teicoplanin, ampicillin-sulbactam, trovafloxacin and RP59500 alone or in combination with rifampin or gentamicin against different strains of vancomycin-intermediate *Staphylococcus aureus* by time-kill curve methods. Antimicrob. Agents Chemother. 43: 717-721.

Hiramatsu, K., Hanaki, H., Ino, T., Yabuta, K., Oguri, T., and Tenover, F.C., 1997. Methicillin-resistant *Staphylococcus aureus* with reduced vancomycin susceptibility. J. Antimicrob. Chemother. 40: 135-136.

Hood, J., Edwards, G.F.S., Cosgrove, B., Curran, E., Morrison BD., and Gemmell, C.G. Vancomycin-intermediate *Staphylococcus aureus* at a Scottish hospital. J. Infect. 40: A11.

Hooper, D.C,2002. Glycylcyclines. Proc. of 41st Interscience Conference on Antimicrob. Ag. and Chemother. Chicago, Illinois, USA, p. 506.

Huang, H.W. 2000. Action of antimicrobial peptides: two-state model. Biochemistry. 39: 8347-8352.

Humble, M.W., Eykyn, S. and Phillips, I. 1980. Staphylococcal bacteremia, fusidic acid, and jaundice. Br. Med. J. 280: 1495-1498.

Inoue, M., Nonoyama, M., Okamoto, R., and Ida, T. 1994. Antimicrobial activity of arbekacin, a new aminoglycoside antibiotic, against methicillin-resistant *Staphylococcus aureus*. Drugs. Exp. Clin. Res. 20: 233-239.

Jamjian, C., Barrett, M.S., and Jones, R.N. 1997. Antimicrobial characteristics of quinupristin/dalfopristin (Syndercid at 30:70 ratio) compared to alternative ratios for *in vitro* testing. Diag. Microbiol. Infect. Dis. 27: 129-138.

Johnson, D.P., and Bannister, G.C. 1986. The outcome of infected arthroplasty of the knee. J. Bone. Joint. Surg. 68: 289-291.

Jones, R.N., Ballow, C.H., and Biedenbach, D.J. *et al.* 1998. Antimicrobial activity of quinupristin-dalfopristin (RP 59500, synercid) tested against over 28,000 recent clinical isolates from 200 medical centers in the United States and Canada. Diagn. Microbiol. Infect. Dis. 31: 437-451.

King, A., and Phillips, I. 2001. The *in vitro* activity of daptomycin against 514 gram-positive aerobic clinical isolates. J. Antimicrob. Chemother. 48: 219-223.

Korzeniowski, O., and Sande, M.A. 1982. Combination antimicrobial therapy for *Staphylococcus aureus* endocarditis in patients addicted to parenteral drugs and in nonaddicts: a prospective study. Ann. Intern. Med. 97: 496-503.

Kotra, L.P. 2000. Daptomycin. Current Opin. Anti-infect. Invest. Drugs. 2: 185-205.

Kureishi, A., Jewesson, P.J., Rubinger, M., Cole, C.D., Reece, D.E., Phillips, G.L., Smith, J.A., and Chow, A.W. 1991. Double-blind comparison of teicoplanin versus vancomycin in febrile neutropenic patients receiving concomitant tobramycin and piperacillin: effect on cyclosporin A-associated nephrotoxicity. Antimicrob. Agents Chemother. 35: 2246-2252.

Lamp, K.C., Ryback, M.J., Bailey, E.M., and Kaatz, G.W. 1992. *In vitro* pharmacodynamics effects of concentration, pH and growth phase on serum bactericidal activities of daptomycin and vancomycin. Antimicrob. Agents Chemother. 36: 2709-2714.

Lawrence, T., Totstein, C, Bean, R.,and Gorzynski, E.A. 1993. *In vitro* activities of ramoplanin, selected glycopeptides and methicillin-resistant *Staphylococcus aureus* strains in New York city hospitals J. Clin. Microbiol. 13: 754-759.

Lawrence, T., Totstein, C, Bean, R., Gorzynski, E.A., and Amsterdam, A. 1993. *In vitro* activities of ramoplanin, selected glycopeptides, fluoroquinolones and other antibiotics against clinical blood stream isolates of gram-positive cocci. Antimicrob. Agents Chemother. 37: 896-900.

Leach, T.S., Kaja, R.W., Eckert, S.M., Schaser, R.J., Todd, W.M., and Hafkin, B. 2000. Linezolid versus vancomycin for the treatment of MRSA infections: results of a randomized phase III trial. 9th International Congress on Infectious Diseases 2000; April 10-113: Buenos Aires, Abstract 95.079, p224.

Leclerq, R. Nantas, L., and Soussy, C.-J., and Duval, J. 1992. Activity of RP 59500, a new parenteral semisynthetic streptogramin, against staphylococci with various mechanisms of resistance to macrolide-lincosamide-streptogramin antibiotics. J. Antimicrob. Chemother. 30 (Suppl A): 95-99.

Li, Z., Wilke, R.J., Pinto. L.A., Rittenhouse, B.E., Rybak, M.J., Pleil, A.M., Crouch, C.W., Hafkin, B., and Glick, H.A. 2001, Comparison of length of hospital stay for patients with known or suspected methicillin-resistant *Staphylococcus aureus* infections treated with linezolid or vancomycin: a randomized, multicenter trial. Pharmacother. 21: 263-274.

Lina, G., Quaglia, A., and Reverdy, M.E., LeClerq, R., Vandenesch, F., and Etienne, J. 1999. Distribution of genes encoding resistance to macrolides, lincosamides, and streptogramins among staphylococci. Antimicrob. Agents Chemother. 43: 1062-1066.

Levine, D.P., Fromm, B.S and Reddy, B.R. 1991. Slow response to vancomycin or vancomycin plus rifampin in methicillin-resistant *Staphylococcus aureus* endocarditis. Ann. Intern. Med. 115: 674-680.

Lina, G., Quaglia, A., Reverdy, M.E., Leclercq, R., Vandenesch, F., and Etienne, J. 1999. Distribution of genes encoding resistance to macrolides, lincosamides and streptogramins among staphylococci. Antimicrob. Agents Chemother. 43: 1062-1066.

Louie, A., Baltch, A.L., Ritz, W.J., Smith, R.P., and Asperilla, M. 1993. Comparison of *in vitro* inhibitory and bactericidal activities of daptomycin (LY 146032) and four reference antibiotics, singly and in combination, against gentamicin-susceptible and high-level gentamicin-resistant enterococci. Chemother. 39: 3302-3309.

Markowitz, N., Quinn, E.L., and Saravolatz, L.D. 1992. Trimethoprim-sulfamethoxazole compared with vancomycin for the treatment of *Staphylococcus aureus* infection. Ann. Intern. Med. 117: 390-398.

Matsuzaki, K. 1998. Magainins as paradigm for the mode of action of pore forming polypeptides. Biochim. Biophys. Acta. 1376: 391-400.

Mellor, J.A., Kingdom, J., Cafferkey, M., and Keane, CT. 1985. Vancomycin toxicity: a prospective study. J. Antimicrob. Chemother. 15: 773-780.

McManus, J. 1999. Vancomycin resistant staphylococcus reported in Hong Kong. Brit. Med. J. 318: 626.

Mi-Na, K., Pai, C.H., Woo, J.H., Ryu, J.S., and Hiramatsu, K. 2000. Vancomycin intermediate *Staphylococcus aureus* in Korea. J. Clin. Microbiol. 38: 3879-3881.

Nadler, H., Dowizicky, M.J., and Feger, C. *et al.* 1999. Quinupristin/dalfopristin: a novel selective-spectrum antibiotic for the treatment of multi-resistant and other gram-positive pathogens. Clin. Microbiol. Newsl. 21: 103-112.

National Committee for Clinical Laboratory Standards. 2000. Methods for Dilution Antimicrobial Susceptibility tests for bacteria that grow aerobically. 5th ed. Approved standard M7-A5. Villanova, Pa. NCCLS.

Neu, H.C., and Labthavikul, P. 1983. *In vitro* activity of teichomycin compared with those of other antibiotics. Antimicrob. Agents Chemother. 24: 425-428.

Neu, H.C., Chin, N-X., and Gu, J-W. 1992. The *in vitro* activity of new streptogramins, RP 59500, RP 57669 and RP 54476, alone and in combination. J. Antimicrob. Chemother. 30 (Suppl A): 83-94.

Nichols, R.L., Graham, D.R., Barriere, S.L., Rodgers, A., Wilson, E.E., Zervos, M., Dunn, D.L., and Kreter, B. 1999. Treatment of hospitalized patients with complicated gram-positive skin and skin structure Infections: two randomized, multicentre studies of quinupristin/dalfopristin versus cefazolin, oxacillin or vancomycin. J. Antimicrob. Chemother. 44: 263-273.

Noviello, S., Ianniello, F., and Esposito S. 2001. *In vitro* activity of LY333328 (oritavancin) againt gram-positive aerobic cocci and synergy with ciprofloxacin against enterococci. J. Antimicrob. Chemother. 48: 283-286.

Panlilio, A.L., Culver, D.H., Gaynes, R.P., Banerjee, S., Henderson, T.S., Tolson, J.S., and Martone, W.J. 1992. Methicillin-resistant *Staphylococcus aureus* in US hospitals, 1975-1991. Infect. Control. Hosp. Epidemiol. 13: 582-586.

Patel, S.S., Balfour, J.A., and Bryson, H.M., 1997. Fosfomycin tromethamine. Drugs 53: 637-656.

Ploy, M.C., Grelaud, C., Martin, C. de Lumely, L., and Denis, F. 1998. First clinical isolate of vancomycin-intermediate *Staphylococcus aureus* in a French hospital. Lancet. 351: 1212.

Portier, H. 1990. A multicenter, open, clinical trial of a new intravenous formulation of fusidic acid in severe staphylococcal infection. J. Antimicrob. Chemother. 25 (Suppl. B): 39S-44S.

Raad, I., Bompart, F., and Hachem, R. 1999. Prospective, randomized dose-ranging open phase II pilot study of quinupristin/dalfopristin versus vancomycin in the treatment of catheter related staphylococcal bacteremia. Eur. J. Clin. Infect. Dis. 18: 199-202.

Rello, J., Paiva, J.A., Baraibar, J., Barcenilla, F., Bodi, M., Castander, D., Correa, H, Diaz, E., Garnacho, J., Llorio, M., Rios, M., Rodriguez, A., and Sole-Violan, J. 2001. International conference for the development of consensus on the diagnosis and treatment of ventilator-associated pneumonia. Chest. 120: 955-970.

Rello, J., Quintana, E., Ausina V., Castella, J., Luquin, M., Net, and A., Prats, G. 1991. Incidence, etiology and outcome of nosocomial pneumonia in mechanically ventilated patients. Chest. 100: 439-444.

Rello, J., Torres, A., Ricart, M., Valles, J., Gonzales, J., Artigus, A., and Rodriguez-Roisin R. 1994. Ventilator-associated pneumonia by *Staphylococcus aureus*. Comparison of methicillin-resistant and methicillin-sensitive episode. Am. J. Respir. Crit. Care. Med. 150: 1545-1549.

Richards, M.J., Edwards, J.R., Culver, D.H., and Gaynes, R.P. 1999. Nosocomial infections in medical intensive care units in the United States. National Nosocomial Infections Surveillance System. Crit. Care Med. 27: 887-892.

Rubinstein, E., Cammarata, S., Oliphant, T., and Wunderink, R. 2001. Linezolid (PNU-100766) versus vancomycin in the treatment of hospitalized patients with nosocomial pneumonia: a randomized, double-blind, multicenter study. Clin. Infect. Dis. 32: 402-412.

Rubinstein, E., Prokocimer, P., and Talbot, G.H. 1999. Safety and tolerability of quinupristin/dalfopristin: Administration guidelines. J. Antimicrob. Chemother. 44: 37-46.

Ryback, M.J., Albrecht, L.M., Boike, S.C. Chandrasekar, PH. 1990. Nephrotoxicity of vancomycin, alone and with an aminoglycoside. J. Antimicrob Chemother. 25: 679-686.

Sambatakou, H., Giamarellou-Bourboulis, E.J., Grecka, P., Chryssouli, Z., and Giamarellou, H. 1998. *In vitro* activity and killing effect of quinupristin/dalfopristin (RP59500) on nosocomial *Staphylococcus aureus* and interactions with rifampicin and ciprofloxacin against methicillin-resistant isolates. J. Antimicrob. Chemother. 41: 349-355.

Schaberg, D.R., Culver, D.H., Gaynes, R.P. 1991. Major trends in microbial etiology of nosocomial infection. Am. J. Med. 91 (Suppl 3B): S72-S75.

Schmitz, F.J., Fluit, A.C., Gondof, M., Beyrau, R., Lindenlauf, E., Verhoef, J., Heinz, H.P., and Jones, M.E. 1999. The prevalence of aminoglycosides resistance and corresponding resistance genes in clinical isolates of staphylococci from 19 European hospitals. J. Antimicrob. Chemother. 43: 253-259.

Schoifet, S.D., and Morrey, B.F. 1990. Treatment of infection after total knee arthroplasty by debridement with retention of the components. J. Joint. Bone. Surg. Am. 72: 1383-1390.

Shresthha, N.K., and Tomford J.W. 2001. Fosfomycin: a review. Infect. Dis. Clin. Pract. 10: 255-260.

Sieradzki, K., and Romasz., A. 1997. Suppression of beta-lactam antibiotic resistance in a methicillin-resistant *Staphylococcus aureus* through synergic action of early cell wall inhibitors and some other antibiotics. J. Antimicrob. Chemother. 39 (Suppl A): S47-S51.

Simon, G.L, Smith, R.H, and Sande, M.A. 1983. Emergence of rifampin-resistant strains of *Staphylococcus aureus* during combination treatment with vancomycin and rifampin: A report of two cases. Rev. Infect. Dis. 5 (Suppl): S507-S508.

Simon, V.C., and Simon, M. 1990. Antibacterial activity of teicoplanin and vancomycin in combination with rifampicin, fusidic acid, or fosfomycin against staphylococci on vein catheters. Scand. J. Infect. Dis. Suppl 72: 14-19.

Small, P.M., and Chambers, H.F. 1990. Vancomycin for *Staphylococcus aureus* endocarditis in intravenous drug users. Antimicrob. Agents. Chemother. 34: 1227-1231.

Smith, S. R., Cheesbrough, J, Spearing, R., and Davies, J.M. 1989. Randomized prospective study comparing vancomycin with teicoplanin in the treatment of infections associated with Hickman catheters. Antimicrob. Agents Chemother. 33: 1193-1197.

Smith, S.M., Eng, R.H.K., Bais, P., Fan-Havard P., and Tecson-Tumang, F. 1990. Epidemiology of ciprofloxacin-resistance among patients with methicillin-resistant *Staphylococcus aureus*. J. Antimicrob. Chemother. 26: 567-572.

Smith, T.L., Pearson, M.L., Wilcox, K.R., Cruz, C., Lancaster, M.V., Robinson-Dunn, B., Tenover, F.C., Zervos, M.J., Band, J.D., White, E., and Jarvis, W.R. 1999. Emergence of vancomycin resistance in *Staphylococcus aureus*. N. Engl. J. Med. 340: 517-523.

Sorrell, T.C., Ackham, D.R., Shanker, S., Foldes, M., and Munro, R. 1982. Vancomycin therapy for methicillin-resistant *Staphylococcus aureus*. Ann. Int. Med. 97: 344-350.

Sorrell, T.C., and Collignon, P.J. 1985. A prospective study of adverse reactions associated with vancomycin therapy. J. Antimicrob. Chemother. 16: 235-241.

Stalker, D.J., Wajszczuk, C.P., and Batts, D.H. 1997. Linezolid safety, tolerance and pharmacokinetics following oral dosing twice daily for 14.5 days. 37th Interscience Conference on Antimicrobial Agents and Chemotherapy Sep 28:Toronto, abstract 23.

Stein, A., Bataille, J.F., Drancourt, M., Curvale, G., Argenson, J.N., Groulier, P., and Raoult, D. 1998. Ambulatory treatment of multi drug-resistant *Staphylococcus*-infected orthopedic implants with high-dose-oral co-trimoxazole (trimethoprim-sulfamethoxazole). Antimicrob. Agents. Chemother. 42: 3086-3091.

Swnaberg, L., and Tuazon, C.U. 1984. Rifampin in the treatment of serious staphylococcal infections. Am. J. Med. Sci. 287: 49-54.

Swaney, S.M., Aoki, H., and Ganoza, M.C., and Shinabarger, D.L. 1998. The oxazolidinone linezolid inhibits initiation of protein synthesis in bacteria. Antimicrob. Agents. Chemother. 42: 3251-3255.

Swaney, S.M., Shinabarger, D.L., Schaadt, R.D., Bock, J.H., Slightom, J.R., and Zurenko, G.E. 1998. Oxazolidione resistance is associated with a mutation in the peptidyl transferase region of 23SrRNA. Proc. of 38th Interscience Conference on Antimicrob. Ag. and Chemother. San Diego, CA, USA (Abstr C-104).

Stevens, D.L., Smith, L.G., and Bruss, J.B. 2000. Randomized comparison of linezolid (PNU-100766) versus oxacillin-dicloxacillin for treatment of complicated skin and soft tissue infections. Antimicrob. Agents. Chemother 44: 3408-3413.

Taburet, A.M., Guibert, J., Kitzids, M.D. Sorenson, H., Acar, J.F., and Singlas, E. 1990. Pharmacokinetics of sodium fusidate after single and repeated infusions and oral administration of new formulation. J. Antimicrob. Chemother. 25 (Suppl B): 23-31.

Tally, F.P., Zeckel, M., Wasilewski, M.M., Carini, C., Berman, C.L., and Drusano, G.L. and Oleson, F.B. 1999. Daptomycin; a novel agent for gram-positive infections. Exp. Opin. Invest. Drugs. 8: 1223-1238.

Tsiodras, S., Gold, H.S., Sakoulas, G., Eliopouos, G.M., Wennersten, C., Venkatatraman, and L. Moellering, R.C. 2001. Linezolid resistance in a clinical isolates of *Staphylococcus aureus*. Lancet. 358: 207-208.

Tsukayama, D.T., Wicklund,B., and Gustilo, R.B. 1991.Suppressive antibiotic therapy in chronic prosthetic joint infections. Orthopedics. 14: 841-844.

Watanakunakorn C, and Guerriero, J.C. 1981. Interaction between vancomycin and rifampin against *Staphylococcus aureus*. Antimicrob. Agents Chemother. 19: 1089-1091.

Watanakunakorn, C. and Tisone, J.C. 1982. Synergism between vancomycin and gentamicin or tobramycin for methicillin-susceptible and methicillin-resistant *Staphylococcus aureus* strains. Antimicrob. Agents Chemother. 22: 903-905.

Welshman, I.R., Stalker, D J., and Wajszczuk, C.P. 1998. Assessment of absolute bioavailability and evaluation of the effect of food on oral bioavailability of linezolid. Antiinfect. Drugs Chemother. 16 (Suppl. 1): 54.

Widmer, A.F. 2001.New developments in diagnosis and treatment of infection in orthopedic implants. Clin. Infect. Dis. 33 (Suppl.): S94-S106.

Widmer, A.F., Gaechter, A., Ochsner, P.E., and Zimmerli, W. 1992. Antimicrobial treament of orthopedic implant-related infections with rifampin combinations. Clin. Infect. Dis. 14: 1251-1253.

Wilcox, M. Tang, T., and Hafkin, B. 2001. Linezolid versus teicoplanin for the treatment of hospitalized patients with gram-positive infections. Proc. of 41st Interscience Conference on Antimicrobial Agents and Chemotherapy Chicago, Illinois, USA, December 16 - 19. Abstract L-1481.

Wilson, A.P., Gruneberg, R.N., and Neu, H.C. 1994. A critical review of the dosage of teicoplanin in Europe and the USA. Int. J. Antimicrob. Agents. 4: S1-S30.

Wilson, M.G., Kelly, K., and Thornhill, T.S. 1990. Infection as a complication of total-knee replacement arthroplasty: risk factors and treatment in sixty even cases. J. Bone Joint Surg Am. 72: 878-883.

Van der Auwera, P., Aoun, M., and Meunier, F. 1991. Randomised study of vancomycin versus teicoplanin for the treatment of gram-positive bacterial infections in immunocompromised hosts. Antimicrob. Agents. Chemother. 35: 451-457.

Voss, A., Milatovic, D., Wallrauch-Schwarz, C., Rosdahl, V.T., and Braveny, I. 1994. Methicillin-resistant *Staphylococcus aureus* in Europe. Eur. J. Clin. Micobiol. Infect. Dis. 13: 50-55.

Yelandi, V., Strodtman, R., and Lentinio, JR. 1988. *In vitro* and *in vivo* studies of trimethoprim-sulfamethoxazole against multiple resistant *Staphylococcus aureus*. J. Antimicrob. Chemother. 22: 873-880.

Zimmerli, W., Widmer, A.F., Blatter, M., Frein R., and Ochsner P.E. for the Foreign-Body Infection (FBI) Group. 1998. Role of rifampin for treatment of orthopedic implant-related staphylococcal infections. A randomized controlled trial. JAMA. 279: 1537-1541.

Zurenko, G.E., Yagi, B. JH., and Schaadt, R.D., Allison, J.W., Kilburn, J.O., Glickman, S.E., Hutchinson, D.K., Barbachyn, M.R., Brickner, S.J. 1996. *In vitro* activities of U-100592 and U-100766 novel oxazolidinone antibacterial agents. Antimicrob. Agents Chemother. 40: 839-845.

From: *MRSA: Current Perspectives*
Edited by: A.C. Fluit and F.-J. Schmitz

Chapter 12

Prevention and Control of Methicillin-Resistant *Staphylococcus aureus* (MRSA)

Uwe Frank

Abstract

This chapter is intended to clarify controversial infection control issues in the management of methicillin-resistant *Staphylococcus aureus* (MRSA). For more than three decades, MRSA strains have been identified as a major source of nosocomial infections and outbreaks in the health-care setting. MRSA presents a challenge to infection control departments in hospitals of varying size attempting to control and eradicate this micro-organism. Controlling the transmission of MRSA is primarily the responsibility of healthcare workers providing direct bedside care to in-patients. Hand-washing and appropriate application of barrier precautions are the two primary transmission prevention measures. The use of single rooms is strongly recommended. If resources are limited, cohorting MRSA colonised/infected patients in one room, ward or area may be a consideration. Unfortunately, all health care facilities, regardless of location or the complexity of care rendered will experience sporadic or endemic MRSA over the next few years.

Introduction

There are many different strains of Methicillin-(oxacillin-) Resistant *Staphylococcus aureus* (MRSA), some of which may be epidemic in character causing serious outbreaks (Emmerson, 1994; Farrell *et al.*, 1998; Goetz *et al.*, 1992; Guiguet *et al.*, 1990; Harbarth *et al.*, 2000; Hitomi *et al.*, 2000; Lingnau *et al.*, 1994; Meier *et al.*, 1996; Moore *et al.*, 1991; Nicolle *et al.*, 1990; Reboli *et al.*, 1990; Romance *et al.*, 991; Schumacher-Perdreau *et al.*, 1994; Tambic *et al.*, 1997; Witte *et al.*, 1994; Zafar *et al.*, 1995). New epidemic strains of MRSA continue to emerge and decline for unknown reasons (Ayliffe, 1997). The main mode of spread is person-to person within a unit or hospital and subsequently to other hospitals or chronic care facilities (Bradley, 1997; Bradley, 1999; Harbarth *et al.*, 1997; Witte *et al.*, 1994). Risk factors for acquiring MRSA infection are length of prior hospitalization, invasive procedures and the number of antibiotic treatments (Guiguet *et al.*, 1990). It has been suggested that early detection, effective infection control measures, and rational use of antibiotics may limit the transmission of these organisms (Ayliffe, 1997; Matsumura *et al.*, 1996; Solberg, 2000).

Generally the following applies:

(1) MRSA is no more virulent than sensitive *Staphylococcus aureus* strains (Humphreys *et al.*, 1997).
(2) The main reservoir is the nasopharynx (Boyce, 2001; Solberg, 2000).
(3) MRSA is primarily transmitted via the hands (Boyce, 1991; Farrell *et al.*, 1998; Solberg, 2000).
(4) Hand hygiene protects from transmission and nasal colonization (Pittet *et al.*, 2000).

Measures Used to Control MRSA

Guidelines for Control of MRSA

Guidelines dealing with the control of MRSA in hospitals (Table 1) have been published in different countries, such as Britain (Combined Working Party, 1998), Germany (Robert-Koch-Institut, 1999), the Netherlands (Working Group on Infection Prevention, 1995), and the United States (Mulligan *et al.*, 1993; Boyce *et al.*, 1994). One major guideline represents a consensus statement developed by the American Hospital Association (AHA) (Boyce *et al.*, 1994).

Table 1. Measures used to control MRSA*

General Control Measures

- Guidelines for control of MRSA
- Infection control programs
- Early detection of MRSA
- Hand hygiene
- Inpatient management: Education, communication
- Standard precautions: Hand washing, gloving, gowning, masks, device handling, appropriate handling of laundry
- Microbiological surveillance and decolonization therapy of MRSA carriers (patients, personnel)

Specific Control Measures

- Contact precautions: Single room/ cohort isolation
- Barrier precautions: Gloves, gowns, masks
- Housekeeping/domestic cleaning/ terminal cleaning

---

*   Recommendations are based on the latest epidemiologic information on the transmission of MRSA in the health care setting and described in the text.

Furthermore, the Hospital Infection Control Practices Advisory Committee (HICPAC) of the Centers for Disease Control and Prevention (CDC) has published general guidelines for isolation precautions in hospitals (Garner *et al.*, 1996). The latter guidelines, however, not only deal with MRSA, but with many different nosocomial pathogens. They are categorized as follows: *Category IA*: Strongly recommended for all hospitals and strongly supported by well-designed experimental or epidemiologic studies. *Category IB:* Strongly recommended for all hospitals and reviewed as effective by experts in the field and a consensus of HICPAC based on strong rationale and suggestive evidence, even though definitive scientific studies have not been done. *Category II:* Suggested for implementation in many hospitals. Recommendations may be supported by suggestive clinical or epidemiologic studies, a strong theoretical rationale, or definitive studies applicable to some, but not all, hospitals. *No recommendation:* Unresolved issue. Practices for which insufficient evidence or consensus regarding efficacy exists. Wherever necessary we will refer to these categories in the following text.

Infection Control Programs

MRSA has become endemic in many institutions. Since there is no standard approach to controlling MRSA, health care professionals must develop specific programs to prevent transmission within their institutions. Control programs should be employed when: (1) MRSA has been newly introduced into a hospital, particularly into the intensive care unit, (2) MRSA causes a high incidence of serious nosocomial infections and, or (3) MRSA accounts for more than 10% of nosocomial staphylococcal isolates (Boyce, 1991). When designing a control program, staff should consider the prevalence of MRSA in their hospital, the prevalence of MRSA in referring facilities, the risk factors in the patient population, the frequency of nosocomial transmission, the current reservoirs and modes of transmission, and the resources available (Herwaldt, 1999). Efforts to control spread of MRSA should be targeted to key hospital areas, such as intensive care units, where the impact of infection is likely to be greatest; general infection control measures should be taken throughout the hospital (Humphreys *et al.*, 1997).

Three general strategies for the control of MRSA have been proposed:

(1) the *Scutari Strategy* (named after the hospital in which Florence Nightingale worked in the Crimean war from 1854-56) based on simple hygienic measures and barrier precautions which may apply to nursing homes, convalescent hospitals and small, non-surgical hospitals in which MRSA patients rarely colonize or infect other patients if quite simple infection control measures such as hand washing are used without isolation; (2) the *Search and Destroy Strategy* with strict isolation of all colonized and infected patients and attempts to eradicate MRSA from the environment, which may apply to hospitals previously known to be free of MRSA but facing recent epidemic situations; and (3) the *SALT Strategy* (Staphylococcus aureus Limitation Technique) with isolation precautions only for non-containable infections and "infectious precautions" for colonized patients, which may be appropriate for epidemic situations, low incidence of infections and limited resources (Spicer, 1984; Vandenbroucke-Grauls *et al.*, 1991).

Early Detection of MRSA

Clinical microbiology laboratories must be proficient in the detection of MRSA. Susceptibility testing of *Staphylococcus aureus* should be performed using oxacillin-salt agar screening plates, broth microdilution methods or agar dilution methods with 2% NaCl, according to the guidelines of the National Committee for Clinical Laboratory Standards (NCCLS, 1993). Although automated susceptibility testing systems detect most strains of MRSA, questionable results should always be tested on oxacillin-salt plates.

Frequent review of microbiology records to identify patients with MRSA represents a minimal surveillance level and is practical in hospitals with on-site laboratories (Boyce, 1991). Clinical laboratory-based surveillance may be efficient in monitoring the prevalence of MRSA in the hospital: Surveillance records of the base rate of colonization or infection in each hospital ward may allow for early recognition of any MRSA problem that requires intervention by the Infection Control Team (Adeyemi-Doro *et al.*, 1997). Modern computer alert systems may assist in performing this kind of surveillance (Pittet *et al.*, 1996).

Prevalence surveys involve periodic culturing of high risk patients at several body sites such as anterior nares, perineum, surgical wounds, and skin lesions, in order to determine if they are colonized with MRSA (Solberg, 2000). Prevalence surveys are used by hospitals in which MRSA is endemic in order to identify colonized patients who have not been detected by clinical cultures. In settings with numerous high risk patients, this approach can be very labor-intensive, and the circumstances under which it may be cost-effective, or exactly which patients should be cultured, has as yet not been determined in studies.

Label charts of patients with MRSA may facilitate prompt isolation of patients colonized with MRSA at the time of re-admission (section Communication). Such patients may account for up to one-third of all MRSA cases admitted to the hospital (Boyce, 1991). Surveillance cultures of high-risk patients, e.g. residents of long-term care facilities and transfer patients, on admission to the hospital have contributed to control several hospital outbreaks. Screening high-risk patients appears to be a cost-effective method of controlling nosocomial MRSA transmission if the organism is prevalent in referring health care facilities (Papia *et al.*, 1999).

One of the most important steps in controlling MRSA in the hospital is typing of strains which allows the confirmation or exclusion of genetic relatedness among isolates associated with a given epidemiological situation. Typing data are needed to discriminate between epidemic, endemic or sporadic strains (Hryniewicz, 1999). Various typing methods have been used for this purpose: (1) Phenotypic methods such as biotyping, phage typing, serotyping, and antibiograms, and (2) molecular techniques such as genomic DNA restriction fragment length polymorphism (RFLP) involving the pulsed-field gel-electrophresis (PFGE) and polymerase chain reaction (PCR) amplification-based methods (Hryniewicz, 1999; Mueller-Premru, 1998). While PCR-amplification based techniques have the advantage of easy handling and rapid performance, PFGE is superior in terms of discriminatory power between strains (Schmitz *et al.*, 1998). PCR-based methods, e.g. random amplification of polymorphic DNA (RAPD), may be useful for preliminary epidemiological investigation as results are available within 8 hours; where a cluster of clonally related isolates is suspected, further investigation by PFGE is recommended (Tambic *et al.*, 1997).

Hand Hygiene

The usual rules of hand hygiene, i.e. hand washing and disinfection procedures, must be observed for activities where there is a risk of contamination, i.e. before and after contact with the patient or his/her immediate environment (Combined Working Party, 1998; Robert Koch-Institut, 1999; Working Group on Infection Prevention, 1995; Mulligan *et al.*, 1993; Boyce *et al.*, 1994). Hospital campaigns promoting hand hygiene procedures may produce sustained improvement in compliance and reduce nosocomial transmission and infection by MRSA (Pittet *et al.*, 2000). Hands should be washed thoroughly using an antiseptic detergent or, alternatively, if physically clean, be treated with an alcoholic hand rub (Garner, 1996).

Various antimicrobial hand washing products, including waterless alcohol-based antiseptic agents, have been recommended by experts as a way to help prevent cross-contamination and spread of MRSA in hospitals. Hand hygiene procedures must be performed, even if disposable gloves have been worn. Hand hygiene procedures must always be performed before leaving the room (even if there has been no direct contact with the patient, e.g. after ward rounds, serving food or airing the room, etc.).

Special care must be taken to ensure that transmission of the pathogens from the colonized or infected part of the body to areas at high risk of infection (e.g. from infected wounds into tracheal secretion) is avoided. Before further handling the patient, the hands must be disinfected carefully each time after manipulation of the colonized region or infected areas of the body.

Inpatient Management

Education of Personnel, Patients, and Visitors

If MRSA is isolated in a patient, the following precautions must be implemented:

A member of the Infection Control Team should be notified. Education of hospital staff is an essential component of any MRSA control program (Cosseron-Zerbib *et al.*, 1998). For success of the control measures, information and motivation of staff are essential (Heuck *et al.*, 1995). Continuing education programs for hospital staff including those who are responsible for decision making regarding patient care should include a thorough review of current infection control practices and recent guidelines.

Patient education is essential to control the transmission of MRSA in the facility. MRSA patients should wash their hands after contact with secretions and before touching other objects (e.g. immediately after coughing). Patients

ATHENS.......

How do I register for an Athens username and password?

- Register online at: https://register.athensams.net/nhs/

Why should I register?

An Athens password gives you access to:

- 7 major healthcare databases, including Medline, Embase, Cinahl and PsychINFO
- 1,200 full text electronic journals
- 50,000 medical images and an interactive 3D anatomy site

Where can I use Athens?

Use Athens resources wherever you have Internet Access at:

- www.athensams.net/myathens

Where can I get training on using Athens resources?

Contact the Libraries on gm.e.mtw-tr.ilibrary@nhs.net, or phone your nearest service:

- Maidstone: 01622 224647   OR:
- Kent & Sussex: 01892 632384   Outreach Training (Community): 01892 633019
- Pembury: 01892 633119

on isolation and their families need to be informed on the reason for isolation, control measures, and expected duration of isolation.

Visitors should be instructed that items are not to be shared between patients unless they can be appropriately cleaned. When visiting patients on "contact precautions" (section Contact precautions: single room/cohort isolation), visitors should be instructed regarding control measures, with special emphasis on hand washing.

Communication

When transferring a patient with MRSA colonization/infection, it is the responsibility of the transferring facility to inform the admitting hospital of the patient's MRSA colonization/infection status and history. On the other hand, when a transferred patient is found to be colonized/infected with MRSA within 48 hours of admission, it is the responsibility of the admitting hospital to inform the transferring facility as soon as possible. Early identification of patients with MRSA at the time of re-admission to the hospital can assist the medical staff to implement infection control measures promptly. This measure, however, requires some indication in the patient's medical record and/or computer file, which is accessed at the time of admission. The system used should maintain patient confidentiality. Personnel who will have direct contact with MRSA patients should be made aware of appropriate infection control measures prior to room entry. Traditionally, this has been accomplished by placing instructional cards on the patient's door and a label on the patient's chart.

Patient Transport Within and Outside of the Hospital

Transport of patients with MRSA to other departments should be kept to a minimum (only, if there is a clear medical indication). Prior arrangements must be made with the department to which the patient is to be transported. The Infection Control Team has to advise on the necessary infection control precautions. Patients visiting specialist departments should be seen or treated at the end of the working session and spend a minimum time in the department. Devices used on the patient must subsequently be wiped with a disinfectant. The hands of the personnel must always be disinfected. Patients visiting specialist departments should be seen immediately and not left in a waiting room with other patients. Medical devices should be sent along with the patient in a tightly shut plastic bag. Before transporting a patient with an infected wound, the dressing should be changed if necessary (the dressing must always remain dry). Before leaving the room, the patient's hands should be disinfected.

In case of nasal colonization the patient must wear a surgical mask during transportation. Bed-ridden patients should be moved to a trolley. If this is not possible, the patient should be transported in his bed. However, the bed linen must be changed first and the bed frame must be disinfected. The staff transporting the patient must be informed, and they must disinfect their hands thoroughly before leaving the room. Gowns or gloves do not need to be worn during transportation. Transporting staff should wear protective gowns to handle the patient (these should subsequently be left in the room). Hands must be disinfected before leaving the ward and after moving the patient at the point of destination. After transporting the patient, the trolley or wheelchair must be wiped with disinfectant.

If the patient is transferred to another hospital, identification of infected or colonized patients is the responsibility of the transferring hospital. Transfer of patients to other wards or departments within the hospital must be kept to a minimum and preferably only after consultation with the Infection Control Team. Inter-hospital transfers should be restricted wherever possible. Medical charts and records must be marked and other hospitals or nursing homes informed. A letter should be mailed giving the relevant clinical details as to whether the patient is infected or colonized.

MRSA patients destined for residential or nursing home care should complete any treatment prescribed for MRSA eradication either before leaving the hospital or under supervision in the home (Cox *et al.*, 1995). If the patient is discharged, the general practitioner involved in the outpatient treatment should be informed of the patient's status and a copy of the decontamination protocol should be given, so that the course of treatment may be completed. If the patient is being discharged to a nursing or convalescent home, the receiving medical staff must be informed in advance. Patients should be informed that there is no risk to healthy family members. If the patient dies, precautions taken by the health care worker performing last offices should be similar to those observed during life. All skin lesions should be covered with impermeable dressings.

All patients known to be infected or colonized must be admitted directly to a single room with appropriate isolation precautions (section Contact precautions: single room/cohort isolation). In situations where there is more than one patient colonized or infected with MRSA, it may not be feasible to place each patient in a private room. In these cases, two or more patients with MRSA may be placed in the same room, area or unit (cohort isolation) (Barakate *et al.*, 1999). Cohorting, however, may be done only on the advice of the Infection Control Team. A patient with MRSA should be discharged from hospital as soon as his clinical condition allows. This measure is highly desirable and simple, unless a colonized patient is awaiting transfer to a facility that does not accept patients with MRSA. If early discharge is impossible, he/she should be isolated with implementation of appropriate infection control measures. If the patient requires treatment in any other hospital or health care facility, the clinician and the Infection Control Team at the receiving hospital must be informed.

Standard Precautions

Standard precautions should be practised for contact with every patient. The term "standard precautions" is defined in the HICPAC-Guideline (Garner, 1996). Standard precautions include recommendations for hand washing, wearing gloves, masks, gowns, handling of patient-care equipment, environmental control, and laundry.

Treatment of MRSA Carriers

To determine the extent of MRSA colonization, swabs must be taken of the nose, throat (carriers of dentures!), the perineum (alternatively the groin), and of all wounds and skin lesions. Other sampling sites to detect MRSA colonization or infection include insertion sites of IV lines, catheter urine samples, and sputum if expectorating. Treatment should be prescribed on an individual basis in consultation with the Medical Microbiologist and/or Infection Control Team. For intranasal carriers, decolonization with topical administration of mupirocin ointment applied to the anterior nares three times a day for five days should be tried (Frank *et al.*, 1989). This agent has played an important role in terminating hospital outbreaks of MRSA (Barrett, 1990; Boyce, 2001). Intranasal mupirocin application, however, has limited effectiveness in eradicating MRSA colonization in patients who carry the organism at multiple body sites. In this instance, skin and wound disinfection appears as an essential complementary measure. Careful attention should be given if topical mupirocin is used to eradicate staphylococci from lesions such as eczema and small sores where resistance to the topical agent can emerge (Brun-Buisson *et al.*, 1994). Therefore, long-term application of mupirocin to open wounds, and prolonged and widespread use of mupirocin in endemic situations should be avoided (Boyce 2001). If mupirocin resistance becomes widespread, this control measure will no longer be useful (Hill *et al.*, 1990; Miller *et al.*, 1996).

As an adjuvant measure - provided the skin is in good condition - once daily whole body wash down and twice weekly hair wash should be carried out with an antiseptic detergent, e.g. with Betaisadona ®-antiseptic lotion (not to be applied to wounds) or Octenisept® solution 50% (diluted 1:1 with Aqua dest.; leave to soak in for 2 min). If Octenisept® is used, the patient's skin should subsequently be washed as usual; liquid soap or other care products may be added. Subsequently the hair should be washed with normal shampoo. Cave: Antiseptic detergents should be used with care on patients with dermatitis and open skin lesions, and must be discontinued if skin irritation occurs. It should be noted that topical antimicrobials are not a substitute for hand hygiene and isolation precautions; they should be used only as a part of a global strategy in which isolation precautions remain of primary importance (Brun-Buisson *et al.*, 1994).

Microbiological Surveillance of and Decolonization Therapy

Swabs of the nasal cavity and the perineum/groin should be taken once a week; sputum and urine samples where applicable. If the swabs are negative, controls should be made at 24 h intervals, until three successive swabs are negative, in which case isolation may be discontinued. Follow-up screening swabs should be taken once a week until the patient is discharged. Relapses may occur and prolonged treatment may be required. Relapses are particularly likely in patients receiving antibiotics. If the patient is readmitted to hospital, swabs must be taken of the nasopharynx, the perineum, of all wounds and any skin lesions. In addition, if an indwelling catheter is inserted, a urine sample must be taken. At the same time, the Infection Control Team must be informed.

A swab of the nose and perineum should be taken from fellow patients sharing the same room; swabs should also be taken of all wounds and skin lesions. If dentures are worn, a throat swab should be taken. If in doubt, the infection control team should be consulted. Mupirocin should be use to treat any fellow patients with nose and/or throat colonization (see above) and general precautions should be implemented.

Selective screening and isolation of MRSA carriers on admission to intensive care units are beneficial compared with no isolation (Chaix *et al.*, 1999). For the following reasons, routine screening of staff and other patients on the ward is not recommended: (1) The time, staff and expense involved is disproportionate to the benefit derived (if at all, only in cases of short-term MRSA colonization); (2) if screening identifies a carrier of MRSA among the staff, this still does not mean, that he/she is responsible for nosocomial transmission. In fact molecular biological typing should be done to ensure that the same strain is involved. Even if found to be identical, it is still unclear whether the member of staff was responsible for colonization, or whether he/she was colonized by the patient. Screening the staff and all other patients on the ward is an option in outbreak situations, but should only be undertaken if all the other measures described above do not suffice to prevent MRSA transmission to fellow patients. When epidemiologic evidence indicates that healthcare workers are the most likely source for the outbreak, infection control should attempt to identify and eliminate this hospital reservoir (e.g. by application of intranasal mupirocin), since these individuals can disseminate MRSA both within the hospital and into the community (Meier *et al.*, 1996; Reboli *et al.*, 1990).

Rational Use of Antibiotics

Excessive use of antibiotics has played a major role in the increase of antimicrobial resistance; it is difficult, however, to establish the role of any particular antibiotic in the increased frequency of MRSA (Ayliffe, 1997; Crowcroft *et al.*, 1999). Third-generation cephalosporins and fluoroquinolones may exert a selective pressure. The control of antimicrobial use, in addition to measures for prevention of nosocomial infections, may be the best approach to control the spread of MRSA.

Systemic antibiotics should not be used to treat colonization with MRSA. Decolonization therapy with systemic antibiotics such as rifampin, cotrimoxazole, or clindamycin used alone or in various combinations has been reported as ineffective and potentially hazardous, because colonization persisted or recurred in more than half of the patients and substantial antimicrobial resistance was noted in MRSA strains after therapy (Strausbaugh *et al.*, 1992).

Decontamination should be exclusively tried by use of intranasal mupirocin ointment (three-times daily for five days) and, if necessary, by concomitant use of antiseptic soap/lotions. Microbiological controls for documentation of MRSA eradication can be started on the 3rd day after end of mupirocin treatment. If mupirocin treatment fails or if there is resistance to mupirocin, treatment with another antiseptic, e.g. Octenisept nose salve or PVP iodine nose salve, is recommended (contact Infection Control Team).

Clearance of MRSA carriage should be attempted before surgery whenever possible. Colonized or infected patients should be operated upon at the end of an operating list. During the operation, skin lesions must be covered with an impermeable dressing and adjacent areas treated with an antiseptic. Prophylactic vancomycin or teicoplanin should be considered for colonized or infected patients undergoing surgery or any invasive procedure. Treatment with systemic antibiotics should only be administered for MRSA infection and not for colonization. Drugs of choice are glycopeptide antibiotics, e.g. vancomycin or teicoplanin. Although susceptibility testing of MRSA against a variety of other antibiotics (e.g. cephalosporins, fluoroquinolones, cotrimoxazole, clindamycin) may give a positive result of sensitivity, these antibiotics will not be effective in-vivo. In the individual case, quinupristin/ dalfopristin (Synercid®) and linezoid (Zyvoxid®) are available as reserve antibiotics.

Specific Measures to Control MRSA

The efficacy of most specific measures used for prevention and control of MRSA has not been established in controlled clinical trials. Therefore, the following recommendations are based mainly on general infection control principles and on review of published articles dealing with the epidemiology and control of MRSA in hospitals and long-term care facilities (Bradley *et al.* 1991, Muder *et al.*, 1991, Strausbaugh *et al.*, 1991 and 1992). Factors to be considered when implementing specific measures to control MRSA include the prevalence of MRSA in and outside the hospital, i.e. in other facilities that refer patients for admission, the incidence of nosocomial transmission within the facility and the likelihood that MRSA will spread among high risk patients, and last but not least, the extent of available infection control resources (Boyce *et al.*, 1994). Thus, we recommend that each healthcare facility develops its own strategy to control MRSA on the basis of local conditions.

Contact Precautions: Single Room / Cohort Isolation

It is strongly recommended that a patient with MRSA be placed in a single room with en suite toilet facilities (Combined Working Party, 1998; Robert-Koch-Institut, 1999; Working Group on Infection Prevention 1995; Mulligan *et al.*, 1993; Boyce *et al.*, 1994; Garner *et al.*, 1996). The door must only be opened to allow entry and exit of essential staff. The patient may not leave the room without consultation with the nurse in charge of the ward. If medically indicated, the patient may by all means be transported. In case of wound colonization only, the patient may leave the room (hand disinfection first), if it can be guaranteed that the dressing always remains dry. Other isolation measures must be implemented, e.g. single room, hand disinfection or wearing gloves and gowns. If resources are limited, and more than one patient with MRSA-positive culture can be found, cohort isolation of MRSA patients in two- or four-bed rooms, or even in separate units or wards, may be necessary (Barakate *et al.*, 1999; Murray-Leisure, 1990). The number of staff nursing the patient should be kept to a minimum (e.g. one person per shift). Cohort nursing appears to have been helpful during some outbreaks (Boyce, 1991). This practice requires that nurses caring for MRSA patients do not care for non-colonized patients. Staff with skin lesions, eczema or superficial skin infections must be excluded from contact with the patient. However, further "social" isolation of the patients should be avoided (Tarzi *et al.*, 2001). The nursing workload may influence nosocomial MRSA transmission; closure of wards to new admissions may be used as a part of MRSA outbreak control (Farrington *et al.*, 2000). Visitors must report to the nurse-in-charge before entering the room. Visitors are not routinely required to wear protective gowns; disinfecting the hands suffices before and after each contact with the patient, and after leaving the room. Household contacts should be provided with information (important: hand disinfection with specific instruction by nursing staff; "Information guidelines for patients and household contacts").

Barrier precautions

Gloves

Among the published guidelines, there is no consensus when gloves should be worn by staff caring for MRSA patients (Mulligan *et al.*, 1993; Boyce *et al.*, 1994). The HICPAC guideline recommends that gloves should be worn whenever entering a room of a patient with MRSA (*Category IB*; Garner, 1996). Our recommendation is that single-use disposable gloves must be worn when providing care in direct contact with the patient, handling contaminated tissue, body fluids (e.g. manipulation of the urinary catheter), secretions (e.g. during endotracheal suction, in oral care), dressings or linen. Gloves must always be worn when washing the patient (whole body wash down). After contact with infected or colonized parts of the body and their secretions, the gloves should be removed, the hands disinfected, even if the same patient requires additional handling or examination. Gloves must always be removed immediately before leaving the bed area (e.g. to prepare medication, infusions etc). This must be followed by careful hand hygiene techniques.

Gowns

Published guidelines do not agree on when gowns should be worn by staff caring for MRSA patients (Mulligan *et al.*, 1993; Boyce *et al.*, 1994). Adequate studies investigating the value of gowns in preventing transmission of MRSA are lacking. The HICPAC guideline, however, recommends that personnel entering a room of an MRSA patient wear gowns if substantial contact with either the patient or with environmental surfaces or equipment items can be anticipated or if wound drainage is not contained by a dressing (*Category IB*; Garner, 1996). Since personnel may easily contaminate their clothing when performing intensive nursing activities (Boyce *et al.*, 1997), we recommend that long-sleeved gowns be worn for all nursing activities involving contact with the patient or his/her environment (e.g. when making the bed, moving the patient, during physiotherapy, X-ray or bedside (invasive) diagnostic procedures). Impermeable single-use aprons must be used when there is risk of contact with body fluids and secretions. Protective gowns and disposable aprons may be used more than once; they should be hung up near the patient. Single-use aprons must be marked on the outside, and should be turned inside out when hung up. Gowns and aprons must always be changed after contamination; otherwise they should be changed once daily on general wards, and three-times daily in intensive care units (i.e. after every shift).

Masks

There is no consensus regarding the use of masks when caring for MRSA patients. Adequate studies investigating the value of masks in preventing transmission of MRSA are lacking. We favour the recommendation that a mask be worn whenever heavy contamination of the air with staphylococci is to be expected, for example, for all procedures that may generate staphylococcal aerosols, e.g. sputum or endotracheal suction; the use of a closed suction system is preferable on all these patients; chest physiotherapy; procedures on patients with exfoliative skin condition; when changing the dressing of an extensively infected wound; when making the bed if the skin and perineum are colonized. The mask used should preferably be of a high efficiency filter type; a surgical mask, however, may be acceptable too. The use of mouth/nose masks is questionable in any other situation, since colonization of the staff takes place above all via the hands. Systematic hand hygiene protects against nasal colonization.

Patient-Care Equipment

The role of the environment as a reservoir for MRSA and the role of environmental control measures are controversial (Boyce *et al.*, 1994). Current guidelines stress the importance of disinfecting equipment used in MRSA patient rooms before being used on other patients. The usual reprocessing procedures for disinfection / sterilization of instruments suffice. Devices such as blood pressure meter, stethoscope, tourniquet, thermometer, etc. should be left in the patient's room and should be wiped to disinfect them (i.e. with 70% alcohol). A minimum of materials should be kept in the patient's room; trolleys and drug cabinets should be restocked every day. If the patient is discharged or isolation is discontinued, open packages and any other materials (e.g. gloves, swabs, etc.) kept in the vicinity of the patient should no longer be used and should be discarded, if they cannot be disinfected otherwise. In as far as possible, medical devices may be disinfected by wiping with 70% alcohol and then reused.

Housekeeping

Routine Domestic Cleaning

The role of the environment in hospital spread of MRSA is unclear. The survival time for MRSA on environmental surfaces varies greatly and is dependent upon numerous environmental factors and also upon the particular MRSA strain. Heavy contamination in areas immediately adjacent to patients' beds has been reported, but widespread environmental contamination seems to be rare (Moore *et al.*, 1991). The nurse-in-charge of the ward must advise the domestic service manager regarding isolation, disinfection and cleaning procedures. The room should be cleaned after all other areas of the ward. Domestic staff must wear a disposable plastic apron and household gloves. Preferably, separate (colour-coded) cleaning equipment should be used. Disposable cloths should be used and discarded after each use. It has been shown that hospital dust plays an important role in the epidemiology of MRSA (Rampling *et al.*, 2001). Therefore, rooms should be cleaned once a day, ensuring that all dust collecting areas are cleaned adequately. Current guidelines stress the importance of surface cleaning in the immediate vicinity of MRSA patients. We recommend that all surfaces near the patient should be wiped with disinfectant once daily on general wards and three-times daily in intensive care units. In principle, specific disinfection must be performed after visible contamination of surfaces and devices. The surface disinfection procedure takes place by usual means and using usual concentrations (according to the "cleaning and disinfection guidelines of the hospital"). After use, plastic (single-use) aprons should be disposed of in the waste bag, and household gloves should be washed and dried. Hands should be disinfected after removing aprons and gloves and after contact with the patient and his/her environment.

The patient's clothes and the bed linen should be changed once a day, and whenever necessary. Used linen must be handled gently at all times. Special care should be taken to avoid shaking the bed clothes, possibility of distributing skin scales. Linen bags must be sealed at the bedside and removed directly to the dirty utility area or the collection point. Subsequently, the hands must be disinfected.

Cutlery and Crockery should be reprocessed by usual processing procedures (food tray system); the tray should be placed in the food trolley immediately. If this is not possible, the tray should remain in the patient's room until the container-wagon is available again. Subsequently the hands must be disinfected.

All waste (including single use-items and dressings) can be disposed of in the patient's room as domestic waste. Waste bags must be sealed before leaving the room. All handling of sealed waste bags should be followed by immediate hand disinfection.

Terminal Cleaning

Terminal cleaning must take place after the patient is discharged or after isolation is discontinued (cf. Isolation protocol, by prior consultation with Infection Control Team). Terminal cleaning should commence only after the patient and his/her possessions have been removed from the room or the area. Any disposable items or equipment, which cannot be disinfected appropriately should be discarded. All waste can be disposed of in bags for household waste. Waste bags must be sealed before leaving the room. Domestic staff must wear a disposable plastic apron and household gloves. Long-sleeved gowns should be worn when processing the bed. Bed linen and pillows should be put in normal, non-infectious laundry bags. Laundry bags must be sealed before leaving the room. Bed rails and mattress should be wiped with an appropriate disinfectant. After disinfecting bed rails (in the room, not in the corridor) all horizontal areas should be disinfected, including the floor (walls and ceilings do not have to be disinfected). All high ledges, window frames and curtain tracks should be dusted, ensuring that all dust collecting areas are cleaned adequately. In cases of massive colonization and long patient-stay, the curtains should be washed. The room may be used again for other patients when all surfaces are clean and dry.

References

Adeyemi-Doro, F.A., Scheel, O., Lyon, D.J., and Cheng, A.F. 1997. Living with methicillin-resistant *Staphylococcus aureus*: a 7-year experience with endemic MRSA in a university hospital. Infect. Control Hosp. Epidemiol. 18: 765-767.

Ayliffe, G.A. 1997. The progressive inercontinental spread of methicillin-resistant *Staphylococcus aureus*. Clin. Infect. Dis. 24 Suppl. 1: S74-S79.

Barakate, M.S., Harris, J.P., West, R.H., Vickery, A.M., Sharp, C.A., Macleod, C., and Benn, R.A. 1999. A prospective survey of current methicillin-resistant *Staphylococcus aureus* control measures. Aust. N. Z. J. Surg. 69: 712-716.

Barrett, S.P. 2001. The value of nasal mupirocin in containing an outbreak of methicillin-resistant *Staphylococcus aureus* in an orthopaedic unit. J. Hosp. Infect.. 15: 137-142.

Boyce, J..M. 2001. MRSA patients: proven methods to treat colonization and infection. J. Hosp. Infect. 48 Suppl A: S9-S14.

Boyce, J..M. 1991. Should we vigorously try to contain and control methicillin-resistant *Staphylococcus aureus*? Infect. Control Hosp. Epidemiol. 12: 46-54.

Boyce, J..M., Jackson, M..M., Pugliese, G., Batt, M.D., Fleming, D., Garner, J.S., Hartstein, A.I., Kauffman, C.A., Simmons, M.,Weinstein, R., O'Boyle Williams, C., and the AHA Technical Panel on Infections within Hospitals. 1994. Methicillin-resistant *Staphylococcus aureus* (MRSA): a briefing for acute care hospitals and nursing facilities. Infect. Control Hosp. Epidemiol. 15: 105-115.

Boyce, J..M., Potter-Bynoe, G., Chenevert, C., and King, T. 1997. Environmental contamination due to methicillin-resistant *Staphylococcus aureus*: Possible infection control implications. Infect. Control Hosp. Epidemiol. 18: 622-627.

Bradley, S.F. 1997. Methicillin-resistant *Staphylococcus aureus* in nursing homes. Epidemiology, prevention and management. Drugs Aging 10: 185-198.

Bradley, S.F. 1999. Methicillin-resistant *Staphylococcus aureus*: long-term care concerns. Am. J. Med. 106: 2S-10S; discussion 48S-52S.

Bradley, S.F., Terpenning, M.S., Ramsey, M.A., Zarins, L.T., Jorgensen, K.A., Sottile, W.S., Schaberg, D.R., and Kauffman, C.A. 1991. Methicillin-resistant *Staphylococcus aureus*: colonization and infection in a long-term care facility. Ann. Intern. Med. 115: 417-422.

Brun-Buisson, C., and Legrand, P. 1994. Can topical and nonabsorbable antimicrobials prevent cross-transmission of resistant strains in ICUs? Infect. Control Hosp. Epidemiol. 15: 447-455.

Chaix, C., Durand-Zaleski, I., Alberti, C., and Brun-Buisson, C. 1999. Control of endemic methicillin-resistant *Staphylococcus aureus*: a cost-benefit analysis in an intensive care unit. JAMA 282: 1745-1751.

Combined Working Party. 1998. Revised guidelines for the control of methicillin-resistant *Staphylococcus aureus*. J. Hosp. Infect. 39: 253-290.

Cox, R.A., Mallaghan, C., Conquest, C., and King, J. 1995. Epidemic methicillin-resistant Staphylococcus aureus: controlling the spread outside hospital. J. Hosp. Infect. 29: 107-119.

Crosseron-Zerbib, M., Roque Afonso, A.M., Naas, T., Durand, P., Meyer, L., Costa, Y., el Helali, N., Huault, G., and Nordmann, P. 1998. A control programme for MRSA (methicillin-resistant *Staphylococcus aureus*) containment in a paediatric intensive care unit: evaluation and impact on infections caused by other micro-organisms. J. Hosp. Infect. 40: 225-235.

Crowcroft, N.S., Ronveaux, O., Monnet, D.L., and Mertens, R. 1999. Methicillin-resistant *Staphylococcus aureus* and antimicrobial use in Belgian hospitals. Infect. Control Hosp. Epidemiol. 20: 31-36.

Emmerson, M. 1994. Nosocomial staphylococcal outbreaks. Scand J. Infect. Dis. Suppl. 93: 47-54.

Farrell, A.M., Shanson, D.C., Ross, J.S., Roberts, N.M., Fry, C., Cream, J.J., Staughton, R.C., and Bunker, C.B. 1998. An outbreak of methicillin-resistant *Staphylococcus aureus* (MRSA) in a dermatology day-care unit. Clin. Exp. Dermatol. 23: 249-253.

Farrington, M., Trundle, C., Redpath, C., and Anderson, L. 2000. Effects on nursing workload of different methicillin-resistant *Staphylococcus aureus* (MRSA) control strategies. J. Hosp. Infect. 46: 118-122.

Frank, U., Lenz, W., Damrath, E., Kappstein, I., and Daschner, F.D. 1989. Nasal carriage of *Staphylococcus aureus* treated with topical pseudomonic acid (mupirocin) in a children's hospital. J. Hosp. Infect. 13: 117-120.

Garner, J.S. 1996. Guideline for isolation precautions in hospitals. The Hospital Infection Control Practices Advisory Committee. Infect. Control Hosp. Epidemiol. 17: 53-80.

Goetz, M.B., Mulligan, M.E., Kwok, R., O'Brien, H., Caballes, C., and Garcia, J.P. 1992. Management and epidemiologic analyses of an outbreak due to methicillin-resistant *Staphylococcus aureus*. Am. J. Med. 92: 607-614.

Guiguet, M., Rekacewicz, C., Leclercq, B., Brun, Y., Escudier, B., and Andremont, A. 1990. Effectiveness of simple measures to control an outbreak of nosocomial methicillin-resistant *Staphylococcus aureus* infections in an intensive care unit. Infect. Control Hosp. Epidemiol. 11: 23-26.

Harbarth, S., Martin, Y., Rohner, P., Henry, N., Auckenthaler, R., and Pittet, D. 2000. Effect of delayed infection control measures on a hospital outbreak of methicillin-resistant *Staphylococcus aureus*. J. Hosp. Infect. 46: 43-49.

Harbarth, S., Romand, J., Frei, r., Auckenthaler, R., and Pittet, D. 1997. Inter- and intrahospital transmission of methicillin-resistant *Staphylococcus aureus*. Schweiz. Med. Wochenschr. 127: 471-478.

Herwaldt, L.A. 1999. Control of methicillin-resistant *Staphylococcus aureus* in the hospital setting. Am. J. Med. 106: 11S-18S; discussion 48S-52S.

Heuck, D., Braulke, C., Lauf, H., and Witte, W. 1995. Analysis and conclusions regarding the epidemic spread of methicillin-resistant S. aureus. Zentralbl. Hyg. Umweltmed. 198: 57-71.

Hill, R.L., Casewell, M.W. 1990. Nasal carriage of MRSA: the role of mupirocin and outlook for resistance. Drugs Exp. Clin. Res. 16: 397-402.

Hitomi, S., Kubota, M., Mori, N., Baba, S., Yano, H., Okuzumi, K., and Kimura, S. 1999. Control of a methicillin-resistant *Staphylococcus aureus* outbreak in a neonatal intensive care unit by unselective use of nasal mupirocin ointment. J. Hosp. Infect. 46: 123-129.

Hryniewicz, W. Epidemiology of MRSA. Infection. 27 Suppl 2: S13-S16.

Humphreys, H., and Duckworth, G. 1997. Methicillin-resistant *Staphylococcus aureus* (MRSA) – a re-appraisal of control measures in the light of changing circumstances. J. Hosp. Infect. 36: 167-170.

Kommission für Krankenhaushygiene und Infektionsprävention am Robert-Koch-Institut (RKI). 1999. Empfehlung zur Prävention und Kontrolle von Methicillin-resistenten *Staphylococcus aureus*-Stämmen (MRSA) in Krankenhäusern und anderen medizinischen Einrichtungen. Bundesgesundheitsbl. 42: 954-958.

Lingnau, W., and Allerberger, F. 1994. Control of an outbreak of methicillin-resistant *Staphylococcus aureus* (MRSA) by hygienic measures in a general intensive care unit. Infection 22 Suppl 2: 135-139.

Matsumura, H., Yoshizawa, N., Narumi, A., Harunari, N., Sugamata, A., and Watanabe, K. 1996. Effective control of methicillin-resistant *Staphylococcus aureus* in a burn unit. Burns. 22: 283-286.

Meier, P.A., Carter, C.D., Wallace, S.E., Hollis, R.J., Pfaller, M.A., and Herwaldt, L.A. 1996. A prolonged outbreak of methicillin-resistant *Staphylococcus aureus* in the burn unit of a tertiary medical center. Infect. Control Hosp. Epidemiol. 17: 798- 802.

Miller, M.A., Dascal, A., Portnoy, J., and Mendelson, J. 1996. Development of mupirocin resistance among methicillin-resistant *Staphylococcus aureus* after widespread use of nasal mupirocin ointment. Infect. Control Hosp. Epidemiol. 17: 811-813.

Moore, E.P., and Williams, E.W. 1991. A maternity hospital outbreak of methicillin-resistant *Staphylococcus aureus*. J. Hosp. Infect. 19: 5-16.

Muder, R.R., Brennen, C., Wagener, M.M., Vickers, R.M., Rihs, J.D., Hancock, G.A., Yee, Y.C., Miller, J.M., and Yu, V.L. 1991. Methicillin-resistant staphylococcal colonization and infection in a long-term care facility. Ann. Intern. Med. 114: 107-112.

Mueller-Premru, M., and Muzlovic, I. 1998. Typing of consecutive methicillin-resistant *Staphylococcus aureus* isolates from intensive care unit patients and staff with pulsed-field gel electrophoresis. Int. J. Antimicrob. Agents. 10: 309-312.

Mulligan, M.E., Murray-Leisure, K.A., Ribner, B.S., Standiford, H.C., John, J.F., Korvick, J.A., Kauffman, C.A., and Yu, V.L. 1993. Methicillin-resistant *Staphylococcus aureus*: A consensus review of the microbiology, pathogenesis, and epidemiology with implications for prevention and management. Am. J. Med. 94: 313-328.

Murray-Leisure, K.A., Geib, S., Graceley, D., Rubin-Slutsky, A.B., Saxena, N., Muller, H.A., and Hamory, B.H. 1990. Control of epidemic methicillin-resistant *Staphylococcus aureus*. Infect. Control Hosp. Epidemiol. 11: 343-350.

National Committee for Clinical Laboratory Standards. 1993. Methods for dilution antimicrobial susceptibility tests for bacteria that grow aerobically, 3rd ed. Approved Standard M7-A3. Villanova, Pa. National Committee for Clinical Laboratory Standards, 1993.

Nicolle, L.E., Dyck, B., Thompson, G., Roman, S., Kabani, A., Plourde, P., Fast, M., and Embil, J. 1999. Regional dissemination and control of epidemic methicillin-resistant *Staphylococcus aureus*. Manitoba Chapter of CHICA-Canada. Infect. Control Hosp. Epidemiol. 20: 202-205.

Papia, G., Louie, M., Tralla, A., Johnson, C., Collins, V., and Simor, A.E. 1999. Screening high-risk patients for methicillin-resistant *Staphylococcus aureus* on admission to the hospital: is it cost effective? Infect. Control Hosp. Epidemiol. 20: 473-477.

Pittet, D., Hugonnet, S., Harbarth, S., Mourouga, P., Sauvan, V., Touveneau, S., and Perneger, T.V. 2000. Effectiveness of a hospital-wide programme to improve compliance with hand hygiene. Infection Control Programme. Lancet. 356: 1307-1312.

Rampling, A., Wiseman, S., Davis, L., Hyett, A.P., Walbridge, A.N., Payne, G.C., and Cornaby, A.J. 2001. Evidence that hospital hygiene is important in the control of methicillin-resistant *Staphylococcus aureus*. J. Hosp. Infect. 49: 109-116.

Reboli, A.C., John, J.F. Jr., Platt, C.G., and Cantey, J.R. 1990. Methicillin-resistant *Staphylococcus aureus* outbreak at a Veterans' Affairs Medical Center: importance of carriage of the organism by hospital personnel. Infect. Control Hosp. Epidemiol. 11: 291-296.

Romance, L., Nicolle, L., Ross, J., and Law, B. 1991. An outbreak of methicillin-resistant *Staphylococcus aureus* in a pediatric hospital- -how it got away and how we caught it. Can. J. Infect. Control 6: 11-13.

Schmitz, F.J., Steiert, M., Tichy, H.V., Hofmann, B., Verhoef, J., Heinz, H.P., Kohrer, K., and Jones, M.E. 1998. Typing of methicillin-resistant *Staphylococcus aureus* isolates from Düsseldorf by six genotypic methods. J. Med. Microbiol. 47: 341-351.

Schumacher-Perdreau, F., Jansen, B., Seifert, H., Peters, G., and Pulverer, G. 1994. Outbreak of methicillin-resistant *Staphylococcus aureus* in a teaching hospital—epidemiological and microbiological surveillance. Zentralbl. Bakteriol. 280: 550-559.

Solberg, C.O. 2000. Spread of *Staphylococcus aureus* in hospitals: causes and prevention. Scand. J. Infect. Dis. 32: 587-595.

Spicer, W.J. 1984. Three strategies in the control of staphylococci including methicillin-resistant *Staphylococcus aureus*. J. Hosp. Infect. 5 Suppl A: 45-49.

Strausbaugh, L.J., Jacobson, C., Sewell, D.L., Potter, S., and Ward, T.T. 1991. Methicillin-resistant *Staphylococcus aureus* in extended-care facilities. Infect. Control Hosp. Epidemiol. 12: 36-45.

Strausbaugh, L.J., Jacobson, C., Sewell, D.L., Potter, S., and Ward, T.T. 1992. Antimicrobial therapy for methicillin-resistant *Staphylococcus aureus* colonization in residents and staff of a Veterans Affairs nursing home care unit. Infect. Control Hosp. Epidemiol. 13: 151-159.

Tambic, A., Power, E.G., Talsania, H., Anthony, R.M., and French, G.L. 1997. Analysis of an outbreak of non-phage-typeable methicillin-resistant *Staphylococcus aureus* by using a randomly amplified polymorphic DNA assay. J. Clin. Microbiol. 35: 3092-3097.

Tarzi, S., Kennedy, P., Stone, S., and Evans, M. 2001. Methicillin-resistant Staphylococcus aureus: psychological impact of hospitalization and isolation in an older adult population. J. Hosp. Infect. 49: 250-254.

Working Group on Infection Prevention: Management policy for methicillin-resistant *Staphylococcus aureus* infection in hospitals. No. 35a. Leiden, The Netherlands. Working Group on Infection Prevention. 1995. p. 1-8.

Vandenbroucke-Grauls, C.M., Frenay, H.M., van Klingeren, B., Savelkoul, T.F., and Verhoef, J. 1991. Control of epidemic methicillin-resistant *Staphylococcus aureus* in a Dutch university hospital. Eur. J. Clin. Microbiol. Infect. Dis. 10: 6-11.

Witte, W., Braulke, C., Heuck, D., and Cuny, C. 1994. Analysis of nosocomial outbreaks with multiply and methicillin-resistant *Staphylococcus aureus* (MRSA) in Germany: implications for hospital hygiene. Infection 22 Suppl. 2: S128-134.

Zafar, A.B., Butler, R.C., Reese, D.J., Gaydos, L.A., and Mennonna, P.A. 1995. Use of 0.3% triclosan (Bacti-Stat) to eradicate an outbreak of methicillin-resistant *Staphylococcus aureus* in a neonatal nursery. Am. J. Infect. Control. 23: 200-208.

From: *MRSA: Current Perspectives*
Edited by: A.C. Fluit and F.-J. Schmitz

Chapter 13

Concluding Remarks

A.C. Fluit and F.J. Schmitz

To Summarize

Staphylococcus aureus is an important human pathogen capable of causing a variety of diseases (Chapter 1). This variety of diseases is made possible by an armamentarium of virulence factors, which include adhesins, immune system evasion factors, enzymes, and toxins. It has frequently been suggested that MRSA are more pathogenic than MSSA, but are they? To answer this question one has to consider two possibilities. First, the mobile genetic element SCC*mec* harboring *mec*A entered only the more virulent strains. Second, SCC*mec* contributes to pathogenicity. No conclusive evidence for the first hypothesis is available. Some studies suggest that MRSA are indeed more virulent, whereas other studies contradict this. No studies have been published on the contribution of SCC*mec* to the virulence of MRSA. Only one possible virulence factor involved in adhesion is encoded on some SCC*mec*, but its role as a virulence factor is still a matter of debate. So, despite a large volume of research it still not clear whether MRSA are more pathogenic than their methicillin-susceptible counterparts. However, one remarkable finding favors the hypothesis that at least some MRSA are more virulent. All cases of necrotizing pneumonia in youngsters are caused by Panton-Valentine Leukotoxin (PVL) encoding *S. aureus*, which were all methicillin-resistant (*See* Wright and Novick, Chapter 9).

The role of Small Colony Variants (SCVs) in virulence is totally unclear (*See* von Eiff and Becker, Chapter 10). However, the possibility that SCVs play a role in long-term colonization and the chronic diseases attributable to *S. aureus* infections cannot be excluded.

With respect to virulence it is important to note that sub-inhibitory concentrations of antibiotics have been shown to affect the expression of virulence factors. These antibiotics include fluoroquinolones and β-lactam antibiotics; both of which are commonly used to treat *S. aureus* (and other) infections. The implications of this remain to be established (*See* Wright and Novick, Chapter 9).

The development of antibiotics around the Second World War appeared to give clinicians near magic tools to combat often fatal *S. aureus* infections. However, this euphoria was short lived. Soon after the introduction of penicillin, the first resistant strains emerged. Fortunately, new antibiotics were also developed and these were effective against *S. aureus*. In particular, celebin, later re-named methicillin (and now replaced by highly related compounds such as oxacillin) was available for the treatment of penicillin-resistant strains. However, history repeated itself. Within an even shorter time, methicillin-resistant strains emerged. This had several consequences.

The emergence of MRSA severely curtailed the options for the empirical treatment of MRSA infections. The mechanism of methicillin resistance was unknown, however it was shown not to be β-lactamase mediated. How could MRSA best be detected? How do these MRSA strains spread and can this spread be curtailed? These questions led to a wealth of data, but four decades later most these questions are still not completely answered and the problems with MRSA are only increasing.

The onslaught of MRSA meant that both the detection and typing of strains became important issues. It had already long been evident that a high level methicillin resistance is (in nearly all cases) dependent on the expression of the *mecA* gene. Unfortunately this does not mean that quick and cheap detection methods are available. Culture methods are generally cheap, but culture and the reliable expression of the *mecA* gene takes at least 48 hours. In addition, culture is a phenotypic method dependent on the expression of all factors necessary for methicillin resistance (*See also* Rohrer *et al.*, Chapter 3). For some strains, culture conditions are less than optimal for the expression of methicillin resistance. Amplification methods, on the other hand, are quick but expensive. Furthermore, amplification is prone to inhibition, especially when the target for amplification is obtained directly from patient samples. Sensitivity of detection is an issue for both types of detection (*See* Brown and Cookson, Chapter 2).

These problems with the detection of MRSA by culture methods are not surprising when one considers the mechanism of methicillin resistance in staphylococci. Methicillin resistance is dependent on the expression of the *mecA* gene, which encodes PBP2a (also known as PBP'). Expression may be constitutive or regulated. Regulation is effected by the *mecI* and *mecR* genes, which encode a transmembrane-β-lactam-sensing transducer (MecR1) and a repressor, MecI. In the absence of a β-lactam, *mecA* transcription is repressed. Methicllin and oxacillin, however, are poor inducers of transcription. This results in low levels of PBP2A and thus a phenotype that is not resistant. In consitutive *mecA* expressing strains, the *mecR/mecI* genes are truncated or mutated, although other mechanisms lead to high-level expression of *mecA*. Methicillin resistance expression is often heterogeneous. Highly resistant subclones arise under antibiotic pressure at a frequency above that of spontaneous mutations. Despite a wealth of data on factors that influence methicillin resistance, the mechanism behind heteroresistance has not been solved (*See* Rohrer *et al.*, Chapter 3).

The resistance problem with MRSA does not stop with their resistance to to β-lactam antibiotics. In fact, resistance not only includes the well-known resistance to antibiotics such as gentamicin, erythomycin, clindamycin, and fluoroquinolnoes, but it is also includes emerging resistance to the recently introduced antimicrobial compound combination of quinupristin/dalfopristin. In addition, resistance against linezolid, another recently introduced antibiotic has been reported (*See* Morrissey and Farrell, Chapter 4). A special problem is caused by vancomycin resistance, because until recently, this was the only antibiotic available that could be given empirically to treat suspected MRSA infections. The great majority of MRSA strains with reduced susceptibility to vancomycin derive this resistance from the overproduction of peptidoglycan in the bacterial cell-wall. Since the detection of these strains cannot be performed with a standard inoculum, these isolates may have been missed in the past (*See* Cui and Hiramatsu, Chapter 8).

Recently, another mechanism of vancomycin resistance was discovered in MRSA strains. Two reports describe the emergence of VRSA strains which harbor the VanA transposon. This transposon is responsible for high-level vancomycin resistance in enterococci. In fact, the *vanA* gene cluster is commonly found among vancomcyin-resistant enterococci. This is especially worrying because, since this gene cluster is located on a transposon, it can potentially be mobilized to other strains of *S. aureus* including MRSA strains (Anonymus, 2002a, 2002b; Centers of Disease Control and Prevention, 2002).

This leads us to a bleak situation in terms of the available treatment options; especially empiric treatment options. However when the resistance profile of the MRSA isolate is known, treatment options other than vancomycin or linezolid often remain. In addition to the resistance profile, the treatment options available depend on the type of infection. Future treatment options include new drugs such as oritavancin and cationic peptides, which are currently undergoing pre-clinical study (*See* Ben-David and Rubinstein, Chapter 11).

It is clear that the implications of MRSA infections are grave. Typing of MRSA to study their epidemiology is therefore important. However, typing methods for MRSA strains also have their limitation. Firstly, the right typing method for the question at hand should be chosen. The determination of the population structure of MRSA requires a different choice of methods than asserting an ongoing outbreak in a hospital. In addition, it is nearly impossible to prove that two isolates are identical, but different isolates still can be related, e.g., the introduction of a lysogenic phage may alter typing results (*See* van Leeuwen, Chapter 5).

The *mecA* gene is encoded on the Staphylococcal Cassette Chromosome *mec* (SCC*mec*). Four distinct types of SCC*mec* have been described. SCC*mec* regions range from 21-67 kb in size. Depending on the type of SCC*mec*, additional antibiotic resistance determinants may also be present. In addition, a number of other orf's may be present, however most of these appear to be non-functional. It has been demonstrated that enzymes encoded in the SCC*mec* region, enable the excision and integration of the SCC*mec* region into and out of the *S. aureus* chromosome. SSC*mec* is integrates into an open reading frame (*orfX*) of unknown function in the *S. aureus* chromosome. The origin of the *mecA* gene is unknown, however it has has been postulated to originate from *Staphylococcus sciuri*. If this is true, it raises the question of how the *mecA* gene became incorporated in four different types of SCC*mec* (*See* Fitzgerald and Musser, Chapter 6; Fluit and Schmitz, Chapter 7). An important new development is the emergence of community-acquired MRSA. Most of the community-acquired strains have a SCC*mec* type IV genotype.

Using typing methods and a genomics approach it was established that at least four waves of MRSA caused problems during the last four decades. (*See* Fitzgerald and Musser, Chapter 6). These waves are reflected in the MRSA population structure. Bacterial populations are built by the exchange of genetic material and the expansion of successful clones. Amongst MRSA strains, several successful clones have appeared; for example, the Iberian, Brazilian, and Peadiatric clones. Some clones appear to be limited to a hospital, whereas others have a pandemic spread. Some MRSA strains also appear to spread more easily than MSSA strains. The reasons for these differences are unclear. The role of horizontal transfer in the spread of MRSA, i.e., in initiating new clonal lineages, has long been disputed, but an increasing body of evidence favors a more important role than long believed. However, the mechanism of transfer is still unknown (*See* Fitzgerald and Musser, Chapter 6; Fluit and Schmitz, Chapter 7). The *mecA* gene can be detected in many different clonal lineages, which are also represent methicillin-susceptible *S. aureu* (*See* Fluit and Schmitz, Chapter 7).

What can we do? It is clear that a reduction in the prevalence of MRSA will be extremely beneficial. For this good hospital hygiene measures are our first line of defense. In the implementation of hospital hygiene measures, it is important to remember that the nasopharynx is the main reservoir and that primary route of transmission is via the hands (*See* Frank, Chapter 12). Problems can be best contained when detection of MRSA is early (*See also* Brown and Cookson, Chapter 2). Three general strategies can be applied to control MRSA. The Search and Destroy policy, the more limited Scutari Strategy, and the SALT Strategy (*Staphylococcus aureus* Limitation Technique). Based on the initial situation, protocols can be drawn up including surveillance, hygienic measures, treatment of carriers, barrier treatment, and decontamination policies of both equipment and rooms (*See* Frank, Chapter 12).

Whatever happens MRSA is likely to stay, but the lessons learned from MRSA should be applied to the appearance of strains, harboring resistance against new antibiotics. It remains to be seen, whether the application of these lessons to linezolid-resistant isolates and VRSA, especially those strains harboring the vancomycin transposon, does not come to late to save this valuable antibiotic for the treatment of MRSA.

References

Anonymus. 2002a. *Staphylococcus aureus* resistant to vancomycin - United States, 2002. Morb. Mortal. Wkly Rep. 51: 565.

Anonymus. 2002b. Vancomycin-resistant *Staphylococcus aureus* - Pennsylvania, 2002. MMWR Morb Mortal Wkly Rep.51: 902.

Centers for Disease Control and Prevention. 2002. Vancomycin resistant *Staphylococcus aureus*-Pennsylvania, 2002. JAMA 288: 2116.

Index